contents

introduction

"Everything should be made
as simple as possible, but not simpler."

ALBERT EINSTEIN

getting a good live sound requires a combination of technical skills, quick-witted creativity, decent-quality equipment, and many other contributory factors – not all of which are always within your control.

When it's done well, and the results are enjoyed by an audience, it's a hugely satisfying experience – live music has the rare capacity to unite or inspire, transform consciousness, even alter reality. Keep reminding yourself of that when you're dragging the tenth speaker cabinet up those four flights of beer-and-vomit-stained nightclub stairs at two o'clock in the morning...

'Sound reinforcement', as it's also called, may not strike everyone as the most excitingly glamourous job description on the planet, but since the 1960s it's come to play a crucial, central role in most areas of live entertainment – starting from pop and rock tours, festivals and nightclubs/discotheques, then to folk and jazz clubs in the 1970s, and even classical concerts by the 1980s. Today's huge dance scene couldn't exist at all without it.

But the reality is, if everything goes according to plan, your job as a live sound engineer, or as part of the sound crew at a large event, will go un-noticed and unappreciated by the majority of the audience. It's only when things go badly wrong with the sound that the average listener sits up and takes any notice of the huge technological effort that's gone into creating their evening's entertainment – and usually in a disparagingly ill-informed way. That's life.

Part of a sound engineer's job is to minimise the chances of something going wrong in the first place. Which is also what this book attempts to do.

Of course, within the industry itself, good live sound engineers are highly sought-after by top touring acts and revered by their peers – some are even reasonably well paid. Many got into the profession almost by default – helping out friends' bands and finding they have a knack or an affinity with the job, and gradually making it a serious career.

Nowadays there are training courses available – though these can sometimes be as disconnected from 'frontline' reality as a driving test is from real motoring. (Industry professionals warn that any course lecturer who is not an experienced hands-on PA operator or mixing engineer is

THE **live** SOUND MANUAL

A BACKBEAT BOOK

First edition 2002

Published by Backbeat Books

600 Harrison Street,

San Francisco, CA94107

www.backbeatbooks.com

An imprint of The Music Player Network United

Entertainment Media Inc.

Published for Backbeat Books by Outline Press Ltd,

Unit 2a Union Court,20-22 Union Road, London SW4 6JP

www.backbeatuk.com

ISBN 0-87930-699-8

Art Director: Nigel Osborne

Design: Paul Cooper

Editor: Paul James

Production: Phil Richardson

Origination and Print by Colorprint (Hong Kong)

08 09 10 11 6 5 4 3 2

unlikely to teach what actually needs to be known.) Academic qualifications by themselves cut little ice in this area.

Everyone in the business will tell you there's still no substitute for learning by doing. The knowledge gained by years of trial, error and success is what makes a top live sound engineer worth their weight in power amps.

It's some of this accumulated inside-knowledge that we hope to pass on in this book – from hard facts that explain how the technology works, and why you need it, to hands-on practical tips to ease or guide the application of that knowledge and make your working life easier – not to mention making life more fun for the performers and audiences you're bringing together.

The PA system – and why less is the new more

The process of 'concert sound reinforcement' and the equipment used for this, is commonly referred to by the abbreviation 'PA'. This derives from the oddly antiquated term 'public address' – which sounds rather quaint these days, bringing to mind images of summer/holiday camps, school sports days, and train station announcements, with their honking or aggressively sharp and notoriously unclear tone.

Modern PA systems, as used at band gigs, clubs and dance events, are a subtler and more sophisticated sort, built and tuned to accurately deliver all types of music, whatever its qualities or complexities.

Fundamentally any PA system – even the most advanced – still generally aims to achieve what the name suggests: to allow an artist (singer, band, DJ, poet – anyone with something to say) to address, and hopefully entertain, an audience of whatever size.

Seems straightforward enough... but the means of doing so has become an insatiable multi-billion-dollar international industry – almost imperceptibly, as far as the general public is concerned. All the regular gig-goer has seen is a steady variation in the number and shape of black boxes around the stage.

Since the 1960s PA systems have grown to be more elaborate, with more mixer channels, greater amplification, and often more speakers being used than for comparable past events. Yet they've also become more efficient, so they can be physically smaller and still deliver the same (or more) power.

Still, even with all the expensive modern hardware, and about 40 years of hard-earned industry know-how, no one can guarantee the sound will always be perfect – which brings us back to the numerous 'contributory factors' we alluded to earlier.

We all know how difficult it can be to get a consistently perfect sound out of a good home hi-fi system, playing a top-quality recording in the stable set-up of a living room, and with just one or two people's tastes to satisfy. When it comes to PA systems, the variables are multiplied enormously.

First of all, a live performance is not like a studio recording – there's no chance for editing or post-production polish. And the instruments, their players, or both, may be having a bad day. As might the sound engineer, or any other members of the production crew (no matter how professional).

And of course public spaces are generally bigger, taller and often more oddly-shaped than (even the largest) living rooms. Many spaces – even those regularly used for musical performances – in fact provide the worst possible acoustic (not to mention aesthetic) environment for music, with walls and floors made of hard, shiny or 'false' (sound-transparent) materials.

In the worst cases this can cause long-term harm – perhaps physical damage to people's hearing (which we'll cover in Appendix C), but also emotional upset, possibly turning people off live music altogether.

It goes without saying that concert sound level (known to non-professionals simply as 'volume') is usually considerably greater than is feasible in most homes. And even though you'd expect large spaces like music venues to be capable of handling high levels of acoustic energy, these high energy levels (together with the rapidly changing patterns of louder, more vigorous sorts of music) create energy-chaos on a scale that defies acoustic science.

In short, you could define the art of getting a good sound from a PA as, 'mastery of high-energy sound in chaotic real-time environments'...

The practicalities are complicated further by the fact that the concert sound industry nowadays suffers from a fairly routine 'overkill' use of technology. From the late-1960s onwards it was considered macho to have a massive PA rig – the idea being that the more watts and speaker

introduction

cabinets you have, the better. (There is a good argument for more 'headroom', which we'll talk about in Chapters 5 & 6, but sadly the result of bigger PAs is often simply to make a 'messy sound' a little louder.)

Added to that we have 101 makers of profitable electronic boxes nowadays producing 505 hardware solutions to fix problems that are caused... well, usually by the other hardware they've previously sold you.

The well-grounded (perhaps only) solution is to return to basics: to cut back to the least gear you can get away with, and try to find solutions that use fewer, not more, boxes.

A return to minimalism has already been shown to help create alive-sounding hi-fi (there haven't been any equalisers or other gimmicks in any seriously satisfying, high-end domestic hi-fi equipment for over 20 years), and the same can be said of experienced engineer's recording studio set-ups.

It can be educational, as well as inspiring, to listen afresh to 1960s and 1970s recordings (particularly of US soul and R&B musicians) made on shoestring budgets with low-tech gear by astute engineers.

What is 'good' sound –
and why is it so hard to get?

Good sound quality connects you emotionally with the music. Good sound quality is not always perfectly clean, but it will (by definition) not hurt your ears or cause real discomfort. Any sound that causes pain is simply not very clever.

Clean sound never seems as loud as it really is – in the same way that a marching brass band in the street doesn't sound quite as loud as it really is. As a measure of this, to cleanly replicate the sound level of this 'mere' acoustic band, you'd need a fairly substantial PA system (say about ten kilowatts of amplification – which is more than you'd find in many medium-sized rock venues).

There's an easy way to spot good loud clean sound: when you hear it you assume you can comfortably talk over it to someone standing next to you – and then you find you can't.

Some people prefer 'dirty' sound systems, because of the adrenaline rush they get (cheap thrills). Dirty PAs also sound louder and fiercer (cheap volume), which 'turns on' the lower, limbic centres of the brain, stimulating mental states beyond conscious control, which humans share with creatures as far down the evolutionary scale as reptiles.

The results are not always pretty – it could be argued that many tragic or unpleasant gig experiences (from individual hearing damage to fights and riots) can be associated with loud PA systems that were painfully distorted.

It should maybe be better known (but isn't in the interests of most audio equipment makers to mention) that once a sound signal is damaged or degraded in some way – or if the original signal is of poor quality to start with (because of sub-standard instruments, vocal ability, microphones, or whatever) – then no amount of processing can recover it, let alone make it 'perfect'. In other words, as the saying goes – 'garbage in, garbage out'.

Most 'improvements' to the sound signal are made at the expense of some useful part of the music. As a basic example: removing hiss also inevitably removes some top-end sparkle from the sound.

In such a subtle subject as sound, 'solutions' always come with a cost. That's why it is so important to get the best sound you can 'at source' (for example good instruments, reasonable-quality, well-positioned mikes etc).

Strangely, you cannot buy any decent PA system straight out of a box, or off-the-shelf. Like a good hi-fi or a well-furnished home, it has to be assembled from diverse supplies of 'good stuff', none of which is ever all available at once (even if you have the cash), but which is built-up over time, and which all has to be matched and mastered.

It's also a fast-moving target, with new equipment and methods regularly appearing, so keeping up-to-date can be a complicated and expensive business.

Even on big-name tours, where you assume money is no impediment, it's an eternal frustration in the PA world that stage-lighting tends to receive a larger slice of a touring budget than the sound system. This is despite the fact that visuals are rarely – or at least surely shouldn't be – the biggest reason people come to live music shows (though sound engineers might grudgingly admit that good lighting can add atmosphere to the proceedings).

Another recurring problem in the PA industry is that flux levels are high. Nothing stands still. Every time you use a PA, it gets changed and enhanced – it's unlikely that every piece of gear is exactly the same from one gig to the next.

Crew turnover is also high, so specialist knowledge is easily evaporated: the pissed-off monitor man who quits (to manage a guitar shop, after too much abuse by bands), may have learned some clever or innovative tricks which aren't getting passed on for everyone's benefit.

PA has always been a heuristic (teach-yourself) craft. The industry is in its fifth decade, yet this is one of the few in-depth, practical books on the subject. As already noted, only a limited number of really pertinent courses exist, and their influence is not widespread. So word-of-mouth is still the main means of skill communication.

It's never been a 'joined-up', co-operative industry either – specialist manufacturers stick to their own schemes and methods, and no company successfully produces all elements of a PA system, at least at pro level. It's partly as a consequence of

this that PA sound quality – a big, multi-dimensional subject to learn under the best of conditions – continues to suffer through misinformation and misuse.

Blinded by science –
and why you should trust your ears anyway

In some areas of the PA industry in recent years there's been wholesale adoption of an over-zealous scientific approach to live sound. This manifests itself in the belief (by some less worldly PA users at least) that manufacturer's specification sheets can somehow provide a reliably accurate map of the sonic capabilities of their gear – that you can in some way 'hear sound with your eyes'.

In fact, the most sophisticated measurements currently available still prove relatively little about what a given system will actually sound like in practice – especially as so many other variables come into play.

A very brief history of PA

The first PA systems date back over 2000 years, and were made of stone... Way back in those pre-electrical power times Roman amphitheatres were designed and constructed in such a way as to make the most of natural acoustics to enhance or 'amplify' public speaking or theatrical events.

In practice, of course, prior to the 20th century if you wanted to be heard in public you either had to have a loud voice or play a naturally loud musical instrument – or get together with others of a like-mind, which partly accounts for the appeal of choral singing and the growth of orchestral music and, later, big-bands (all of which can be startlingly loud without any amplification).

The megaphone – a large cone held to the mouth to help project and target the voice – was about the only man-made 'PA' device available right up until the late 19th century, when the development of the telephone also led to primitive sound systems (and of course to

radio, recording and talking movies). In fact the earliest PAs were built with help from Bell Telephone Labs (in the US, about 1915) – and probably sounded much like loud telephones.

This ability to convert sound into electrical signals and magnify it using valves (tubes) revolutionised public life in the 20th century. As well as allowing political leaders to address and influence huge crowds (not always a positive thing, as it turned out), the burgeoning technology was also employed on-stage by musicians and singers, which in turn allowed a new breed of non-operatically trained 'crooners' to achieve huge success. Few people would probably ever have heard of (or even heard at all) the likes of Bing Crosby or Frank Sinatra without microphones and PA systems – and the same could be said of most successful popular music acts since then.

The rise of (relatively) rabble-rousing rock'n'roll in the 1950s created the desire for better and

louder concert amplification – but it was slow to develop (which is hard to believe in today's shamelessly commercialised, opportunistic age). Few PA options were available for the hordes of proto-rockers following in the wake of Elvis, Little Richard, Eddie Cochran et al: up-and-coming singers would sometimes have to misuse domestic tape recorders to amplify their vocals.

Even by the mid-1960s, international megastars like The Beatles still only had a few Vox speaker cabinets and the latest 'state-of-the-art' 100-watt amps to try to entertain a sports stadium-full of screaming fans.

But it was this kind of experience that gave some technically-astute entrepreneurs a kick, and showed them the potential for larger concert sound systems. The following 25 years saw an intensive growth spurt in this area, particularly in Britain and the US, with PA technology developing almost beyond recognition, to become the huge industry it is today.

People who've successfully approached PA from a more artistic, creative, often self-taught background would never think of buying or using equipment on the basis of fancy-coloured literature. This message needs to be spread: pretty catalogues are to be treated as snapshot approximations of reality, dressed-up to look like what sales people think will help make a sale.

When you have the real cab or mixer in front of you, plug in, tune up, listen – and throw the glossy pictures away (or at least file them for future reference...).

From its earliest days, PA equipment was designed and manufactured by people very closely associated with live shows. More than a few successful manufacturing companies were originally begun by former 'roadies' and soundmen (many of whom had worked for/with bands that were shaping musical history), and the inventions and developments of such companies were driven largely by what they saw was needed by the industry they knew – often in spite of 'scientific' advice.

(Charlie Watkins, founder of pioneering British PA company WEM, was warned by a physicist back in the mid-1960s that creating a PA of over 300 watts would be dangerous to life...)

Nowadays it's not uncommon for large conglomerates to swallow up smaller specialist companies. Which is fine if they utilise the knowledge base, and use corporate strength to keep costs down and quality high. Sadly this isn't always the case. Corners are sometimes cut and quality goals redefined to fit what is cheap and easy to make, or what suits lowest-common-denominator tastes, rather than what is truly best or innovative.

In some cases gear designers have simply become removed – physically and mentally – from the day-to-day requirements of the working sound engineer. The best ones, on the other hand, combine their computer-aided-design skills with an intimate understanding of live sound.

In short, as in all industries, newer isn't necessarily better, and the bigger names don't always produce the best gear. When making a purchasing choice, don't dismiss the independent makers just because they don't take out as many full-page advertisements in the music magazines.

Equally, be ready for a mega-corporation to make something that's really good too – that can

happen every now and again. Let your ears be the ultimate judge.

Another area where you have to be guided by the sound rather than anything else is the age-old analog-versus-digital debate. It's easy to be persuaded these days that 'digital is always best, for everything'. It can certainly seem cheaper, and is more efficient and cost-effective for many tasks – such as control circuitry, and memorising/recalling of settings.

But low-budget digital gear (especially the older stuff), though it may be adequate for domestic or hobby use, can often struggle in high-level professional live sound environments. And you may find there is a tried-and tested analog alternative available that's worth a slightly higher outlay.

Of course it's perfectly possible nowadays for fully digital audio to sound very good, but to achieve this will have required enormous design and development effort from the manufacturer – so in fact it may end up not costing much less than comparable analog gear. In the end it comes down to what you think sounds best.

Whichever you go for, be sure you're not just paying for added gimmicks that aren't really improving anything substantially, and in fact may even make the end result worse.

Power and glory –
and why the two don't always match up

There are several common mistakes and misconceptions prevalent among less-experienced PA users, especially when it comes to PA power. One prime example of this would be feeding, let's say, a loud drum kit through a smallish PA. A drum kit's acoustic power is far more than you'd think – even a ten kilowatt PA may be of little use if the drummer plays heavily.

There is no point at all in putting any instrument through a PA if the PA's own sound level capability (very roughly indicated by the power that the amps can deliver) is not big enough to reinforce the sound.

It's the same with the brass and horn families – especially trumpets, which are unfeasibly loud when blown hard (and don't often need amplifying at all).

At small gigs it's far better to use the PA for just the weaker instruments (acoustic guitar, piano, vocals). In general, the less you put through a PA the better it will sound – for one thing there are fewer instruments competing for the limited PA power, and there's less to go wrong.

A common (but perfectly understandable) misbelief is that a 20kW PA is twice as loud as a 10kW one. For a start, PAs of different power (especially from different manufacturers, despite what quality claims they may make) don't use exactly the same parts, or in the same way, so they can't readily be compared – except through listening tests and careful system-matching.

As a fairly extreme example, a 10kW PA with inefficient 'sealed-box' cabs can actually be quieter than a 1kW PA using efficient horn-loaded cabs (see Chapters 5, 6 and 7 for more on power and speakers).

To summarise, here is the 'bad news' about relative PA power – even assuming the PAs being compared are the same in every other respect: a 1kW (1,000-watt) PA system is barely twice as loud as a 100-watt set-up – but it can cost around four times as much.

To double loudness levels again, you'd probably need to pay another four times as much – though the slightly silver lining here is that for this money you'll probably acquire yourself a 10Kw PA rig.

In other words, a 10kW PA is in reality only around four times as loud as a 100W PA (not 100 times louder, as the numbers alone suggest), while perhaps costing a hefty 16 times as much.

The secret of this seemingly bizarre arithmetic is that the wattage scale operates logarithmically – which means it rises with a constant ratio of about ten times to achieve each loudness doubling.

Again, in the real world, even this figure isn't consistent – it's most accurate at mid frequencies, but in the low bass, above a certain sound level, the wattage increase needed to double the loudness is in fact closer to being double. This fact, so vital to PA, is rarely explicitly mentioned in any academic book.

All of this at least gives you an idea just how much extra power (and money) is needed to make any headway in increasing a PA's overall loudness. It also explains the sheer size and weight of some professional PA systems – and the reason why PA is big business for manufacturers. And, conversely, it makes you realise why a 500kW (half a megawatt) PA, such as you might hear at a festival, is not necessarily as loud as you might imagine.

And here's another oddity that's often overlooked. Despite the conception that modern PAs might be 'too big', many problems are in fact caused by not using enough power (amplifier power, to be precise).

In PA, it's a long established practice that amplifier power-delivery ratings should be higher than the speakers' average power-handling ratings. The norm is between two and five times as high (see Chapters 5 & 6).

When underpowered, PAs soon hit clip (overload), which creates a vicious, hard sound and can damage ears and perhaps the PA. No amount of equalising or compression or other tweaking will greatly alter the bare fact that you don't have enough power. (In fact, over-EQ'ing can create as many problems as it solves, as we'll see in Chapters 3 & 4.)

It's also important to remember that power is not simply a number on a knob – the amount of drive a power amplifier needs to persuade it to deliver its maximum load varies with the make/model, the controls and internal settings, and operating condition – including the speakers attached to it.

Also, speakers themselves don't 'have power' as such – the phrase 'powerful speakers' is a misnomer. Instead, speakers have a 'power-handling capability'. In fact they waste between 50 and 99.7 per cent of the energy that's delivered to them. That's why they get very hot (and occasionally burn-out), and why we need them to be able to handle high power in the first place.

Valuable energy (electrical and human) is also wasted in more subtle ways. One cause is incorrect polarity (incompatible wiring at any stage of the audio chain, which reduces or perverts the resulting signal – more of which in Chapters 1 & 7).

Another is poor gear placement – wrong positioning of microphones and arraying of speaker cabs (see Chapters 1, 6, 7, 8, 9). These and other oversights or miscalculations can fritter away some or most of the power you've paid for, and create ugly sound into the bargain.

It's rather like a boat with two engines, attached to separate propellers, each pointing in different directions – the two can largely cancel each other out, creating some strange vibrations in the process. And it's amazing how often you see (and hear) this, when it can be very easily fixed just by 'lining up' one or two vital elements in the system.

Two final, irritating, and sadly familiar practices need addressing here before we start. Firstly – red flashing 'overload' or 'clip' LEDs on an amp or mixing console are not a virility symbol, despite what some misguided sound engineers seem to think.

Usually in pro-audio gear a red light means hard distortion is just starting to occur, or could well be imminent. When PA systems are driven into distortion in one way or another they create masses of spurious, indiscriminate notes (or harmonics).

Some of these are not only not what was musically intended, but are also actively anti-musical. (Some particular types of overdriven harmonics can, as it happens, enhance the music – but that doesn't excuse mindless, indiscriminate, everything-in-the-red engineering.)

In general, analog devices such as microphones, speakers, transformers and valves (tubes) are better at producing nicer 'natural' harmonics. They also break down more gracefully when pushed hard.

By comparison, solid-state (transistor and chip-based) electronics, whether analogue or digital, may produce less distortion under ideal conditions, but any small harmonics they do create are more likely to contain 'non-musical' qualities.

Again, this can be used creatively – depending on the taste and sensibilities of the audience – but be aware that, to many ears, uncontrolled digital distortion is more unpleasant than analogue overdrive.

Loud can be good. It can be a major part of the excitement and adrenaline of a live gig. But at ultra-high sound levels, especially when distortion of any kind is present, most people will suffer ear pain or general discomfort.

If humanitarian considerations don't hit home, consider the economic ones: as a sound engineer you're paid to provide an entertainment – if audiences are put off coming because they no longer find it fun, you're simply turning away their future investment in your industry and your own career. Engineering a purposely painful sound is not macho, just stupid.

The other enemy – again one that's rarely mentioned – is bland sound: live sound engineers should rejoice in the fact that they don't have to worry about churning out an anodyne mix that 'sounds good on the car radio' or won't frighten the MTV viewers. Why not be creative with your live mixing.

Using this book

In the end it's all about delivering music to an audience – which by definition ought to be an enjoyable and rewarding experience all round. To reach that goal inevitably involves some grasping of occasionally complex technology, which this book doesn't shy away from. But keep in mind that it's all a means to an end.

If you're already a working engineer, hopefully there will be some tips and technical insights in these pages that will make your job easier.

If you're just starting to dabble in the world of live sound and PA, there's plenty here to take you forward, get you thinking, experimenting and problem-solving.

Before you start on the journey, a quick note about how to approach this book. It's big, it can get technical, but it doesn't need to be daunting.

If you're completely new to the subject, start by scanning through all the 'Quick View' intros to each chapter – if you just read these, in order, at one sitting, you'll get a good lay-person's grounding in the component parts of a PA system and how it's set up for use. Then you can dip into the chapters themselves as and when you feel the need.

For instance, if you're a musician or DJ, and you already know all there is to know about microphones and turntables, you might want to skip Chapter 1 on Signal Sources. (On the other hand, you might just learn something useful – maybe a handy tip or two about DI boxes or mike splitters…)

If you're convinced you're already well-enough versed in all the gear involved in PA work, and just want some hands-on advice on wiring, operating and troubleshooting a system, you could go straight to Chapter 7 and start reading from there. The other chapters can be used for reference, should you need a technical refresher.

Enough with the introduction – time to get to the show on the road…

CHAPTER 1

signal sources

microphones, DI boxes etc

Quick view

The first step in the process of 'sound reinforcement' is to take the original sound (voice/sampler/ guitar/record, or whatever) and translate it into a suitable electrical signal which the rest of the PA system can then manipulate (if necessary), amplify, and eventually feed to the speakers (which turn electrical impulses back into sound). This first step is achieved in one of two main ways: using a microphone, or by 'direct injection' via a cable into a direct or DI box. These essentially constitute the 'signal sources' in a PA system. (Note: explanations of unfamiliar terms can be found in the Glossary.)

microphones (mikes)

The microphone's job is to pick up a sound as accurately as possible and transmit it to the next part of the audio chain. There are many kinds of microphone, although you can do maybe 80 per cent of all miking with just a few basic types. Through experience you'll come across a diversity of more unusual mikes that are particularly appropriate for special jobs – we'll mention some of these a bit later.

Mikes used on stage have a lot in common with those used in recording studios (which you may or may not be familiar with) – although those with more rugged construction are obviously favoured for live work, as the gigging/touring environment can be brutal on all equipment.

Broken microphones are an occupational hazard for PA operators, and some otherwise ideal studio mikes (large-diaphragm Neumann mikes, say, or fragile valve-containing ones) are just too easy to damage, and expensive to mend.

Microphones aren't readily repairable – you might find professional PA users who learn to make their own cables, re-cone speaker units, and carry out basic servicing to amplifiers, but with microphones even 'bench repairs' in a controlled environment are fiddly, let alone 'field' repairs on the road or at a gig. Damaged mikes must usually be sent away to be fixed, which could take days or weeks. That's why the sturdier the gear is to start with, the better.

Many widely used stage mikes are amazingly old designs – some models date back several decades, which is a long time in a technology-led industry like music. They're still in production simply because they do their job perfectly.

Microphones come in all shapes and sizes, and you can't always tell much about their use from their appearance – although mikes intended for vocals often have more substantial protective 'grilles' to help distance their inner workings from (to put it politely) 'moisture'. You'll get little indication from the seemingly random model numbers either – if you don't already know them, you'll soon become very familiar with names like SM57, MD421, D12, M88...

Widely used stage mikes usually come from either one or two American companies or are of German/ Austrian origin, with the odd Japanese and British model making an impression from time to time. Bear in mind that many of the big-name microphone brands (including Shure, Beyerdynamic, Sennheiser, AKG, Audio-Technica et al) also produce budget lines, which will be less suitable for high-quality live work. (As a very rough guide, under £100/$150 is pretty much budget range for a microphone; top quality professional models can cost up to several hundred more.)

Mike types

There are two main basic kinds of microphone used on-stage. These differ in how they work, but not necessarily what they look like.

Dynamic mikes
Used for most jobs – perhaps accounting for up to 70 per cent of all mikes at a gig. They're also called 'moving-coil' mikes, because of the internal mechanism – a coil moving over a magnet – which they use to convert sound waves to electrical impulses. The name 'dynamic' comes from the fact that they work somewhat like tiny dynamos (or, strictly speaking, alternators) in the way they generate current, and not because they necessarily result in any particularly 'dynamic' type of sound.

Dynamic mikes are preferred for stage work because they are generally sturdier, more affordable, and largely free of complications (no separate power source is needed, for instance) – you just plug and play. Some main exceptions to this rule – where dynamic mikes are not ideal – would include miking instruments with high treble (toppy) content (cymbals, for example) or very low bass, or else where a strictly 'flat', accurate frequency response is essential (rather than perhaps a warmer, more 'musical' one).

Capacitor mikes
These are the other predominant type – commonly referred to as 'condensers' (a term which actually dates back to primitive 18th century ideas of 'condensing' electricity). In electronic terms a capacitor is a component that can store and release energy; a capacitor mike creatively employs the same principles, but adapted for capturing sound.

Stage-worthy capacitor/condenser mikes need to be of sturdy build, though they're still often more vulnerable to rough treatment than dynamic mikes

Soundwaves cause fluctuations between a moving diaphragm and a fixed plate, and the resulting change in capacitance appears as a very delicate fluctuating voltage. This has to be amplified within the mike itself (which therefore requires power), but the result is a signal even more robust and also about three times higher in level than that from a dynamic microphone.

Stage-worthy capacitor/condenser mikes will need to be of sturdy build, though they're often still more vulnerable than dynamic mikes to rough treatment (either physical or sonic), and to damp or blowy conditions (fitting a 'windshield' can help to an extent – we'll come back to this later).

Capacitor/condenser mikes, as we've said, always require a power source. Usually this is supplied in the form of 'phantom power', which comes either directly from the mixing desk or from an active 'mike-splitter' (there's more about these later), cleverly avoiding the need to use special cables or plugs or batteries. It's sent 'transparently' – without affecting the signal – up standard three-wire signal lines (in other words balanced lines, which we'll explain in a minute).

Capacitor (sometimes abbreviated to 'cap') mikes are appropriate where an 'uncoloured' response is required over a wide frequency range, and especially where an extended and fairly flat top-end response is needed. (Note: 'flat' in this context just means 'even', or not biased in any particular frequencies – it has no negative connotations of 'dullness'.) This could make them ideal for miking-up a full range cab connected to a wide-frequency instrument like a synthesiser, or they're also often used with percussion and high strings, such as violins, to pick up that extra 'presence' and 'zing'.

Another advantage of a capacitor/condenser mike signal is it has what's called a lower source impedance (see sidepanel below), which means you can use longer mike cables with less signal loss or degradation.

There are a few other types of mike you may come across from time to time on-stage:

Electret mike

Usually a low-budget capacitor type. Don't be deceived by the styling, which can look 'professional' or might even resemble one of the well-known pro-standard 'condenser' mikes.

One sign of an electret is if the mike has a battery in it: the electret capsule doesn't need power in itself, so it doesn't need a high voltage (like the typical +48v used for phantom powering), but there's still a head-amp inside the mike that needs a few volts, hence the battery.

Another feature of an electret is that it often won't have 'balanced' connections (see sidepanel on p14 looking at 'balanced & unbalanced').

Ribbon mike

This relies on a piece of very fine metal foil to generate a signal, rather than plates or a coil, so it can be particularly fragile. You're unlikely to use one of these for live work, except perhaps for sedate classical work or low-key indoor vocal performances, which are more suited to the delicacy of this kind of mike.

Boundary mike

Also known as the pressure zone microphone (PZM), this is easily identifiable by its almost completely flat appearance: it's basically a mike capsule attached face-down onto a (square metal) plate. Often used for picking up ambient sound at live shows.

You may sometimes come across 'nuisance mikes' – basically toys or fakes: mikes that are all style and contain sub-standard guts. They may either be poor imitative 'copies', or a genuinely expensive second-hand mike which at some stage has had the internal mike capsule replaced by one from a 'toy' microphone (say from a domestic tape recorder) before the mike was sold on. Anyone who then buys that mike on the basis of looks alone may regret not trying it out first.

Semi-pro mikes

In-between the two extremes (pro or toy) are the

Impedance & sensitivity

Most devices in a circuit present some kind of electrical resistance (whether weak or strong) to the current passing through it. In AC systems (which includes audio signals) this resistance is known as **impedance** (or the symbol 'Z').

You'll hear references to two different types of impedance in PA – 'source' and 'load' – depending on the role of the device in question in the system. A microphone is a 'source', as it delivers a signal to the 'load', which could be a mixer, or an amp, or a speaker. (Many units in a system will of course have both a 'load' impedance at their inputs, and a

'source' impedance at their outputs).

To create an ideal 'match', the load impedance should be at least ten times higher than the source impedance that feeds it – this creates the maximum voltage transfer between the two devices. For example, a professional, low-impedance microphone will usually have a source impedance of less than 300 ohms, and should be fed into a mixer input (for example) with an impedance of around 3k ohms (which may be labelled 'low impedance').

If a high impedance mike (10k ohms+) is plugged into a low impedance input, some signal loss will occur. (High-Z mikes

are also more vulnerable to interference and noise when using long cables.)
(See also 'Impedances in context' in Ch5.)

Sensitivity refers to the amount of input voltage required to obtain a certain output level of sound. Low sensitivity means more volts are needed.

Capacitor/condenser mikes have high sensitivity, so they don't need so much amplification to bring them up to a decent level – unlike, say, ribbon mikes, which have very low sensitivity. Moving coil/dynamic mikes are a bit more sensitive than ribbon mikes, but will still need considerably more amplification than condensers.

signal sources

mass-produced semi-pro mikes – these may look like professional mikes, and may even have an XLR connection, but they'll usually have unbalanced wiring, to keep the cost down. For the same reason they may also have a high impedance output – with the result that, if you use cables that are over a few yards/metres in length, the top end (high frequencies) of the signal quickly start to be lost.

If in doubt, either test the mike yourself with different cables (noting the treble differences between using a short two-yard/metre cable and say a longish 32-yard/metre one), or ask the manufacturer – quoting the exact model/version in question.

This is a good place to point out a rarely mentioned hazard – if you plug an unbalanced mike into a balanced input on a mixer with phantom power applied, it can damage the mike.

If you're unsure whether a particular mike is balanced or unbalanced, first try asking the maker's or distributor's technical department. They can tell you any identifying numbers or features to look for.

A practical test is to plug the mike into a balanced input (making sure the phantom power is *off* first), and hear what it sounds like – if the sensitivity is poor (see sidepanel p13), and there's added hum and buzz, it's probably an unbalanced mike (either that or a faulty balanced one).

If, for whatever reason, you have no option but to use an unbalanced mike, it's possible to get an accessory transformer (usually a small box or canister) that will convert it to balanced operation, and also to low impedance, if it's high. This is almost the same as using a decent mike in the first place, but with more messy wiring and slightly lower overall performance.

Headset/head-worn microphones

Headset systems – where a head-worn attachment positions a miniature mike in front of the mouth – are fashionable items with the more gymnastic performers who need their hands free while they're singing.

In actual fact there are a limited number of headset mikes that are high enough quality to make good vocal mikes – some are even used by performers who mime (because they say their dance movements are too intricate to sing over...). No names, please.

They can look a bit silly, but one bonus is that head-worn mikes have removed the need for singers to clasp all-too-obviously phallic objects and point them down their open throats – if they don't want to.

Headset mikes can also suit or help some singers and acoustic players who have problems with straying 'off mike' – accidentally drifting away from their microphone at inappropriate moments.

The related downside of head-worn mikes is they make conventional microphone technique largely impossible, as there's little the singer can do from note-to-note to alter mike position and distance from the mouth.

Head-worn mikes need to be professional models, with balanced outputs and beltingly-high SPL handling capability – not the inferior ones made for telephonists, which can look very similar. Those with a capacitor capsule may have (or require) a belt-mount battery pack for powering. This will be essential if, like many headset mikes, the link is by radio/wireless (there's more about this later in the chapter). Some head-worn mikes incorporate a single headphone for monitoring.

Pro standards

Other than being rugged, stage mikes have to conform to a few basic professional norms – unless you enjoy sorting out problems. These are mainly as follows:

Balanced v unbalanced

Unbalanced connections (as found in domestic and non-professional systems) use cables with one core wire to carry the signal plus a braided metal 'screen' or 'shield' wrapped around it (underneath the rubber or PVC covering). This screen operates as both a barrier against external noise interference and a return path for the signal – and inevitably fails to do either entirely efficiently.

Balanced wiring involves three separate wires (and usually XLR connectors). Two core wires carry the signal: each has opposite polarity, or 'split phase', meaning one is positive, the other negative (hot & cold), which helps to cancel out any noise that may leak into the cables on-route. (The signals are then put back into phase when they reach their destination.) There's also a braided screen, which operates solely as a noise-shield in this case. The result is a cleaner and much more reliable signal.

- The rear end should have a three-pin (male) XLR connection (also known as a Cannon connector), so that any industry-standard three-pin XLR-M to-XLR-F (male to female) cable can be plugged in. Old or oddball mikes have some rather individualistic connectors that may look vaguely similar to XLRs, but aren't. They won't be as tough and, if they break, no one will have spares; and cables won't be interchangeable. So don't even think of using them – even if the mixer/stagebox end of their dedicated cable is fitted with an XLR plug.
- The mike should be wired for balanced operation. This allows it to connect to professional gear's mike inputs without wiring problems or hums and buzzes.
- The mike should be low impedance ('low-Z'). Its 'source' impedance may typically vary between 150 and 500 ohms – see sidepanel for more on impedances.

Whatever the value, a low-Z mike's impedance is always low enough (compared to a high impedance mike) that it doesn't greatly affect the sound when the cabling to the mixer is long, even over 100 yards/metres. A mike splitter (see later in this chapter) further helps to isolate the mike's sound from the effects of long cables.

Directional sensitivity

The directional response of a particular microphone model is usually described by referring to its 'polar' pattern – so-called because it's shown on a grid of concentric circles, not unlike looking down at the north pole on a globe of the earth. The microphone is imagined in the middle of this grid, and the response pattern shown drawn around it. The grid lines move away from the mike where the sensitivity is higher.

Most mikes used in PA are 'pointable' – they have a more-or-less 'uni-directional' response, to help avoid extraneous noise leaking into the mike from elsewhere on the stage.

On a polar diagram, the typical pattern for this type looks roughly heart-shaped – hence the common name 'cardioid'. If you imagine the mike capsule is pointing into the indent at the top of a (cartoon-style) heart, the pattern shows that the mike is helpfully most sensitive from the front, where the desired sound source is, slightly less

from the sides (still enough to allow the odd bit of microphone sharing by a vocalist), and relatively deaf to sounds at its rear end, where the cable exits (and where unwanted and potentially problematic noises lurk – such as monitors, and the audience).

The other main polar patterns found among stage mikes are: omni-directional – meaning all-around; and super/hyper-cardioid – which has an even narrower focus than a cardioid, with less spread-out frontal sensitivity. Omni mikes are used for picking up ambient sound, including natural (room) reverb and the audience (for a live recording, for example). Hyper-cardioids are best for ignoring instruments on either side, but are more fussy as to placement because they have a weak spot: the pattern can show a 'rear lobe', which indicates increased response at the back – so their rear end would also have to be kept from pointing at loud unwanted sounds.

A microphone's response pattern changes as soon as someone picks it up – even if it's designed to be hand-held

Mikes with a 'figure-eight' response pattern are equally sensitive to sounds from the front and back. This becomes useful for stage work when you have two vocalists sharing the same mike, singing face-to-face.

It's important to be aware that a microphone's response shape changes as soon as someone picks it up – even though most mikes are designed to be hand-held. This is soon learned when a nervous singer grasps the mike and has to deal with wild gestures from the monitor mix person as their monitor speaker (let alone the out-front PA) starts to squeal and howl.

Generally, the problem happens if any of the slots, grilles or apertures on the mike's body are covered up – this can change some peaks and dips in the mike's frequency response, and in many situations it is enough to trigger acoustic feedback.

Some mikes are designed to be particularly

signal sources

good at noise-cancelling – reducing high levels of ambient noise – which may be useful for getting separation when miking parts of loud drum kits, or for use with bands that have very loud monitoring.

A few mikes have a switchable response pattern, so by flicking a switch they can be converted from, say, a cardioid to a figure-of-eight pattern mike.

Proximity effect

A feature of most directional mikes is that the bass end of the signal is boosted when the mike is brought very close to the source. The obvious use of this proximity effect is for singers who want that intimate, smoochy, late-night vocal tone, or the bass-rich 'Elvis in Las Vegas' croon. This is widely experimented with and used by most creative vocalists, who make it part of their microphone technique.

The snag is that breath noises and 'explosive' consonants, like 'p' (which cause popping sounds) are also boosted, making careful technique all the more important.

Some microphones have a switchable setting that can minimise the proximity effect. A few mikes have a special design that prevents the effect (though the mike remains directional) – sometimes these types can permit higher levels of gain before feedback.

Handling noise

Handling noise can include the rustling when a mike (or its lead) is stroked, and the thudding sound when a mike is clipped onto its stand. Good mikes will keep these noises to a minimum,

only audible when the mixer channel used has a high gain setting.

Low handling noise is even significant for mikes that aren't being directly held, as sound vibrations from the stage-boards can travel up microphone stands.

Vibration transmission is also a concern with mikes mounted directly onto instruments. As with DJ sources (which we'll come to later in this chapter), isolation is a good thing. A mike's isolation can be increased with a 'shock-mount' holder or adaptor.

Another possible cause of noise is an on/off switch. Not all mikes have these fitted – their usefulness (for instance, giving the performer personal control over muting) can be outweighed by the complications they can introduce (for instance, giving performers control over muting... which might result in accidental or inappropriate use).

On-off controls can also cause a loud, alarming pop when switched – something which must be ascertained before use, and avoided while the volume levels are up.

This is why mikes with switches are not liked by PA people – except maybe for 'talkback' mikes so the engineer can communicate with the people on-stage.

Reflections

Even though a mike may be directional, when setting its position and angle you have to consider the reflective qualities of nearby objects.

In general, small and (especially) shiny objects/areas are enough to reflect high frequencies; it takes larger objects – several square feet in area – to reflect mid frequencies, while bass will only be affected by wall-sized areas.

Decibels (dB)

Decibels are not actual amounts. They are 'ratiometric' – a way of describing a ratio between two values, or their relative levels. They're used to compare voltages, watts, sound pressure levels etc, each identified by a suffix (eg dBSPL).

Once a reference is given, a logarithmic scale is applied, and comparative figures

can be assessed. For instance, 0 dBSPL is accepted as the lower threshold of hearing – and in this scale each addition of 6dB represents a doubling of the sound pressure level, while adding 20dB means a tenfold increase in the level. This shows why dB tables can be misleading – for instance, 90dBSPL is actually ten times 'louder' than 70dBSPL. Loud music

is normally somewhere in this region, though excessive sound systems have been known to peak much higher, at 120dB or more – which is considered to be the threshold of pain, and hearing damage.

For other uses of the decibel as a measurement of ratio, see the Glossary towards the back of the book.

Reflections cause colorations – possibly unwelcome and complex tonal changes.

Maximum sound level handling

It's not mentioned enough by microphone makers that the sound levels encountered by mikes positioned close to powerful acoustic instruments (such as kick drums and loud brass) can be high enough to push many lesser mikes into overload – sometimes way above 125dB-SPL RMS or 140dB peak (see Appendix C).

When a mike overloads it's generally 'graceful' – less unpleasant-sounding than an overloaded amp or a digital circuit. But overloading any mike is still not a good idea, assuming you don't want to hurt your audience's ears. Prolonged overload might also ultimately damage the capsule (the mike's delicate guts).

Many mikes used for PA are chosen precisely because they can stand very high sound levels. With some high-end capacitor mikes an attenuator can be fitted (as a screw-fitting modular section) to increase the maximum level handling. Usually this is by 20dB (a ten-fold increase – see p16 panel on decibel scales), say from 130 to 150dB-SPL, which should be enough to handle almost any instrument – though not necessarily, if for instance you poke the mike up inside the bell of a loud trumpet.

Multiple miking & polarity

Sometimes miking a particular instrument requires two or more microphones – for instance above and below a snare drum (more of which later), or to create a deliberate stereo effect, or because the instrument is very large (a piano, for instance), or perhaps it's just the only way to get the right sound – mixing say a 'crisp' and a 'dull' mike signal. Sometimes the second mike may just be picking up spill, which could be used creatively in the mix.

In any multi-mike situation, the mikes must usually be 'phased' in a particular way (individually as well as compared to each other – which means wired with the same polarity, with positive and negative connections matching up. (There's more about this in Chapter 7). This will usually be the case anyway, but if there are any hidden wiring polarity/phase 'flipovers' (transpositions) in one of the mikes, cables or even the mixer channels, then the mikes may be 'out-of-phase'.

To check for this – when using, for example, two mikes – the two signals should first be compared for loudness separately at the mixing desk, then combined together at identical volumes. If the overall level (and particularly the bass end) seems to drop, then one is 'out of phase'. The cause may be the wiring inside the mike, or the cabling used.

If the overall level seems to drop, particularly the bass, one of the two mikes is out-of-phase

The true polarity of these items should also be checked later (see Chapter 7 and the Glossary for more on the subtle difference between phasing and polarity). For now, all that's needed while miking up is to remember that mike polarity/phasing can be changed either by using a switch on the mixer or, more securely, by using a 'polarity transposition lead'. This is a short in-line (male to female) XLR adaptor cable with the positive and negative signal wires crossed over.

Whenever the optimum polarity is uncertain, try flipping the 'phase invert' switch ('polarity changeover' is another name for this) on just one of the two mixer channels concerned (see Chapter 2) and listen for the position that gives you the best result.

Every now and again, the out-of-phase (reverse polarity) position may sound better, or may achieve things you can't get any other way. One practical example of this is miking a pair of violinists playing side-by-side: their individual mike signals can be physically connected together using a Y adaptor XLR cable (see chapter 7) – this has two XLR cable sockets that come together and share the connections inside an XLR cable plug. The polarity of one mike is reversed by passing it through a polarity changeover lead.

This arrangement makes the two violinists equally audible, while cancelling out most external sounds to a useful extent. This same method also works well for vocal duets.

Another example of using a polarity (phase) reversal switch (or a reversal cable) with two mikes is when miking a snare drum. The lower mike (inside) must be set so its polarity is reversed, or else the 'thwack' will be lost by cancellation. That's because the main signal received underneath the skin has the opposite polarity.

The situation is different when a third mike is added to the 'miking-up hole' that's in the side of some modern snares. The best polarity for this mike is not so obvious, and may be experimented with to find the ideal sound.

Experimenting with the polarity of some of the mikes on a drum kit can increase the rejection of unwanted sounds ('spill') on some of the sources.

Miking outdoors

At outdoor gigs, if there's any wind above a slight breeze, windshields (windscreens) will be needed for some mikes to keep the noise down. And remember that rainwater will easily damage or destroy a mike, as well as upsetting its acoustic performance. A simple and surprisingly acceptable solution is to fit a condom over the mikes (it makes a good talking point too).

Mike stands

Microphone stands need to be rugged, stable, quick to put up and put away, and corrosion proof (nice shiny chrome can hide 'speed-rust' steel). They also need to be easy to adjust and versatile, useable in as many situations as possible, suiting different types (and sizes) of performer.

Remember sometimes, for speed, the artist may need to make adjustments to the mike stand him/herself, and when they are busy performing, having to fiddle with complex knobs and levers is not especially helpful.

Straight, vertical stands are suitable for singers, horn players, or any situation where the closeness of the stand itself won't impede the performer. Straight stands generally keep their balance quite well, so only require small feet.

Some of the sturdiest ones are those with heavy, often round bases, which anchor the stand on the stage. Most mike stands can be adjusted for height (some require only one hand to do so, and may operate by a kind of pneumatic system).

For many miking situations you will need stands with a 'boom' arm – basically an extendable attachment that can be swivelled and angled in any direction. This means most of the stand can be kept well out of the way, while the mike can be positioned freely, for instance sideways over a grand piano or pointing at a hand drum.

For many miking situations you will need a mike stand with an extendable boom arm

Boom stands are also useful for artists who sing and play instruments, so the vocal mike stand won't impede their playing.

The disadvantage of boom stands is they can topple over if you're using a heavy microphone at full extension. To compensate, some boom stands have very long and splayed-out 'feet' for extra balance – but these can quickly clutter up a stage, or cause a tripping or snagging hazard, if not positioned with some care.

There are also specialised 'floor stands' which comprise a heavy base with a stubby upright stem – useful for miking-up bass drums, or the bottom of pianos, or a bass-cab cone next to the floor.

Different mikes require different attachments for securing to a mike stand, but adapters can be bought to convert unsuitable fittings to make them compatible.

A selection of these should be carried in case an unusual microphone/stand combination is encountered at a gig.

Cables

Professional quality cables have XLR-F (female) to XLR-M (male) plugs at their ends, just like line cables. The difference is that the cable must be even sturdier, able to stand up to possible misuse by performers, such as twisting, twirling or tugging.

signal sources

Rubber cables (or more accurately synthetic butyl rubber) are the most rugged sort, but are usually limited to black or a few drab colours. Cables sheathed with plastics, such as PVC, can be softer and more flexible, and brightly coloured, which can be useful either just for show or for colour-coding (different-coloured leads make it easier to track different signals quickly). But these leads generally won't last so long or stay in good condition when intensively used and/or abused on the road.

Even if the outside sheath is tough, the shielding (braided wire screen) inside ordinary microphone cables degrades with use. At first you may get hums or buzzes if a damaged section of shield (a shield 'hernia') rests over a mains cables on stage. Eventually the shield breaks entirely and the mike will hum all the time.

(To overcome this, a new screening/shield material called 'Suprasafe' has been introduced, made from specially conductive nylon, which handles severe twisting and similar abuse without getting torn, ripped or warped. It also happens to be lighter than standard cable.)

Pro audio-type cable drums (with brakes to stop over-run) are useful for managing especially long cables. They are also one way to store a number of shorter mike cables, plugging the XLR ends together, male-to-female, and reeling them in.

Wireless (radio) mikes & pickups

When they were introduced in the mid-1980s, 'radio' mikes and wireless guitars radically changed the nature (and potential intimacy) of live performances, enhancing artist's freedom to leap across large stages, walk into audiences, prance along catwalks etc.

But any sort of radio transmissions have to be tightly regulated, because it's very easy to cause interference to other broadcasters, and the airwaves could quickly become chaotic.

For broadcasting over a short range with low-powered radio mikes and wireless pickups, a variety of highly specific and narrow bands of radio frequencies have been allocated in different countries.

The FM (frequency modulated) signal they transmit is similar in quality to a live VHF FM radio broadcast (with almost full audio bandwidth, typically up to 16kHz), and is usually received in the stage wings by one or more aerials/antennae. The receiver converts the signal back to audio at

Using radio/wireless mikes

Although there is a modicum of international agreement, licensing wireless PA/stage gear can be a nightmare for those working across national borders – even in the supposed 'common European market'.

Even when working local shows, PA users often choose to *hire* radio mikes, leaving all the licensing paperwork to the supplier.

Certain frequencies are de-regulated in various countries, so gear using these frequencies can be used without paying for a licence, provided the radio mike system complies national standards that ensure it won't interfere with, say, with any of the emergency services or other users.

Most VHF and UHF frequencies require a licence. But some radio/wireless systems (possibly ones bought at a suspiciously low price) may broadcast on frequencies that aren't legally available. These won't be officially licensable, and are used at your own risk. They might, for example, block a vital ambulance or marine transmission that could cost lives. If operated well away from any other radio users, harm is unlikely – though be wary that such equipment can be confiscated and those deemed as 'the user' could be fined or even prosecuted.

Whatever the system, the frequencies may need to be changed in a hurry, if interference occurs just before the show or even during a set. A good system, and wise user, will anticipate this.

Tips from the top

Some advice to help you avoid panic when wireless mikes act up:

● First check with DJs, MCs and other acts who might be using a radio mike of their own. Find out what channel they are on.
● Keep a 'crib list', of the frequencies you're licensed to use – tape it to the receiver.
● Before the show, leave the receiver channel on and open as much as possible, even while the mikes are switched off. Then you'll stand a good chance of hearing if anyone else is using your frequency before it can become a problem.
● Have a standard 'cabled' mike as a standby, already sound -checked and ready to go, in the wings. Sod's law dictates that your wireless mike will probably now work fine.

signal sources

line level. It is then handled like a DI box source (more of which a bit later).

Equipment operating on these frequencies should be officially authorised for stage use – in the UK by the Department Of Trade & Industry (DTI), and in the US by the Federal Communications Commission (FCC). In some cases the equipment can be operated for free, and in others on purchase of a licence/license (which can be annual or even for a number of days).

Radio transmissions have to be tightly regulated, because it's very easy to interfere with other broadcasters

---

Wireless gear is available in a choice of operational wavebands – either VHF or UHF (Very and Ultra-High Frequency). The main difference is UHF has shorter waves and is more readily absorbed by (and so contained by) buildings – which means in dense urban areas and multi-gig venues/events, UHF sets can offer the best chance of finding a spare, 'clean' channel, because there's less competition from other radio frequency users. UHF systems are also typically more expensive.

Capture effect, diversity and range

A main benefit of using FM (rather than simpler AM radio transmission) in all stage wireless gear is 'capture effect'. This means the strongest signal (hopefully that of the mike transmitter) will largely 'wipe-out' any weaker radio signals, so there is relatively little worry about interference – even when there is no signal in the mike (in other words FM can broadcast 'silence').

The downside of capture effect is that if a radio mike's signal isn't the strongest, or if it falls below the level of prevailing RF (radio-frequency) noise, the wanted signal will drop out and be completely replaced by an unwanted stray signal.

The high radio frequencies used, particularly with UHF, are also easily reflected by metal and solid objects, which can cause reception anomalies – such as a second, delayed signal or total signal loss, due to the level dropping below the 'capture ratio'.

One solution is 'diversity'. This means having two aerials/antennae feeding same-frequency inputs on the receiver, which switches automatically (and usually noiselessly) between one and the other signal, seeking the strongest of the two. This solves most problems with reflections and signal cancellations. More sophisticated UHF systems have 'frequency agility', meaning they can automatically change their frequency if there's any competition on the one being used. (This is also sometimes called 'true diversity'.)

Diversity is less likely to be required with lower-frequency VHF, or where there are fewer hard surfaces or obstructions nearby – as in rural areas, and for open air performances (under a wood or fibreglass-framed covering, as long as there is little in the way of reflective metals, brick or concrete in the performing area).

The range of radio mikes is commonly designed to be at least 100 metres (330 feet). Few stages are this wide or deep, but as this range applies only under ideal conditions, it provides some safety margin. If the performers' aerial isn't polarised ideally (meaning oriented fairly vertically, like the receiver's aerial should be), then the wanted signal level can be reduced, as can the range (distance) over which capture can be maintained.

Some 'intelligent' wireless systems can be programmed to recognise an identifying tone, so the receiver knows it has the right signal, and other signals can then be ignored. This overcomes capture problems, and improves reception quality in difficult environments – for example an outdoor festival where different PA companies might be using the same frequency for different stages less than 100 yards/metres apart.

Transmitters and batteries

The transmitter (the bit that sends the wireless signal) can either be inside the mike body, or 'belt-mounted', as can wireless pickups. The belt-pack, which needs to be of rugged construction for

signal sources

regular use, is usually mounted on the side or rear of the artist's waist or hips. (An unusual social feature of beltpacks is that a somewhat intimate personal zone has to be entered when PA crew help fit, adjust or diagnose them. So a special professional etiquette – akin to a tailor's or make-up artist's – may apply.

All radio mikes and beltpacks depend on batteries for power, which of course tend to run down (solar trickle-charging has yet to really happen). Radio mikes in turn can stop quite abruptly when the power source fades – suddenly not hearing yourself can be quite a psychological blow to a performer's confidence.

Modern radio mikes claim to work for typically between ten and 12 hours, or longer with the latest power-saving models – but it could be much less if consumable batteries are out-of-date or have been partly discharged, either through over-use or lack of use, or, if they're re-chargeable, they were not fully re-charged. (Some receivers are fitted with a battery charger for one or more transmitters.)

On older and low-budget radio mikes, battery metering and 'about-to-run-out' warning lights are often primitive, if supplied at all. But the latest up-market models have battery 'fuel gauges' on both the mikes and, more importantly, at the receiver, so the engineer can monitor them.

Sometimes just moving the receiver a few feet to one side can make all the difference

---

Still, seasoned users don't take any chances and use fresh batteries every day. If using expensive throw-away batteries, buy and use a handheld battery condition meter – low-cost types showing general condition, with coloured bands of say red and green, are fine.

Experiment or learn from experience about how many hours of use remain with given readings (for a given transmitter pack). Never assume a battery is full – use the condition meter to check that new or freshly charged batteries are not in fact partly discharged. (Part-used batteries needn't be thrown away – they can be saved and used for something less critical, in the office or home.)

It's rarely mentioned that re-chargeable batteries will stop working very suddenly – whereas consumable batteries fade, giving you more warning. Rechargeables also require management if they're to work smoothly and last longer. Each type of battery technology requires different treatment. Read the maker's instructions. The most common NICAD type batteries should be flattened/fully discharged, before being re-charged. Some PA professionals insist that rechargeable batteries and stage radio wares don't really mix and that the combination is best avoided altogether.

Wireless mike system features

As for the microphone itself, wireless/radio mikes made by the conventional mike manufacturers are usually based on an existing model. They may look similar, other than perhaps having a longer body and an absence of cable running from the end.

Other makers may use generic capsules (the mike guts), or those taken from well-loved mikes, and fit them into their own mike casings.

In all instances, it's best to use a system built to professional standards. This should include:

- An RF input socket, allowing either a direct aerial/antenna connection, or connection via an RF coaxial (co-ax) cable
- Readily selectable radio channels (frequencies).
- A/B diversity indicator showing which aerial/antenna (of two) is in use
- LEDs to indicate when RF (carrier signal) and audio are present, to help with troubleshooting. Some high-end systems have optional monitoring that runs on a computer. Advantages include signal strength being recorded during a walk test before use, allowing you to work out the best aerial/antenna site.
- A sensitivity switch, so RF signals that are too strong and will cause distortion can be reduced.
- A balanced output through an XLR – many systems provide 'consumer grade' unbalanced outputs on low-budget jack sockets, creating more problems than they are worth. Avoid these, or see if you can have the output adapted

signal sources

to be balanced, possibly using an add-on balancing transformer in a box.

- 'Squelch' control – works rather like a noise gate (which we'll come to in Chapter 3). When the signal level is below a certain threshold, showing that there's nothing useful being received, this mutes the signal, preventing any noise appearing on the channel

Wireless practicalities

To find the best reception, you should experiment with placement of the receiver and positioning of aerials/antennae. As mentioned already, the more advanced wireless gear allows the received RF level to be recorded during a walk test, so PA crew can work out the best receiver site.

Avoid obstructions, especially anything hard or metal, between the transmitter and receiver

Sometimes just moving the receiver a few feet to one side can make all the difference.

Avoid obstructions between the transmitter and receiver (especially anything metal or hard). The receiver should be situated at least 4ft (1.3m) off the floor, and as close as possible to the artist – particularly if the conditions are difficult, with lots of strong neighbouring interference.

If this isn't possible, then at least the receiver's aerial(s)/antenna(e) should be sited close – perhaps hanging above the performers' heads, and connected to distant receiver(s) with good quality co-ax cable. Up to 100ft (30m) is OK (at least on VHF) without significant signal loss.

All outboard aerials/antennae must be dipole types – in other words symmetrical pairs (a bit like insect antennae).

In cases of reception problems, compare the polarisation (orientation) of the receiver aerial(s) against the mike's own. VHF systems in particular need to have a similar orientation – if, for example, they are at right angles to each other, the signal reception will be weakened. For stage mobility, this means a generally vertical or slant position is needed. UHF systems aren't as fussy about this.

A lot of radio/wireless mike equipment tends to be lightly built, like domestic consumer gear – partly because designers have become obsessed with keeping down the weight of beltpacks. This often means that receiver sets are cased in fragile plastic, suitable for desktop use but not very roadworthy.

Pro-standard receivers come in rack-mounting metal cases, which are not only stronger but also won't fly through the air when someone trips over a cable. They tend to be equipped with springy 'rubber duck' antennae, as opposed to rigidly attached telescopic aerials which are easily snapped off.

For reliable 'day-in, day-out' stage work, aerials/ antennae should be plug-in types, so they can be changed easily if bent or broken. It's sensible to keep spare antennae, belt-pack parts, and battery packs in your accessories case, and to check all elements regularly for damaged connectors, dirty contacts, etc.

Finally, artists and PA engineers alike need to be aware that merely placing/cupping a hand over/around a UHF radio mike's tiny 'stub' antenna at the base of the mike greatly reduces or even kills the signal.

Instrument pickups

These are not usually the responsibility of the PA provider/engineer – generally, if they're going to use a pickup rather than just a microphone, musicians will already have chosen and fitted a particular type of pickup for their instrument. (In recent years, for instance, high quality electro-acoustic guitars have become available, which are basically acoustic six-stringers with built-in pickups.)

For higher budget tours, it would be the production company's responsibility to organise fitting pickups to say, a top artiste's piano, or perhaps for a brass section.

Occasionally the PA provider may be asked to help, but all parties would probably rely on specialist assistance.

Guide to mike placement

Other than opera singers and orchestras, most live performers will need to use microphones in order to be heard over any reasonable distance. Even electric guitarists, who have their own amplification, will need a boost through the PA if the venue is anything beyond what you might call 'intimate'.

By close-miking (see sidepanel below) or DI'ing (which we'll come to in a moment) every individual sound on-stage, the engineer has (theoretically) more control over the relative balance and overall volume of all the instruments and voices – though of course it does make mixing more complex, and potentially troublesome.

The following guide provides just a starting point to the art of mike placement – and it is an art, arguably the most important skill a sound engineer can possess. With the help of people who have 'good ears', experimentation can be rewarding. There's no law about mike placement – you won't get arrested if you get it wrong. Just try not to do it again…

As for models of mike used for particular jobs: observe what other bands and engineers use (you need to develop keen eyesight and recognition abilities as well as listening skills). Make notes of sounds you like and how they are achieved. If we were pushed, and had to recommend, say, five or six mikes to start your collection with, you wouldn't go far wrong by experimenting with accepted classics like the Shure SM58 (or Beta 58), SM87, Sennheiser MD421, Beyerdynamic M88, and AKG C451, or 414. You'll be able to cover most jobs with these.

Learning from experience is obviously the best education, but here are some general tips on miking, passed on by several well-travelled sound engineers:

Vocals (often abbreviated to Vox on mixers)

The classic set-up is to arrange the mike on the stand so the mike's 'tail' end is pointing into the wedge monitor.

As every singer has different abilities, techniques and projection, you have to be prepared for unexpected deviations from this

---

Close-miking

As soon as you use mikes live in the same room as PA speakers, feedback (howl) can cause problems. This classic situation is described technically as 'not being able to get enough gain (hence sound level) before feedback'.

The problem is avoided in five ways, preferably in this order:
- by use of directional (as opposed to omni-directional) microphones
- by careful positioning of the mikes – in particular, by the vital practice of close-miking: placing the microphone as near as is sensibly possible to the person/ equipment creating the sound
- by using mikes with a fairly flat (even) frequency response, whenever possible; and finally, as a last resort…
- using EQ (expanded on in Chapters 2, 4, 8 & 9).

Close-miking maximises the direct signal that the mike picks up, at the expense of the peripheral sounds you don't want. Because the wanted sound level is also higher because of close placement, the PA system's gain (amplification) for this channel can be lower – which should mean more volume can be used before any feedback that might occur.

An inevitable feature of close-miking is that small changes in the mike position can have a disproportionately large effect on the level. This affects any players or singers whose mike isn't physically attached to them or the instrument they're wearing. Signal compression (more of this in Chapter 3) may be used to help to deal with this, but skilled vocalists learn to try to keep a constant distance from the microphone.

On the other hand, they may also deliberately and intelligently vary the mike-to-mouth distance to change the volume and timbre of their vocals (known as 'microphone technique').

An instrument's actual sound can also change markedly with subtle re-positioning of the mike, so it's useful to be able to listen (or have someone else sound-checking) while you experiment with mike placement.

An unexpected result of close-miking in general is that most instruments sound 'wrong', or at least different to the way they do when heard from a 'normal' distance. That's why corrective EQ may be required (there's more on this in Chapters 2, 3 & 8).

The exception to the rule about close-miking is with classical concert orchestras, which tend to produce adequate volume that needs no PA (until the concert hall or lawn expands to a park or stadium setting).

The inherent high volume and instrumental balance in a well-arranged orchestral piece means overhead, ambient miking may be sufficient – although close-miking can be used to bring out particular instruments as and when required.

signal sources

norm. You'll also need to experiment with things like mike-to-mouth distance.

Guitar – electric (Gtr, El gtr)

Usually this involves miking the guitarist's speaker cabinet or combo amp (either of which may be referred to by their speaker combinations – for instance '2x12', or 'two-by-twelve', meaning a cab with two 12-inch diameter speakers).

Place the mike by the drive unit/speaker that's nearest the floor. The drive units that are higher up off the floor typically suffer from cancelling sound reflections, which cause a hollow sound.

Point the mike at the cone halfway between the dust cap (central dome) and the surround (outer edge). If it's too near the centre, the sound gets harder, while aiming the mike too near the edge of the speaker picks up a softer sound.

Guitar – acoustic (AG, Agtr)

Where a pickup isn't used, mount a mike (on a boom stand) a bit below the soundhole (in the front of the body), pointing up into it. The tricky part is to get close enough to maintain tonal balance and level while allowing the guitarist's hands to move. Try tie-clip mikes (like newsreaders wear on TV) attached to the soundhole, or 'bug' pickups (wired or wireless).

Otherwise, capacitor/condenser mikes are recommended as they seem less sensitive to spacing than dynamic ones. They also better suit the subtler sound of an acoustic guitar, whose frequency range goes higher and lower than its electric cousin.

Bass guitar (Bass, B)

If miking a bass guitarist's speaker cab, the mike is placed close-up and near the edge of one of the drive-unit cones that (preferably the one nearest the floor). If the front of the cab has a grille or mesh, this may need to be removed to get close enough, and also prevent picking up the rattling and buzzing of a typical grille.

Mikes used for this job need to be able to withstand very high sound pressure levels, as well as handle low frequencies, and are often the same as on kick drums.

If you're also using DI for the bass (more of which shortly), this can produce a notably 'round', low-mid-heavy tone, so the miked signal can add some useful top-end 'twang' to help the bass cut through.

Drums

Drum kits are practically always miked up as a series of individual drums, rather than just one instrument. If mike availability is severely limited, it is possible to use just a 'stereo pair' (or even one stereo mike), positioned with care.

(It was good enough for Led Zeppelin, after all: legend has it that's what they used to record John Bonham's drum kit on the classic 'When The Levee Breaks' – a drum part that's been much revered, and sampled, for many years.)

The trick is to achieve the best overall balance, clarity and feel – and results can vary depending on the acoustics of the venue.

If you are going for this minimalist approach, try starting with the mike(s) a few feet in front of, and pointing towards, the centre of the kit, then experiment with placement till you achieve the best result.

Mikes used for bass or drums need to handle high sound levels

Bear in mind you'll pick up lots of spill from other instruments, which limits and muddies your options at the mixing desk. A screen around the drums can help a bit (though it looks unsightly).

With close-miking of individual drums, you have more separation and control, and the

---

Tip from the top
Educating hitters

Many drummers sadly don't know how to tune a drum kit, or else they seem to like a 'cardboard box' sound. Or their toms sort of go 'spht spht' and they expect the sound engineer to make them go 'Bhamm! Bhaam! Bhoom!'. Yet they won't re-tune or fit new heads, or remove the bottom head (which can cause messy reverberations). If the drummer resists your best advice, it may be necessary to try getting the rest of the band on your side and exert a bit of friendly pressure that way.

ambience of the room is less important (though this can be mixed in, if desirable, by using 'ambient' mikes positioned further away from the kit). New problems present themselves, though – for instance close-positioned drum mikes are at risk of being hit, and need to be placed with some feel for the drummer's working space.

Close-positioned drum mikes are at risk of being hit by drumsticks

Some drums have pickups built in, but these generally don't sound as good as using mikes. Some mike makers do produce accessory clips to help mount ordinary mikes onto the different drum parts.

Electronic kits (or, more accurately nowadays, digital-modelling/virtual kits) remove normal miking problems, though their sheer versatility (in some cases EQ, FX and even room ambience can be altered at source) can add more layers of complexity to the mixer's job, unless the drummer is skilled in programming the kit, and keeps the engineer well informed as to possible changes in sound and level.

Kick/Bass drum (Kick, Kik, KD, BD)
Place a cushion/pillow/blanket inside the drum, if this is possible, nestling against the rear head (the one that's struck) to prevent excessive reverberation or 'ringing'. The mike will either be on a floor stand, pointing into the drum, or resting on the cushions inside. Experiment for the best sound from particular kits.

Nowadays there are several microphones designed specifically with low-frequency, heavy-duty bass drum work in mind (these are also often suitable for bass guitar cab miking), but there are no strict rules about what mikes are best for which application. If it sounds good, it is good – though bear in mind that some mikes are not built to withstand the constant battering (sonic or physical) they may get on tour.

A classic example of this is the Beyerdynamic M88, introduced in the 1960s as a general-purpose moving-coil/dynamic recording studio microphone, but soon adopted by live engineers because of its particular ability to reproduce a great bass drum sound. The drawback was that it hadn't been designed for this job, and couldn't survive such abuse for long.

Eventually, in response, the company redesigned a sturdier version – though some engineers still enjoy the sound of them being driven into overdrive, which can prove expensive (and explains why they tend to be used on 'high-budget' tours).

Snare (Sn)
The mike used has to be fairly small so it doesn't impede the player, or get hit. Mount the mike (on a boom stand) above and opposite the playing area, with just the mike's front hanging over the drum. A second mike may be placed underneath – though bear in mind what was said earlier about the possible complications of 'multiple miking'. Again, moving coil/dynamic mikes are generally considered best for snare drums.

Toms/tom-toms
There may be one, two, three, four, five, six... or none, if the drummer is feeling minimalist. The standard arrangement (on a five-piece 'rock' kit) is two smaller toms – either attached to the bass drum ('mounted' toms) or on a separate rack unit – plus the 'floor tom', a larger, deeper drum standing on its own adjustable legs. Variations are endless, though. Again, small mike size can help – sometimes 'tie-clip' attachments are used for maximum proximity with minimal obtrusion.

Hi-Hat, cymbals, other percussion
There are two approaches here: close-miking, which gives more separation and therefore more control over each individual cymbal sound; or overhead miking, which is preferred when mikes are in short supply, but also because it can give a good overall ambient sound, adding a natural presence.

When using a close-miking technique on a cymbal (crash or ride), position the mike on a boom stand (placed as far away from the drummer as possible) above and to the side of the cymbal. Hi-hats should be miked over the domed area – but far enough away so the mike won't be hit by a passing drumstick.

signal sources

(As a rule, always get the drummer to play every drum and cymbal in full flow – as if performing, not just a tap-tap – so you know whether the mikes are in the best position for a good sound and their own safety.)

Overhead miking of cymbals is often preferred because it avoids a forest of mikes around the kit and an excessive use of mixer channels. If you only have two or three mikes to cover all the cymbals, use one on the hi-hats, and one or two as overheads to capture the crashes and rides (cymbals are very loud and and travel well, so their sound can be picked up accurately from a reasonable distance). Overheads (O/H) can also be used to supplement the miking of particular cymbals.

If you have two mikes to spare for overheads, and a pair of free channels, some engineers feel the best stereo image can be obtained by spacing the mikes out, and keeping them parallel to one another (not angled inwards). Others use a 'crossed pair', where the mikes' heads are close (coincident), but angled at 90 degrees away from each other (adaptors are available for mounting two mikes in this way). This offers purist stereo, but increases the unwanted leakage from other sounds on the stage.

If leakage is bad, mikes will have to be moved closer to the sources, as well as using gating and high-pass filtering (see Chapters 2 & 3).

Capacitor mikes are mostly preferred for cymbals, to get a wide and flat (smooth) high frequency response, but some dynamic models are also used.

Organ (Org)

For the low end, you'd use a similar mike to the kind used for a kick drum or bass guitar. If miking up a 'Leslie' cabinet (a box containing a revolving speaker, often used in conjunction with Hammond organs), the sound is in the same range as vocals, and is also quite rough – so a mike that gives too much detail and perfect realism isn't really wanted. A basic dynamic vocal mike will be about right.

Synths/keyboards (Keys)

These are always DI'd for best results, since any cab and amp used by the player will rarely contribute anything beneficial to the sound (that's really just for monitoring by the keyboardist).

Requests to mike-up any electronic keyboard should be met with diplomatic tactics.

Piano

The lid should be completely taken off, or else reflections from it will disturb the sound. An omni-directional mike, as might be used to mike an isolated piano in a studio, will probably be too vulnerable to leakage/spill from surrounding instruments.

In practice, capacitor mikes with a flat response are generally used on-stage – these have the advantage of higher sensitivity, which compensates for not being able to mike-up very closely.

A crossed pair offers purist stereo, but increases sound leakage

One mike can be mounted on a floor stand, close to the piano's soundboard. Another, to pick up higher notes, may be placed inside, and moved around to get a reasonably even response over all the keys. (And see 'using multiple mikes').

Various manufacturers make specialised piano mikes and pickups, including sensitive ribbon microphones.

Horns/brass/woodwind

This includes saxophone, trombone, trumpet, clarinet etc. Members of these families can be amazingly loud at their mouths (the bell), so suitable mikes are those that can handle high sound pressure levels (SPLs –see earlier about max SPL handling).

Some standard vocal dynamic mikes are suitable, though they could sound a bit hard until you 'EQ out' their presence peak. Bell mounting-clips are available for some mikes, while more mobile brass players may use clip-on radio mikes.

Strings

Delicate ribbon mikes are applicable for this job, as well as capacitor/condenser mikes, with their higher sensitivity. It's probably also worth using isolation (shock) mounts to prevent other sounds entering as vibrations up the microphone stands.

Note: miniature mikes attached to instruments or performers – variously known as tie-clip, lapel or lavalier mikes – are invaluable for dealing with some of the instruments above; partly because they're unobtrusive and partly because they enable mike-to-instrument spacing to be maintained, making for more stable sound levels and tonal quality. They also help the performer to be able to move freely, which increases musical confidence.

The stage set even looks better without a forest of mike stands around the drum kit and elsewhere. The one drawback is that some tie-clip mikes are rather semi-professional products, and may not fully meet professional audio requirements.

DI / direct boxes

DI stands for direct injection. It's not especially direct, it just means acquiring the sound (from an 'electric' instrument) in its electronic signal format, rather than waiting for it to come out of the speaker then miking it and re-amplifying it all over again.

Theoretically, direct injection would mean plugging a guitar or synth straight into the mixer's line input. But while you might get away with this in a home studio, it wouldn't be practical with a PA system.

Firstly, even the most professional muso's signals aren't usually balanced (unlike most PA mikes) – which means the various connections will interact and create hums and buzzes, and it gets worse with every added connection. Balanced connections isolate against this.

Also, if there was an electrical power fault in the musician's gear, the mixer would take a 'direct hit'. Plus the high-frequency (top-end) output of 'electric' (and even some electronic) musical instruments is rapidly degraded with more than a few feet of cable.

The solution is to insert a signal isolating transformer, such as a DI box or a mike splitter (which we'll come to next). This simultaneously provides balancing and isolation. It can also (to some extent) 'buffer' the signal, strengthening it so it can pass down longer cables to the FOH mixer without degradation.

So why not DI everything? Well, the DI'd sound isn't always ideal for all occasions. Some of

the musical qualities of electrified instruments (like guitars) are produced in part by their instrument amp and speaker. Six-string electric guitars sound very middly and clean when DI'd, which is not always what's wanted. With bass guitars, it's common for a DI'd signal to be taken in conjunction with a (more spiky) miked-up signal, and the two mixed together to get the best sound.

Types of DI/direct box

The simplest DI box is a transformer in a box, with input and output sockets. But it should be a premium grade transformer – which accounts for the cost (and weight) of a good DI box.

For fuss-free DI use, the transformer also needs to be rated to handle fairly high levels – maybe 10 to 15dB (x3 to x5) than the highest stage mike levels.

DI boxes that have developed into trusted professional tools are additionally provided with switches to set the right input level to best match the mixer's 'window'. Most also use active 'buffer' circuitry to help the transformer along and, as we've said, support driving long cable runs.

Good DI boxes are rugged, because they tend to get trampled on

Some DI boxes are passive, requiring no power – these types have no active electronics, so they tend not to go wrong. Some are battery-powered: flat batteries in the lead guitarist's DI can upset a show. Best stick to properly designed units, which have a reserve battery that can be switched in, and a 'battery low' flashing warning LED, or similar. Some units have a 'sleep mode' that saves the batteries when there is no signal for a while. Others are turned off when the input plug is withdrawn. In all cases, the battery holder should be accessible quickly and without undoing screws.

AC-powered DI boxes are generally good, though of course they add the need for another electrical outlet – as well as potentially another thing to go wrong. If it comes fitted with a consumer-grade 'wall wart' sort of power supply (plastic-cased transformer with puny wiring) this should be re-cased in a sturdy metal box, with thickly sheathed rubber 'stage-proof' AC and DC connecting cables.

Phantom-powered DI boxes are probably the best option – again, 'plug and play'. One problem: if the mixer is switched off, or absent, you can't test the DI. So back-up battery or AC-power is a good idea. Also, note that phantom powering may not be able to work if you use the DI's 'groundlift' option (see later).

Some DI boxes have separate inputs for different levels

Good DI boxes will be rugged, as they are often placed on the floor and tend to get trampled on, kicked and dragged about. The sockets should be easy to replace, as they are open to damage when inserted plugs are trodden on.

Typical DI box features:
- Link out – a second quarter-inch jack socket, to allow 'daisy-chaining', so the signal can carry on wherever it was going, while the DI box 'taps into the signal line'.
- Pad/gain – this drops the input level by a preset amount (-30dB is typical), as on a mixer. Always used if taking signals from a backline (instrument) or slave amp's output. May even be needed with a heavily overdriven line output, although some units have a -10dB pad or gain setting, used for reducing line levels. With both pads switched out, the sensitivity is then right for instrument level, say -10 to -15dBu (about 0.25 to 0.20V). Units with -40dB attenuation will handle 100 volts of signal from a really high power amp, reducing this to typically +2dBu or 1 volt.
- Polarity or phase-reverse – flips the DI's balanced outputs. Useful to reverse any cancellation when there's a miked-up feed, but only when you don't already have such a switch on the mixer, or when adjustment from the

stage end is needed because of some complication.
- LP (low-pass filter) – this removes mid-to-high treble (largely noise/hiss) if you're DI'ing guitar or bass, particularly from instrument amps that have tubes/valves inside, or use cheap integrated circuits.
- Earth/ground lift – usually breaks the earth/ground continuity between the two sides. Some types leave a high value resistance between the sides, to hedge the bets. Try using this switch if you hear hums or buzzing when you connect the DI'd source. Some types use 'auto-lifting' or 'ground compensation diodes', meaning the two sides are joined if there's a small voltage difference above 1/2V, or else they're isolated. The idea is that a larger voltage, which might pose a hazard, bypasses the lifting effect. The 1/2V threshold is safe enough, but may not always be helpful, since the 'signal' that causes the louder buzzes can be above 1/2V without being at all dangerous.
- Some DIs have separate inputs for different levels – this cleverly avoids the need for an attenuation switch, which is probably more likely to break than sockets.
- Direct buffered output – this is an indirect feature, not for PA use. It's an output before the transformer isolation, which can be used to help the instrument drive its own local electronics.

DI boxes in use

Some DI boxes go a bit lower than others in the 'grunt' regions (low-end), and some go a bit higher in the top end – although the missing frequencies would only be applicable to synths making sizzly 'tsst' noises, or the higher harmonics of brushes and cymbals, say.

Some units also pick up hum more than others. As they contain signal transformers, it's a good idea to keep DI boxes away from strong magnetic fields, particularly any wiring carrying high current (the DI's own power wire doesn't, but lighting and other power cables do), and also away from other AC-powered gear.

If there is no choice – say you're rack-mounting some DIs with other gear – then try to leave a gap of a few rack units, or at least experiment with the other equipment's order around the DIs, to try to achieve the least hum while it's all switched on (using headphones to do this can help speed up the process). Removing input connections in turn should help suggest the prime suspect.

Cabling for DI and other line sources

In the real world, these are usually exactly the same cables (preferably XLR-to-XLR) as are used for mikes. You could use slightly less high-quality cables for line levels, as they are less prone (in theory) to interference and don't have to stand so much abuse (again, in theory). But the outweighing operational benefits of only requiring supplies of one type of cable should be obvious enough.

Stage-boxes & snakes

The stage-box is where you plug in all the music signals on-stage – all the mike cables, the line sources, and the DI boxes. Attached to the stage-box is a (usually long) 'multicore' (a thick, sheathed bunch of cables, also known as a snake, loom, or just 'multi') which carries all those signals to the mixing desk, wherever it might be in the venue. (There's more on multicores/snakes in Chapter 7.) Seen from the FOH mixer's viewpoint, a stage-box is the 'breakout point' for signals controlled by the desk.

Sometimes the stage-box is built into a trunk associated with the multicore reel. Or it may be a separate box that has to be mated with the multicore cables, usually through a chunky (but still delicate) circular multipin plug, which must be screwed or locked into place (like a lightbulb).

A workable stage-box comprises a panel with XLR female sockets for the stage cables to be plugged into. Male sockets may be provided if the snake/multi includes the 'returns'. All sockets should be clearly labelled (remember you may be working in low light) either with numbers or space for a replaceable tape 'scribble strip', where signals can be easily identified.

Minimum stage-box requirements are: enough female XLRs for the mixer channels you are using; one or two spare inputs; two spare returns; and at least two male XLRs to connect to the PA wings. This last assumes the cabs are very basic types that use passive crossovers, or else that the active crossover is mounted on the stage, or that it is 'split between' the PA wings.

In most well-equipped situations, you will need: six male XLRs (returns) to connect two three-way crossover outputs to the wings (or eight XLRs for two four-ways, etc.) Make sure you buy a snake/multicore & stage-box combination with enough returns for future expansion

Mike splitters

At times you'll need to feed more than one mixer with the signals coming from the stage – this could simply be to supply a monitor mix (which we'll discuss in Chapter 2), but it could also be for broadcasting or recording a live performance.

While it's obviously good for an act to get this sort of extra attention, such signal-sharing can cause technical problems for a sound engineer. You might innocently imagine that since the electricity that these 'guests' will draw from the artist's/band's signal sources is fairly miniscule, then freely donating it can do no harm. In fact, if someone connects stuff straight across your mikes, DIs or decks, without due precautions (which we'll come to in a moment), it can cause all kinds of havoc: introducing hums, buzzes or RF interference (which may be minor or impossibly

Signal-sharing can cause problems for a sound engineer

loud); making mike levels drop or jump randomly, as your guests alter their console's input settings or connections; trigger feedback, as the mike levels dance about; or even alter sound quality and balance, so the signals have to be re-EQ'd.

Microphones (except capacitor and electret types) are affected more by this than other signal sources. Generally mikes are designed to sound right with one 'mixer's worth' of load (1200 ohms). As you add extra mixers, the sonic tuning of many mikes is thrown off.

Of course, even without sharing with 'guests', every PA mike is already loaded by 600 ohms – the nett effect of the FOH and monitor consoles (1200 ohms each divided by two = 600).

The solution is to use a splitter box to safely isolate the two (or more) separate systems, and control the impedance that the mike has to deal with. There are two main types – passive and active.

Passive mike splitters

These comprise 'raw' isolating transformers, one per stage channel, in a box, with suitable XLR or multi-pin connections. The PA's own feed will likely be 'straight through'. A transformer is also connected across the line. It has at least one, and typically up to three outputs. One will usually be used within the PA, to connect to the monitor console. This will be necessary if there are any hum or buzzing problems – and in some cases may be best done anyway to play safe and save set-up time.

Each of the transformer's output, is a '1:1' (same level or 'unity gain') reproduction of the signal seen at the input side to the transformer. Each of your 'guests' plugs into one of these 'splits' or 'splitter feeds'. The benefit of this is that the windings are isolated – properly wired, this should reduce hum, buzzes and other ugly, interactive noises by enough dBs (typically 30 to 50) to make them practically inaudible.

Transformers are invaluable – as rugged audio isolators, they're not readily replaced

What the passive splitter cannot do is to impart more energy – so it cannot 'fix', and has no effect on, the level losses or related tonal changes. It's still useful, though – it can...

- prevent ground/earth loops between different audio systems. Stops any part causing hum to another, at least by this route.
- provide some protection against shock, fire or electrocution, if any of the plugged-together systems becomes 'live' (at AC supply voltage), due to a fault or miswired connection.

These features are built upon in the evolution into active splitter boxes.

As we said, in splitter transformer language, 1:1 means 'unity gain' – the output is the same size as the input. A three-way split would be written down as 1:1:1:1 and called 'one to one to one to one' – or, more sensibly, 'one in, three one-to-ones out'.

Generally, a superior splitter transformer will be larger and heavier than a standard model. Good transformers are also usually proportionately expensive, because a large part of the cost is the amount of pricey materials used.

Transformers are invaluable for some jobs – as rugged audio isolators, they're not readily replaced. But they're not perfect. They can cause HF resonances (peaks) when connected to some lines and mikes. This effect is unpredictable without in-depth analysis. Any peaks might also 'ring', which can make the sound uncomfortable to the ears. Transformers also add their own distortion, particularly at low frequencies and high levels. Fortunately it is sonically fairly benign and the onset is gentle, though musical detail is impaired.

Partly, their imperfections can be reduced by supporting them with active electronics.

Active mike splitters

Active splitters use powered electronics. Some also use transformers to their best advantage. Inside, each mike's signal is 'actively buffered' by a mike amp (pre-amp) that is also built to drive a heavy load. The effect of the buffer amp is that the mike sees just a single mixer's worth of loading, regardless of what load is placed on the buffer's output. The amp also commonly has some gain – a small, judicious amount, not enough to cause a headroom problem (there's more on this in Chapter 2).

The overall effect is in part as if the mixer's 'front-end' has been moved up onto the stage, near the mikes – and before we split the signal off to different destinations (see Chapter 6).

Why do this? With dynamic mikes, by reducing the distance of the mike cable before the signal is buffered and amplified, the signal (sound) quality is enhanced. The mike's sound is less likely to be upset by resonance caused by long cabling, and direct meetings with transformers, while raised signal levels offer greater noise immunity on the way to the mixers.

The buffer's output then feeds (splits or fans-out to) several outputs. The PA feeds usually all employ 'electronically' balanced outputs. These can offer higher sound quality than using transformers, and avoid unnecessary extra cost. They'll work fine so long as the different equipment is all working on the same electricity power supply, so that large voltage differences – above about 10 volts – can't develop between the connected equipment.

This latter condition might cause reduced performance. One symptom could be distortion, or else the balanced connections may lose their hum-and-buzz-suppressing capacity. Usually there are three PA feeds – to the FOH and monitor mixers, and one spare, which can be used for cross-patching.

It's because of this that the more up-market, robust splitter units are fitted with typically at least one 1:1 balanced output transformer. In some units the other outputs can also be 'transformerised' if need be.

The transformer's 'galvanic isolation' property overcomes almost all noise ('common mode') voltage differences between audio systems working on different mains supplies (or supply phases). This enables a PA system to work without risk of intractable hums and buzzes in practically any environment.

Phantom powering with splitters

When a splitter is used, the phantom power needed for capacitor mikes can't be supplied by the mixer (usually it would be from the monitor mixer – see Chapter 2) , as it cannot pass through the splitter. Instead, the splitter must supply phantom power itself.

Phantom power, as we've said, shouldn't be connected except when required. So the splitter has to have switches to apply phantom power to individual channels – which may have to be altered if a mike is changed from dynamic to capacitor during a set, say.

DJ gear

Turntables/decks

Turntables can either be driven by rubber idler wheels, rubber belts synchronised to the mains frequency, or by DC motors which can be varied or 'speed locked' (for stable speed) and also heavily braked when stopped (so-called 'direct-drive').

Of these, direct-drive is the only serious mass-market contender for the DJ market today. Old idler-wheel decks (such as Garrard broadcast decks) provided quite fast starts, but lack the expensive electronics required for smooth speed control; while belt-drive is made for delicate hi-fi reproduction, and is not capable of fast starts or stops, nor rapid speed control recovery when the platter is spun back and forth then released (as in 'scratching').

A decent DJ turntable should have certain features, some of which may be directly relevant to the sound that's fed through any attached PA. For instance, cueing lights, used to illuminate the area of vinyl the stylus is hovering over, sometimes have dimmer switches – if so, always use a Variac (a variable transformer) or a resistor-based one, as most 'smart electronic' dimmers can cause dreadful interference noises.

The tone arm might be involved in producing feedback – especially if the arm is hollow

The tone-arm (the bit with the cartridge and stylus at the end) should be weight-adjustable – many DJs in fact deliberately over-do the weight adjustment (or 'tracking'), often adding coins for extra weight to keep the stylus planted in the record's grooves. This reduces tone-arm judder when scratching, and makes it less likely to jump if nudged or vibrated. It also trashes the records, of course, over time, gouging out the treble information, and producing a distorted, undetailed sound – but then some people don't seem to mind.

Alternatively, you could see if the vibration problem can be avoided outright by moving or stabilising something. For example, the DJ stand may be on a wooden riser that's being jostled by the crowd – moving this or pushing back the crowd barrier, if possible, would stop the problem. It would also help to use shock and vibration isolation platforms and methods (more of which shortly).

The tone-arm might also be involved in producing acoustic feedback, often around about 350Hz (in the midrange) – especially if the arm is hollow, and if the PA speakers are placed very nearby. The feedback can start as a handling noise, but can build to a groan or howl, just as with mikes. Closing the turntable lid, if fitted, can make the problem worse. Ideally the speakers should be moved away. Equally, moving the decks' position in relation to the speakers, such as away from the usual midpoint between speakers in the stage wings, can help a lot.

To test and see whether the system will feedback, and then manage to reduce the causes, first lower the stylus onto a stationary record; then turn the PA up until you hear feedback just about to start. If the right changes are being made, the PA's level should be able to be gradually raised before the problem comes again. A more permanent solution is to change the arm for a non-hollow one.

Vinyl-playing accessories

The DJ's armoury will also include some or all of the following: stylus balance, alignment tool, spirit ('bubble') level, distilled water and tiny brush (for cleaning stylus), and magnifier to examine the cartridge, spare styli and cartridges, spare bulbs for the cueing light, and a BPM meter for beat-matching. (These will mostly be provided and used by the DJ, though it may be useful for PA crew working with DJs to be familiar with their function.)

CD players for DJ'ing

Vinyl is still the preferred medium for most DJs, as it's the most satisfying to handle, not just in terms of scratching but for its general ease of use, visibility of tracks, and availability of archive recordings. But in recent times CD player manufacturers have been moving into the DJ market more successfully, with top-loading CD machines that allow beat-matching (through digital 'pitch-bend'), fast search dials ('jog' and 'shuttle'), cue and jump, and even digital scratch emulation, without skipping

As with CD players employed to provide backing tracks for bands and other acts, most of these machines will have S/PDIF outputs for a digital signal to be fed to a decent D-to-A converter for higher sound quality.

Anti-vibration platforms

It's important for turntables to be kept level and stable on any given surface. Failure to do so can result not only in uneven vinyl wear but also unpredictable sound quality. Some turntables have built-in spirit levels (if you don't have your own) to check how flat they are, and they can be adjusted by twiddling threaded mounting feet.

But all sound replay sources that involve spinning or similar movement need to be isolated from vibrations and shocks to give their best performance. This includes turntables, CD players, DAT machines, DVD, SA-CD, computer hard disks. CD and other digital source machines can appear to be vibration proof, but sonic quality suffers long before skipping or blanking occur.

(Note that MiniDisc players are a partial exception – in exchange for giving a slightly lower sound quality (less detail) than other sources, MiniDisc handles shocks by storing and then correcting the playback, so the musical stream can be kept steady.)

Moving the turntable away from the speakers can reduce feedback

A few manufacturers produce dedicated anti-vibration units for pro DJs, which could equally be used by PA operators. DIY isolation can be performed by fitting shockmounts and shock-absorbing 'buffer' feet to existing units.

An alternative, more down-to-earth way is to mount the decks etc on a board placed on top of four old inner tyre tubes – the inflation can be altered (bring your footpump too) to get the best isolation at problem frequencies (usually around 40Hz for decks).

An emergency method, using materials to hand at a gig, involves rolls of gaffer tape and rubber bands. Place several rubber bands over each roll, to cover the central hole from several angles. The deck's feet are then seated in this springy recess.

More elaborate-looking isolation hardware, some of it 'scientifically designed', has been developed by/for the home hi-fi industry, and some of it is rugged and practical enough be applied to stage needs. Mostly it has screw-adjusted feet which allow levelling. (Check out hi-fi shops, music shows/fairs/exhibitions and specialist magazines.)

the mixer
front-of-house
& monitor desks...

Quick view

For any show involving more than one or two performers, working without a mixer would get very messy. The mixer is the control centre. It receives all the individual sound feeds from the stage and merges them together, balancing their relative volumes, changing the tone (equalisation) where necessary, filtering out unwanted sounds and frequencies, and adding effects as required (depending on what's available in the FX rack). The signals can also be arranged into groups for easier manipulation, before being passed on down the chain to the amps and speakers. At larger gigs a separate monitor mixer is used to provide personalised mixes for the performers on-stage.

In PA systems there are two main uses for mixers (also called mixing desks, consoles, or boards). One is for bringing together the various signals from the instruments and voices on-stage and blending them into the best possible front-of-house (FOH) sound for the audience to enjoy. The other use is for stage monitoring – blending those same signals (instruments and vocals) personally for each musician, so they can clearly hear themselves and/or the other performers on-stage.

Front-of-house mixers

Mixers are often described by the number of input channels they have – which gives a rough idea of the amount of individual signals a mixer can deal with. This is normally an even number (despite the fact that most PA signals don't actually come as 'stereo pairs'). 12 channels is about the smallest for a basic PA mixer, then you usually find 16, 24, 32, 40 and 48 channels. These are the 'mass market' sizes – specialised PA mixer manufacturers offer more, typically from a standard size of 64 up to as many as 200-odd channels.

Why so many? The basic number of channels needed to fully mike even a small number of musicians – say a three or four-piece band – is in the region of ten to 30. This is often largely due to close-miking of individual parts of the drum kit (see Chapter 1 for more on close-miking). Or there might be lots of keyboards and samplers on-stage, or 'part players' (backing vocalists, string or horn players who appear for just one song, say).

A couple of channels may also be used for 'mood music' before the gig starts – not to

33

mention any required for pre-recorded tracks used in the set, which could involve anything from one simple additional instrument track to a complex 16-track backing. It all eats up channels on the desk.

The number of knobs on even a small mixer can be daunting... don't panic – everybody feels that way at first

---

The other reason for using a mega-channel desk is when there are several acts appearing on the same bill, with limited change-over time between each one – for instance at a festival or award show. A mixer with plenty of extra channels (or even two such mixers) gives you the luxury of being able to set up, soundcheck and leave intact more than one set of mixer connections, 'patches' (wiring set-ups) and configurations at any time, regardless of the number of performers involved in each act.

Using two independent mixers (where the mixing is alternated or 'flip-flopped' between them as the acts change over) may have the disadvantage of requiring two separate multicores/snakes, which many PA engineers would prefer to avoid – because rigging one is hassle enough. So for instance a single 96-channel mixer might be preferred over two 48-channel boards.

This may all seem over-the-top for those working at minor-league gigs and venues, but there's a fundamental point that's true at all levels. While some of the hugest, mega-channel mixers might only be used in over-bloated set-ups by ego-maniac artists, it's nonetheless best to have more channels than you first think you'll need

(rather as with hard disk space on a computer).

Choose perhaps double, if not three times as many channels as your minimum requirement suggests. Extra 'jobs' and additional signals will arrive unexpectedly and soon gobble up space. Don't forget that some channels may suddenly die on you too – and if you don't make allowances in advance, you'll soon end up short of channels again.

Even PA companies on a tight budget might consider having two separate smaller mixers, as it offers more flexibility. They can be used on their own for more intimate gigs. Then at multi-act gigs they can still be used side-by-side and 'flip-flopped' as described above. It also increases confidence if you know you've got a standby should the mixer go wrong or act up.

Most larger companies will own a selection of different-sized desks. Using a desk of about the right size for the job is a good idea – if only to avoid the embarrassment of arriving with something that's physically too big for the venue.

Mixer features & controls

The number of knobs on even a small mixer can be daunting. But don't panic – everybody feels that way at first: look at it in sections and a pattern will start to become clear.

In most cases all the input channel 'strips' (the vertical lines of knobs, usually running from the channel input socket at the top down to the channel fader/slider at the bottom) are all the same; or, at most, there are one or two variations that soon become visually obvious. And once you have cracked the controls on one channel – and similar controls on the other channel-like sections called 'groups' – there isn't a lot more to grasp.

Here are most of the controls and features you'll meet on a PA mixer channel strip, whatever the mixer's size, make, age or cost – though the positioning sequence may vary. (Note that some features on particularly large desks may not be relevant to you or your circumstances – mixers

What the numbers mean
Professional mixers for PA (or recording) are officially described using a three-figure format: for instance 16/6/2 (pronounced "sixteen into six into two").

The first digit is the number of input channels. The last digit is the number of main output channels (this is almost always either 2 (for stereo), 1 (mono) or 3 (stereo plus a centre channel, which

we'll come to later). The middle digit is the number of sub-groups – usually between 4 and 16 (groups and sub-groups are explored in detail later in this chapter – see p45).

are designed to be as flexible as possible, and you'll learn with experience which parts you can safely ignore for certain jobs.)

Input sockets
This is where your signal sources (mikes, DIs etc) plug in. On fully professional mixers over a certain price level you'll find that each channel has a single XLR connector (there's more about connectors in Chapter 9) operating as the input, and this is switchable between line use (from a DI, for example) and mike use – sometimes called 'A/B'. Alternatively there may be two XLRs – one for line inputs and a dedicated one for microphones.

On lower budget mixers there's no switching, and the dedicated line input may be reduced to a jack (quarter-inch/'phone' plug) socket – this saves a few pennies, though may prove to be a false economy if you add up the time wasted later trying to rectify bad connections.

A mixer needs to be above a certain size and sturdiness to cope with a typical 'snake'

To be fair, one reason why jack sockets feature on even quite expensive mixers is that they can provide switching that happens without the engineer having to do anything extra. In particular, jacks are used for line inputs because the jack socket's integral switching can be used to bypass the mike amp, or perform similarly deft re-configuring for higher level line signals. The downside is, unless used and cleaned weekly, this switching can cause intermittency problems.

On larger mixers there may also be a direct connection for a multicore/snake cable. Note that a mixer needs to be above a certain size and sturdiness to cope with the typical weight and stiffness of a large 'snake' without tipping over. More commonly, with smaller desks, the mixer's end of the snake is 'fanned out' to XLR and/or jack plug 'tails' or 'tail ends'. This not only gives flexibility (for quick patching in emergencies,

say), it also physically relieves the mixer of the bending/sagging strain of the heavy multicore/snake.

Stereo inputs/channels
Some inputs accept stereo plugs. By feeding two signals into a stereo quarter-inch phone/jack plug (via a Y adapter if necessary) you can use one these special inputs instead of taking up two channels for, let's say, a stereo synth part, or overhead percussion mikes. So you save a channel, and ultimately may not need as large a mixing desk – but bear in mind you'll have less control over the individual left and right signals, as some controls will be ganged (linked together so one control adjusts both signals).

Mike/line switch (also 'mic/line')
If there's just a single input socket per channel, this switch selects the appropriate sensitivity for the source. The mike input setting also has a lower impedance that will load down some line sources, and might cause them to distort prematurely at higher levels.

If the mike/line switch is set to the wrong position, the outcome won't be permanently damaging, but will either cause overload/distortion (if you feed a line source into an input switched to 'mike'), or there won't be enough level (if you do the opposite).

Phantom powering
This refers to the power (usually 48 volts) that's required for capacitor ('condenser') microphones to work. This is provided on all professional mixers, though in modern PA work the job of supplying this power is mainly in the hands of the monitor console operator – so see later in this chapter under 'dedicated monitor mixers'.

Polarity switch (phase invert)
This is used to flip a signal's polarity (which has the effect of altering the signal phase by 180 degrees, so its wave pattern is exactly opposite to what it was – like a photographic negative). If used in isolation – if there's only one source – don't expect any particular audible effect. It becomes important when combining two signals from one instrument, such as a mike and DI signal, say from a bass, or multiple mikes (see Chapter 1 for more on this).

the mixer

If two virtually identical signals are mutually 'out of phase', they can cancel each other out. This process may also be used creatively with pairs of mikes to reduce ambient (surrounding) sound.

Pad (switch/button)

This is an 'attenuator' – it reduces incoming signal level by a fixed amount. You would use this switch/button to reduce over-loud signals from, for instance, some high-output mikes (eg kick drum, capacitor mikes), and from high-level line feeds from hard-driven DI boxes, DJ mixers, and the like.

In some cases the pad switch is the only mike/ line selecter you have

Sometimes the pad switch is used to enable line signals to be fed into a 'mike' input. This is OK in emergencies, but in some cases the lower impedance (greater loading) of the pad and mike input, as seen by the signal source, won't help sound quality.

In some cases the pad really is the only mike/line selector you have. Your ears have to judge whether it's successful or not.

Signal present

This is generally a green or yellow LED (light-emitting diode) which lights up to show there is some reasonable signal level on a channel. It won't light if the signal is below a certain low level.

(Input) gain control (or 'trim')

This control is used to set the correct signal level (not too high or too low) through most of the input channel part of the mixer. Rather than a 0 to 10 dial, the amount of gain being applied is usually shown in dB. Typically, higher settings (over 40dB, which produces one hundred times more level – see Appendix C) are only used for microphones.

If you're using these on a line source, something is amiss – although gains as high as 35dB can at times be required on line signals. (This is largely due to the annoying, 'domestic'-standard level of -10 ('minus ten') dBu at which fashionable but low-grade, plastic-cased FX equipment operates. This '-10dBu' operating level problem is one of several quite unnecessary sound and headroom-degrading nuisances the PA business has to deal with.)

When a microphone is used in front of a loud instrument, you won't need to apply much gain – typically only 10 to 25dB (x3 to x18). Sometimes with line (DI'd) inputs (and in extreme cases with mikes, no gain at all is needed – which is referred to as '0dB' (x1) or 'unity gain' (sometimes indicated by a 'U').

With quiet instruments (like strings) and insensitive mikes (such as ribbon types), or if more distant miking is used, the maximum gain available and 'lowness of noise' (hiss, buzz, hum) are mixer specifications that suddenly become rather important.

A gain of 60dB, or a tad more, may be needed in such circumstances. Some cheaper mixers do not give this gain range, in which case you'd need quite expensive outboard (external) mike amps offering adequate gain to complete the job.

Peak/clip/overload indicator

In its most basic form, this is simply a single LED that glows to signify an over-high level. It's almost always red, and may be at the top or bottom of the channel strip. More sophisticated mixers offer more detailed peak metering – either several LEDs in a line (or bar), or some other form of graded meter.

(On some mixers, the 'peak' and 'signal present' LED may be combined with a small LED bar-graph meter that shows some intermediate signal levels.)

This LED should, at most, just flash red on the loudest signal peaks in the whole set. Higher levels (where the LED is constantly red) may not cause audibly high distortion straight away, but will be pushing the limits, and have the potential to degrade the sound quality irretrievably.

The pros and cons of different types of metering are covered in more depth later in this chapter – see 'Metering'. For more on what to do if the peak light goes red, see 'Gain Structure' in a few pages time (p46).

EQ (equalisation)

Music is made up of multiple complex waveforms which react with the surroundings (and our ears) in suitably complex ways. EQ is an attempt to exert some measure of artificial control over the sounds we hear by the boosting and cutting of various audible frequencies.

On a mixer this is controlled by a series of at least three knobs (though there are also dedicated outboard rackmount EQ units, which we'll cover in Chapter 3)

But before you even think about recklessly turning any EQ knobs, a few words of caution. Be sparing. Bear in mind that EQ can be costly to sound quality, especially if boosted by any more than a small amount. Even if (by some miracle) you manage to perfectly and tidily compensate for some arbitrary acoustic or electronics-caused degradation, and affect nothing else, the EQ still comes at some cost to sound quality.

Only use EQ as a last resort... even then, use it gently at first

---

For example, EQ is indiscriminate: boosting increases all the noises and harmonics in the boosted frequency area. If these are already making the sound hard or harsh, or are otherwise obtrusive, boosting will make their effect more audible – even if it also makes the kick drum sound right.

It's always best to fix problems as far as possible without using EQ (serious hi-fi listeners have used gear without tone controls for years). Always consider if the sound could be improved by better microphone placement, or making changes to the equipment or performances of the artists, or if there's anything about the

acoustics of the venue that could be easily improved.

Use EQ only as a last resort. Even then, use it gently at first: +2dB here, -3dB there. And failing that, at least try to restrict 'stronger' settings to safe frequencies – ones you know won't blow-up bass or top cabs. If you really have to add EQ in the more 'dangerous' areas – generally meaning below 60Hz or above 10kHz – apply no more than +4dB boost.

It's also not a good idea to make sharply contrasting settings in the same or adjacent frequency areas – for instance, if you're boosting at 250Hz, don't cut at 200 or 300Hz. They will largely cancel each other out, and at the same time leave a more messy frequency response than you began with.

There are other sound reasons for going easy on EQ. For one thing, boosting EQ eats up headroom (the amount of extra level the system has in reserve, to be used only when needed). This is more likely to be an important issue when EQ is used later on in the PA's signal path, so it's dealt with there (see Chapter 4). For now, just be aware that boosting EQ (like borrowing money) isn't free.

EQ can also be deceptive. Our ears quickly adjust to new levels. EQ is like a drug for which the user quickly acquires a tolerance, and a habit, and so craves more. Except that this escalation can be rudely stopped by the lack of headroom and weird, possibly painful sound (unless of course the mix-engineer is too stoned to notice – which is not unheard of).

There's yet another danger with EQ: all uneven frequency responses (at least in all analog gear) come with phase shifts. If the problem being EQ'd is purely a defect of the sound source and mike, and the EQ is carefully set to exactly cancel the problem, then there should theoretically be no net phase shift. In practice there will be varying amounts.

Excessive phase shift is most directly audible at low frequencies, where it may be heard as a delay

---

Frequency and pitch

In music, what we hear as pitch is partly 'frequency sensation'. Frequency is measured in Hertz. The higher the frequency of a soundwave, the higher the pitch we hear. For instance, 440Hz is the fundamental frequency of the note A (above middle C) at concert pitch. When a sound's frequency is doubled, the pitch goes up by one octave, and when the frequency is halved, the pitch drops an octave. So the A an octave higher up has a fundamental frequency of 880Hz (ie double), and the A an octave lower has a fundamental frequency of 220Hz.

in the bassier parts of the sound. Where a given sound source is being captured by more than one mike, it is indirectly audible as reductions in a source's level. The volume actually drops as you boost the EQ, because the phase shift compared to the same sound in another channel (eg DI'd) is enough to cause partial cancellation. (There's more on this in the 'HP Filter section later.)

Mix engineers have to learn to be disciplined if they want to maintain the best overall sound. In practice EQ often overcorrects parts in the sound that didn't actually need to be boosted or cut. This is particularly true of highly un-selective EQs, like simple 'shelving' bass & treble type equalisation. As we're about to see, some EQs can be more accurate than others

Basic bass & treble EQ

Known by engineers as low and high frequency (LF & HF) shelving EQ, these knobs are really just a refinement of the bass and treble boost/cut 'tone' controls found on old or low-budget audio equipment. A simple circuit that controls boost and cut for both frequency areas was invented by Peter Baxandall in the UK in the early 1950s – you'll still sometimes hear of high frequency (HF) and low frequency (LF) shelving EQ called 'Baxandall controls'.

A small amount of boost on each control creates a slightly warm and tizzy sound

---

Using this basic kind of EQ, pre-set symmetrical boost and cut settings can be dialled in either side of a central 'no effect' position on the knob. Shelving EQ always produces the most boost at or towards the frequency extremes, with the boost/cut reducing smoothly as it nears the 'no change' (0dB) region – for both LF and HF this is around the central midrange (1 to 2kHz).

A small amount of boost on each control creates a slightly warm and tizzy sound. It does this better than putting a 'smile' on a graphic EQ

(where the faders are physically aligned in a curve with a dip in the mid-range – see Chapters 3 & 4 for more about graphic equalisers).

In its favour, LF and HF shelving EQ makes sweeping, progressive types of tonal corrections smoothly and naturally without risking the kind of interaction that arises between the graphic EQ's many frequency bands.

(A graphic EQ is a much more powerful tool, and really best kept for more intricate responses – again, see Chapter 3 & 4. On the other hand, some engineers have noticed that it creates phasiness – which makes sense if the 'broad brush' of EQ isn't likely to accurately match actual deficiencies.)

Sweep (bandpass or 'bell') EQ

In this type of EQ, a given band or 'width' of frequencies (as if bordered within a window frame) can be cut or boosted, while the 'window' itself can be moved, or 'swept' up or down to offer a choice of frequency bands to be manipulated. The 'window' isn't straight-sided, because things in nature don't change at right angles – they slope off gradually.

So the frequency response shape you get with any 'area' boost is 'bell-shaped' (with cut, you get an upside-down bell).

On many mid-market mixers you'll find a mixture of shelving and sweep EQ: often there'll be fixed, shelving-type Treble (HF) and Bass (LF) knobs, while the Mid will have sweepable control, sometimes two controls, dealing with low-mid (LM) and high-mid (HM). Their ranges should overlap generously, by at least one octave. So if the LM covers 400Hz to 1.6kHz, the HM should sweep down to 800Hz (half of 1.6kHz, 800Hz = 1 octave below it). (See sidepanel, p37.)

When using this sort of EQ it's useful to know it will appear to have little effect if there is no particular sonic content in the selected frequency area. Avoid boosting too much until the right area has been found.

Parametric EQ (PEQ)

This is an even more sophisticated type of equalisation which adds another control, known as 'Q', to the above-mentioned sweep EQ (sometimes sweep EQs are called 'quasi-parametrics').

The Q governs how narrowly or broadly the

frequencies immediately either side of the named frequency are boosted or cut – which allows more accurate tonal modification of the signal.

Most higher-quality mixing desks have full parametric EQ controls built-in, but you may also find external 'outboard' EQs are used (there's more about these in Chapter 3).

A sound quality issue that's rarely mentioned is that when high settings of the Q knob are used (above $Q=2$ up to $Q=10$ or more), the EQ'd part of the sound is spread out in time, rather as an electronic echo. On bass sounds this gives a springy, 'boingy' effect that may even be artistically desirable.

Technically speaking, high Q settings cause increasing signal decay periods, which develop (with higher Q) into ringing, then resonance, and finally into self-oscillation. But in any properly designed mixer the final stages shouldn't ever happen, as the PA speakers could be damaged (and an audience upset) by the high-level tone signals that would be produced.

Unlike the headroom problems of EQ boost mentioned already, high Q settings do not normally in themselves cause problems with distortion or overdriving the PA. But you'll need to be aware that the slim range of frequencies boosted by any high Q settings – which may not sound very loud – could nonetheless be 'secretly' overdriving some parts of the PA. It's sensible to keep a special check on the peak LEDs of channels using high-Q EQ settings.

Be aware that high Q settings could be secretly overdriving the PA

With parametric EQ's, Q normally varies between 0.1 and 10 (1 is normal). But it's worth noting that on many modern mixers a related figure called bandwidth is used, and this is measured in octaves on a scale that goes in the opposite direction. In other words that same Q variation might vary in bandwidth between 10 and 0.1 octaves. (There's more on PEQs in the section on 'Monitor Mixers', later in this chapter.)

Outboard EQ

For special or 'problem' vocals or instruments, you can always connect an outboard EQ unit. This could be a more sophisticated device, and/or one that's actually nicer sounding (for the job at hand) than the EQ provided in the console – if so, it'll probably be a renowned studio type (Neve, Focusrite, Massenberg etc). We'll look at outboard EQ (and other outboard gear) in more detail in Chapter 3.

EQ on/off

This simply and positively disconnects all the aforementioned sorts of EQ. It's useful for acid test comparisons – like answering the question, "Have I achieved anything yet by five minutes of twiddling?". It can also help in emergencies, such as trying to quickly turn down (say) bass EQ that's causing feedback.

Filtering

Filters are required to 'narrow the window of frequencies' – so that, for instance, a mike placed in front of an instrument on a noisy stage will only receive that instrument's sounds, and the other sounds on stage will be 'filtered out'.

Although not a true EQ, they are in a sense the most important EQ on the board, because in allowing relatively benign and yet selective broad-brush removal of unwanted tonal areas, they achieve what shelving EQ has long been rather misused to attain.

A basic rule of sound engineering is: don't receive signals at any frequencies that you don't need for the instrument at hand. A competent modern mix engineer will be using filters on almost all signal sources. Exceptions might be 'ambient mikes' (which are intended to pick up everything), and wide-range analog synths. Some DI feeds don't need filtering because it's applied at the DI box.

High-pass filter (HPF)

This mixer feature enables you to get rid of unwanted frequencies below an instrument or vocal's range (in other words it keeps or 'passes' the frequencies higher than a settable cut-off). An HPF can very efficiently 'clean up' the mix of a higher-pitched instrument, as well as greatly reduce vocal 'popping' and handling noises on cardioid mikes. It can also help remove unwanted the mixer

breath and air sounds (wheezes and breezes), so it's a saviour if a mike's windshield has been lost.

The frequency below which the cutting-off occurs will be either adjustable, as with sweep EQ, or, less usefully, switchable. A typical useful range is 10-350Hz.

If there's no on/off switch on this filter, simply setting the HPF control down at the lowest frequency will have to do, even though this will add some extra phase shift and signal delay to the instruments' low end – this effect is subtle, but worth mentioning as it is also cumulative and not correctable. For this reason it's best to switch off or avoid using high-pass filtering that you don't really need.

The efficiency or 'selectivity' of an HPF is indicated by its slope, described in units of -6dB/octave. As with crossover filters, at least 12dB per octave is usually needed to get any discrimination. Otherwise the change from 'pass through' to 'stop' is so gentle there's a large, unwanted gray area well outside the required frequency range. What '-12dB per octave' says is, 'signal voltage quarters for each halving in frequency' – in other words, twice as fast as 6dB per octave.

Better quality mixers spend a few more pennies per channel on a -18dB/octave filter slope. Here, the signal falls off twice as fast again, with each octave of decreasing frequency. This is also four times as fast as -6dB per octave, with unwanted signals reducing by a factor of about eight times per octave (ie per frequency halving).

As with highly selective parametric settings, the penalty on some less well-designed mixers (no matter how good they sound to start with) can be some ringing, also described as an 'under-damped' sound. This may or may not sound right, or useful. To suit varying needs, some 'high-end' mixers may offer some choice of slope (-6, -12, -18) or even the actual damping (Q) settings. This sort of variation is particularly accessible with some types of digital EQ (more about these in Chapter 3).

With practice you'll find HP filtering can be used very subtly. As an illustration, it's possible to have bass levels increase when two HP filters are in use, with one turned upwards in frequency. This strange effect occurs because all ordinary filters (even good digital ones) have phase shift

that's quite strong in the initial cut-off region. Depending on conditions, this may be enough to cause bass cancellation, until one or other input is reduced in level or its HP filter setting is altered – which explains the effect just noted.

One complication is that the bass that's being increased may be from the unwanted ambience as much as from the wanted signal. Another is that the effect, or its full extent, will only become apparent when listening to the groups (see later) and/or the final mixdown.

What all of this means is that you should ideally set the HP filter's frequency back quite a bit (downwards in frequency, to half the number of Hz), from where its effect is just starting to be heard (when the sound starts to thin). Secondly, it can be best to avoid using a high-pass filter when it's really not necessary – but always use it when it's needed. Only practice and experience can help you tell the difference.

Low-pass filter (LPF)
These work like high-pass filters, except they allow low frequencies below the cut-off setting to pass unchanged, and filter out those above. So they're most useful on bassier instruments.

A good example is for reducing the spill into a kick drum mike caused by other surrounding loud sounds. But go steady – as some bass instrument sounds can at times contain significant components up to 5kHz, you can't always set the cut-off frequency as sweepingly low as you'd like. Some cautious experimental adjustments will be needed at first.

An LPF can also be used to remove hisses and sizzles on DI'd feeds from instrument cabs

An LPF can also be used to remove hisses and sizzles on DI'd feeds from instrument cabs (guitar, bass, keyboards etc). Another use is to remove 'presence' on instruments that sound unnaturally 'toppy' when close-miked – one particular example is bagpipes.

If there's a problem with feedback, fine-tuning of high-pass filters can sometimes alter the effect and detune it, without much loss of audible bass.

Frequency-tunable high and low-pass filters may have individual on/off switches, or an overall on/off switch. This is helpful if you want to do an 'A/B' check on what's being gained by using the filter, or if you needed to switch a given filter out on a particular song, then back in again, without re-tuning anything.

Other features & controls

Pan

The pan control (or 'pan pot') allows each sound to be 'positioned' between both sides (L & R) of the PA – so it's either 'in the middle' (same level of signal going to both sides of PA) or panned left or right (directed more to one side of the PA than the other).

The word 'pan' derives from the visual and theatrical term 'panorama'. Conventionally the instruments are panned to match the visuals, so for a bass player on the right, the bass guitar sound may be panned a little to that side – just make sure the audience on the other side can still hear it. Pan can also be used as an effect by manually wiggling it to move a sound from side to side of the PA.

Another trick is to split the signal and put echo or reverb on, say, the left side and keep the 'dry' signal on the right. But don't go overboard – remember that people on both sides have paid to hear the artist, not just your special effects.

Also, be aware that when the PA speakers are spaced over 50 feet or so apart, people on one side will hear the other side with something of an echo when the sound on their side is dimmed (by panning too far).

Left-centre-right (LCR) panning

Some mixers take into account the fact that central 'speaker clusters' may require a 'centre' (mono) mix and have 'left-centre-right' panning. It's mainly used for theatre – about the only time you'd use it for music is if you have a mono cluster (or any grouping of PA cabs) behind the FOH mix position/tower, at an outdoor event (see Chapter 6 for more on speakers).

Sub-group assign

This is a set of 'latching' push-buttons (meaning they stay down when first pressed) which you use to choose which sub-groups the channel will be sent to. Usually the groups are assigned in (stereo, L/R) pairs, with all odd numbers (1,3,5 etc) being 'left' and all evens 'right'.

VCA assign

VCA means 'voltage controlled amplifier' – VCAs provide a different method of ganging channels together, which has its own pros and cons. VCA assign is almost like assigning sub-groups (as described above). A VCA allows one fader to control lots of separate signals in a compact area all at once by electronic means. When VCAs are used in this way they allow you to create 'virtual' sub-groups.

Be very wary of VCAs, because good-sounding ones are rare

One difference is that each VCA fader controls two (stereo) channels at once and, related to this, all the signals being grouped go straight to stereo – they don't exist as a discrete (accessible, real) group before this. So rather than routing a piano, say, to groups 1 & 2 (L/R), then to the final L/R outputs, it would be simply routed to VCA group 1. This saves two layers of mixing.

But be very wary of VCAs, because good-sounding ones are rare – even on expensive mixers they have been found to have a detrimental effect on the sound. Also, because VCA groups are 'virtual' they aren't useful if you want a discrete audio group for sending to some effects processors.

Inserts

These are special two-way sockets that allow a signal processor (effects unit) to be 'patched' (connected) into a mixer's channel. Pro standard mixers will have these access points on each channel, where the signal path is broken and the channel signal routed in its entirety into an external device, then routed back again.

On lower-budget mixers (if they have inserts at all) the signal path is automatically broken by the insertion of a quarter-inch plug. On better ones, the insert in/out is bypassable (linked around) with a switch – with which there should also be an insert on/off indicator LED.

Inserts are the norm for connecting processors into a channel – commonly EQ, reverb, compressors, limiters and gates. The frame carrying this equipment is naturally called the 'inserts rack' (more about this in Chapter 3).

Auxiliary sends (aux sends)

Aux sends differ from insert sends in that the signals on (say) the Aux 1 sends, from all the channels, are gathered onto a 'buss' – in other words, mixed down to a single 'Aux 1 send' signal. The level of this is set by an 'aux master' control (normally a knob) positioned in the sub-group or main master section.

Aux sends are mainly used for FX (effects) that are blended in with the main signal, and may even be recycled – such as echo and reverb. These and any related devices are carried in the 'FX rack' (see Chapter 3 for more on the FX rack).

On old-fashioned mixers 'post' sends may even be marked 'echo' or 'reverb'

There are two types of aux send: the first (or first section) of the aux sends is nearly always 'pre-fader' (meaning the signal ignores or bypasses the channel level fader control). These may be used for monitor mixes (which we'll come to later in this chapter). The second (or later) of the aux sends is generally 'post-fader' (meaning accessing the signal after the channel fader). These are used for effects – in fact on old-fashioned mixers these 'post' sends may even be marked 'echo' or 'reverb'.

On more elaborate mixers, some or all aux sends may be assignable between pre and post, so you can have, say, three pre and one post, or maybe vice-versa.

Sends may also have their own mute (silence) buttons. On the largest modern PA consoles there can be up to 36 aux sends, but the principals remain the same – and in 90 per cent of shows only a few aux sends are actually used.

Using pre-fader auxiliary sends

On smaller mixers where the auxiliaries aren't switchable (assignable), the pre-fader aux send will usually be Aux 1 (if there are only two auxes), or Aux 1 & Aux 2 if there are four auxes in all, and so on.

For FOH work these are used to feed an FX unit which you don't want to follow the channel fader's setting. A typical use is with reverb – as the channel fader is reduced, only the direct (dry) signal is reduced, so the vocal will change to pure reverb.

Using the pre-fader auxes for monitor mixes is dealt with later in this chapter.

Using post-fader auxiliary sends

'Post-fade' send is usually Aux 2 if there are two auxiliaries, or Aux 3 & Aux 4 if there are four auxes, etc. The signal set-up by the post (post-fader) aux control knob appears at the corresponding 'post aux send' output socket on the back of the mixer. This output is connected to the desired FX unit. The signal from the FX unit then returns to the mixer, plugged into the 'post aux return' input socket. Normally, this is paired with the send socket.

The returned, effected (wet) signal can then be mixed back with the clean straight-through (dry) signal. But it doesn't have to be. When the mixer has plenty of channels, it's normal to return into a channel input. This gives you the luxury of a channel fader, rather than a cramped knob, to control the level. It also allows the returned signal to be treated like any other as far as mixing goes.

If there's a 'dry versus effect' mix knob on the FX unit itself, turn this hard over to 'effect'. You can then just use the mixer to control the amount of effect used: to make the signal drier (less effect), simply turn down the aux send knob (sending to the reverb), and turn it up to get more of the effect.

Note: if the FX unit distorts, first turn down the aux send; or check whether the FX unit has a 'mike' input or 'high sensitivity' setting that's being used by mistake. Also check to see if there's a switch marked '-10/+4dB' (or similar) round the back. Set this to +4 position, or whatever is the higher level.

Aux send on/off (button)

Some mixers have switches so the FX send and return levels can be pre-set, then switched in. A danger with some mixers is that this on/off switch is either adjacent to, or in some way tied-in with, the pre/post setting. Hitting the wrong button and changing post-fader to pre-fader will usually cause some embarrassingly high levels and loud FX – definitely the wrong effect.

Direct output

This feature is found on some mixers, generally as a jack/quarter-inch socket. It's mainly meant for recording, but can sometimes provide another (post or even pre-fader) aux send – and can possibly be used for monitor mixing too (more of which later). Note that 'inserting' into it can sometimes break the signal path to the sub-groups – unless the plug, cabling and socket are wired to provide a return path. It can also be used to provide a feed to another mixer – possibly to avoid using a splitter (see Chapter 1).

FX return EQ

Sometimes more elaborate mixers may have EQ on the FX returns, which enables you to tweak the sound of the effect without altering the direct sound. But the EQ provided here is often a cut-down version of what's on the channels. The usual approach, when EQ is required, is to return through a channel, so full EQ is available.

Special returns

Some mixers have specialised return circuits. Mixers with stereo channels, for example, will usually also have some stereo returns. These can save space and reduce the need to balance two lots of control settings.

PFL switch (pre-fader listen, 'solo')

Use this switch to bring the right amount of level into the channel. Allows the channel's signal level immediately before the fader to be monitored – usually over cans. Even if the fader is right down, the signal can still be accessed.

Some mixers have a special button in the master section which resets the PFL buttons to become 'solo in place' (explained in a moment). Then any channel with its PFL selected is heard over the headphones at its after-fader level, and as panned.

AFL (after-fade listen – or 'post-fader' listen)

This switch lets you hear the signal level in the channel after the fader has had its effect.

Mute

A button that 'disables' (stops) the channels' output passing onwards to other parts of the mixer – usually the subgroups. On most mixers an adjacent red LED indicates that the mute is in operation.

Solo in place

A potentially dangerous button that mutes every channel or group other than the channel whose solo button is activated. This signal alone is sent to the mixers L & R outputs.

If the solo'd channel has no signal, the whole mixer will appear to be dead. Note that the channel's PFL switch acts to provide what's called plain 'solo' – hearing just the selected channel and no other.

Channel fader (linear fader/slider)

Nearly always at the bottom end of a channel strip, this is a sliding knob/button control used for moment-by-moment level (volume) balancing of each particular sound source.

It's used in conjunction with, but shouldn't be confused with the gain control (described earlier), usually found at the top of the mixer channel. (For details of the way the two work together, see 'Gain Structure' later in this chapter – p46.)

Faders themselves should be ultra-smooth and light in movement. Better quality mixers have faders with a longer 'stroke' (greater travel) for ease of finer adjustment.

What mixer makers rarely mention is that the 'conductive plastic' (CP) type of fader is the only trouble-free, good-sounding technology. Up-market consoles use faders which may be typically guaranteed to keep working for 10 million movements.

Ordinary 'carbon' faders (whose signal-level-reducing 'resistive' element is a carbon-based film that wears away quickly) represent far cheaper 'throwaway technology' which also degrades sonics by adding noise and distortion.

Some intermediate types, such as 'moulded carbon' or 'cermet' can almost rival CP faders for low noise, if not for sound.

Master section

Main outputs (stereo outs)

This section of the mixer/board is where the main faders live. Normally, gain settings are 'normalised' so that these faders are at 0dB when the system is just about to enter clip. If not, the mixer or PA's gain-structure needs looking at (we'll do this very soon). There might also be a mono output fader – useful if any central PA speakers are in use.

Faders themselves should be ultra-smooth and light in movement

---

Matrix

This is mainly a feature for theatre use. For live music mixing, it allows you to create more final outputs from the mixer – for example to feed a mono signal (L&R mixed to mono) to a central PA cluster, or feeds for delayed PA stacks (L or R), and feeds for recording (L&R).

Solo to main out (button)

In the UK this is jokingly called the 'P45 button' (a P45 is a government tax form you get when you leave work, or get fired). If depressed, this button kills everything over the PA, except one selected feed. And in the process perhaps cause a mix engineer's dismissal...

It's really made for studio work, though it can be useful when setting up the PA at the soundcheck. On better-designed mixers it has a transparent flip-top cover, so the button's status can't be missed, and so you can't press it accidentally.

Master PFL

This may have a 'solo' button that makes the feed to your headphones 'solo-in-place' – so you hear only the channel you have selected. In pre-fader position you get a loud signal with no pan information; if set post-fader, you get a quieter signal also with the pan setting.

Headphone output (cans socket)

You'll need some 'cans' for unobtrusive monitoring, fitted with the same quarter-inch (usually) stereo ('A- gauge', normal TRS tip) plug as home hi-fi headphones. The trouble is, mixer makers don't supply headphones and don't even specify beside the socket what impedance the cans need to be.

Headphones can be low impedance (approx 3-30 ohms), medium (30 to 300 ohms), or high impedance (300 to 3000+). Using the wrong ones won't normally cause damage, but it can cause problems – for instance not enough level, even with everything turned up full, or perhaps too much level (and possibly distortion) even at low settings.

Control room section

Mix consoles that can 'double-up' for studio work will usually have a 'control room' section, which may be part of the master section. For PA purposes, this can be used as an added or auxiliary stereo output. It can also be used to supply tape, DAT, or CD-R machines etc for recording the gig. Or it could drive a pair of extra monitor wedges or PA cabs, so the FOH engineer can hear their mix (or PFL feeds) over speakers.

Utilities module & talkback

Again usually on studio-oriented mixers – facilities in this section typically include a basic signal 'tone' generator which can switch between frequencies whenever some pure test tones might be needed for lining-up heads on a tape recorder.

This isn't going to be relevant for PA, though some better-equipped mixers may include a pink noise source as well, which can be useful (see Chapter 9). Switches will be arranged to route these test signals to channels, groups, or direct to the stereo outputs.

This section may also feature a talkback (intercom) system, to enable communication between engineer and performers (or stage crew). It typically comprises a 'stub' (small, apparently sawn-off mike fixed onto the desk) or gooseneck mike (with a thin flexible stem), or at least an XLR socket for plugging in one of these; plus a built-in talkback speaker, volume, talk/listen switch, etc.

It should have latching and momentary switch positions (so the talk/listen switch can be kept permanently on, hands-free, or released when not held down) to make both rapid 'ping-pong' and longer one-sided communications easier.

By routing the talkback mike to the monitor sends, the FOH mix engineer can talk – or secretively whisper – to individual musicians through their monitors.

Groups/sub-groups

Let's say you've set the drum sound by balancing the various drum mikes, which are fed into perhaps six, seven or eight channels on the desk. But now you find the overall volume of the whole drum kit needs to be adjusted – it can be a panic, since it's no easy task to keep all the faders at their relative positions (keeping the balance of the kit) while moving them up and down.

Having sub-groups solves the problem, allowing the level of any sources you want to be grouped together and governed as one.

In general there's a button or switch – usually just above each channel fader – to route (or direct) the signal (which has just been panned into a L&R pair, as described in the 'pan' section above) to one pair of the eight (or however many) sub-group faders. Then there's another switch adjacent to these to route all the subgroups onwards to the usual pair of output faders.

Having selected, for example, groups 1 & 2 for all the drum kit's channels, the first sub-group fader on the right of the mixer now lets us control the whole kit's level. Other possible candidates for sub-grouping are backing vocals, horn and brass sections, choirs, and players using a stack of keyboards/synths.

Sub-groups can have their own aux sends and insert points – so, for example, reverb or compression could be applied overall to a group of vocalists, rather than them using independent effects and settings. (Refer back to the 'Aux Sends' section for more detail on this.)

Sub-groups usually also have their own pan pots, and also PFL switches. Except these are usually actually 'AFL' (compare to channel PFL section above).

Sub-groups were once a sign of a pro-level high-quality desk, but now they even appear on quite basic mixers. An older budget mixer without sub-groups is not always a lot of fun to use for PA if you have more than one signal source per instrument.

Mute groups

As well as having individual channel mutes, it's common to have mute buttons for sub-groups (as well as for the main stereo, mono, and other outputs).

Mute groups are a handy way of 'assigning' a desired selection of mute buttons on individual channels and/or groups to a single (other) 'mute group' button. This lets you, for example, mute all the drums with the hit of one button. It's effectively like 'sub-grouping' the mute controls – and the actual grouping of the signals for muting can be different from the existing groups.

For instance, drums and keyboards may have their own, separate sub-groups, but they can be muted as one, so they all cut out together. Or say the bass guitar and drums are 'locked' together as a sub-group for musical purposes, but when the bass player does a solo (sometimes allowed) all the drum sources can be muted, then brought back at a button-press afterwards.

The actual grouping of the signals for muting can be different from the existing groups

On desks with some degree of computerisation, such facilities and settings can be programmable – stored as a 'snapshot' and recalled at the push of a button.

Line-only mixers

Be wary of 'line mixers' – these can look similar to full-spec mixers, but have no facilities for microphones – for instance no phantom power, 'pad', or high enough input gain. There may not be a gain control at all, as all inputs are assumed to be at 'line level' (commonly +4dBu @ 1V in PA systems). These mixers are suitable for DI'd or DJ feeds, so they make sense where mike level inputs aren't needed. They tend to cost less too.

On the subject of low-cost, you may find some small budget mixers with rotary faders (knobs instead of sliders for adjusting channel level) – these are fine for wine bars, but not ideal for

FOH live use, as they're not going to be easy to control quickly.

Gain structure

This describes the system of (and means of assessing) changing levels throughout a mixer's signal path, which should never be unnecessarily low or high.

Well-designed mixers that come with a proper technical manual should have a gain structure chart that demonstrates levels and settings through the mixing chain that will maintain good headroom and 'dynamic range'.

As an example, let's say the gain (or input 'trim' at the top of the mixer) is set high in one section, or else there's quite a few dB of EQ boost, and it's causing signal levels to be so high that there's occasional overload and sonic break-up. Now say that this high level is reduced (attenuated) at the next stage, often by the channel fader, so the signal level appears normal.

This is an example of bad gain structure. The overloading shouldn't happen in the first place. But as modern mixers are so adept at handling different levels at every stage, and are not fitted with multiple clip indicators (overload is usually only read at a particular point on each channel), the root cause of the 'bad sound on peaks' problem can remain hidden.

Symptom of bad gain structure
There are several possible signs of bad gain structure, which include:

- Severe distortion on peaks (or all the time).
- Hiss and poor signal level, even with everything at maximum.
- Overload LEDs lighting when signal levels are not exceptional.

- Knobs turned up or down by unusual amounts, for regular signal sources.

What to do about it
Start at the beginning of the mixer's signal chain:

If the 'signal present' LED is unlit, there is not enough signal reaching the mixer input. If possible, increase the signal at source, perhaps by better positioning of a microphone, or turning up the volume on an electric/electronic instrument.

If the 'peak' or 'clip' LED or meter goes red, reduce either the gain control knob (*not* the channel fader) and/or the amount of EQ boost being used.

If the meter goes red, reduce either the gain and/or the amount of EQ boost being used

(In better mixer designs the 'peak' LED lights up long before anything overloads, say 6dB (or more) below clip. This helps keep some headroom – rather like the effect of adjusting a speedometer to read 10mph higher than it should, so you drive more slowly.)

Use the 'pre-fade listen' (PFL) to check for individual channel sonic break-up and distortion (using headphones). If overload/distortion occur even with an 0dB setting, or the level is too high to have the fader at the 'U' position, use the pad to knock it back.

Once a channel's input gain is set, and you're happy you haven't over-EQ'd the signal, small level adjustments in the mix are made using the fader at the bottom of the channel strip. Though

Measuring levels
'Peaks' are high-points, or tips, of individual sound waveforms. The averaged-out levels over time are known as RMS (root-mean-square). When RMS and peak values are compared, it's called peak-to-mean ratio (PMR), or 'crest factor'. PMR

shows how 'spikey' or lively-sounding the music is.
To take individual instruments: a bass guitar has high RMS levels compared to peaks that aren't a lot higher, while a snare drum is an example of the opposite – high peaks, with a low RMS level.
When it comes to overall mixes, the

PMR of dreary musical 'wallpaper' is about 4 to 6dB, while with highly compressed commercial radio it's maybe 8 to 10dB. But live 'electric' music can be 10 to 20dB or more – up to 30dB with some jazz and big-bands, classical orchestras and some types of ethnic/world music.

THE live SOUND MANUAL

bear in mind you may still have to alter the gain during a show, maybe because of changes in mike distance from the sound source, or even a forced change in the mike used.

Once all the input gains are set-up – so the balance of sounds is good – look at the faders. These will likely be in different, random positions. But it's actually best if they are all set roughly in a straight line, and about three-quarters of the way up (around the point marked '0dB') – which allows space below for a smooth fade, and some space above for slight increases.

Being able to use this setting under normal conditions also suggests the mixer's 'gain structure' is about right. The next job – a routine one – is to re-adjust (in a compensatory way) the gain setting of each channel that has out-of-line faders, so the level stays the same while repositioning the fader. So you'd listen to that source, make a mental note of how loud it was in the mix, adjust the fader to the 0dB area, then concentrate on resetting the gain to get back the level you remember.

This may be called 'gain normalisation'. It sounds more intricate than it is, partly as it doesn't have to be absolutely accurate, since in many live shows the sound levels and balance requirements vary as you're doing it.

Metering

Meters are a more detailed way of monitoring signal level than the single 'peak' LEDs on channels (which just show if a set maximum level has been exceeded). Even so, metering still only reads signal levels at certain points in a mixer. So be aware that it might be 'blind' to some situations – for example, too much input gain, a bassy signal, but bass EQ turned to max cut.

Metering comes in two main forms: LEDs (light-emitting diodes) arranged in bars, or sometimes a curve; and mechanical 'swing-needle' meters. Both sorts can give a good visual feel for signal dynamics, so long as they're designed correctly. You certainly can't tell how helpful they will be until you use them for real.

Although vastly better for real use than the black-on-grey LCDs (liquid crystal displays) used on disposable 'toy' products (the ones in plastic cases), LED meters (and similarly bright,

fluorescent displays) can still be hard to read in either full sunshine or bright stage lighting. A black hood (or cowling) around them would help, but is rarely found.

Old-style swing-needle meters will usually need to have illuminated (back-lit) scales. Mini incandescent lamps are traditionally used for this, and two are needed to light most meters clearly – these lamps will work for ages, then fail at annoying random intervals as they start to wear out. Some kinds are more fiddly to replace than others, depending on the type of mini lamp employed and the ease of access into the meter's housing. (Some better-designed consoles may have a stowage location in which to carry spares.)

The most basic provision of metering on a mixer is two – one at each of the main stereo outputs. If the mixer has groups, each of these may have a meter too. That's typically from four

Any meter that can read both peak and average levels simultaneously is highly useful

to 16 more meters – already quite a lot of readouts to accommodate, especially as they're usually large displays, ranged along the top of the mixer (on a so-called 'meter bridge').

This is why only more up-market mixers will have meters for all the inputs as well. Even here, to save on expense as well as visual overload, the sub-group meters may be made assignable to the input channels needing them most.

With most mixers above low-budget types, there should be a meter for the PFL and AFL (it may be marked as either). This allows a quick visual check. On lesser mixers there is no dedicated metering for this, but when the PFL button on a given channel is used the PFL's indication is shown on the main L/R meters.

If the mixer were also set to 'solo in place' (on a particular group) at that point, then you'd see the after-fader level (AFL, or post-fader) and pan information (the left and right, L/R levels) on the L/R meters.

What meters measure

VU meters read average levels. They have a linear scale in 'volume units', which work like dBs. Here, '0VU' (zero volume units) is officially +4dBu.

Even well-made VU meters (which meet British 'BS' and international 'IEC' or 'UL' standards) give a reading which isn't real RMS – VU meters tend to under-read all peak levels – by unknown amounts, but at least it will be consistent, and close enough for PA purposes. So with bland 'piped' muzak, a VU meter will under-read the real peaks by about 5dB, on live music it'll under-read by maybe 15dB, and with some jazz or big-bands by maybe over 20dB.

A VU meter indicates roughly how loud the sound will appear

This is the downside of the VU meter – it cannot help you accurately see or know the peak levels that are causing overload, giving you little idea as to how big they really are. The usefulness of the VU (the reason it's been around for over 60 years) is that it successfully indicates roughly how loud the sound will appear – so it's useful for mixing, as a second opinion, assisting the engineer's ears.

Peak-reading meters read the peaks that cause mixer (and PA) overloading. There are several formats: the original and probably still the best is the BBC's own PPM (peak programme meter), which has an unusual scale, numbered 1 to 7. Still, PPMs are rarely (if ever) seen on PA consoles, likely because of their relatively high cost and large size.

Which sort of meter is best?

It's a question that cannot be definitively answered – you really need both VU and peak. As space is rather limited on most mixers, any meter that can read both peak and average levels simultaneously when required is highly useful.

Bar-LED meters (where lights are arranged in vertical or occasionally horizontal graph-like lines) may read levels directly, and you can switch between peak or average readings.

The peak-reading 'ballistics' (the durations of 'attack' and 'release' for meaningful peak readings) may not meet particular or exact standards, but they should at least have been 'tuned' so peak levels sustained enough to be heard are displayed long enough to be read. This means peaks lasting over ten thousandths of a second (for instance) should be stretched to last over 1/3rd second to make it easier to see.

Or there may be LED 'dots' that show the peak level riding over the average level, which is shown by the usual solid bar. This is seen on some tape decks. But again the meter's ballistics must be suitable (fast peak capture and hold) for professional use.

Some VU meters have a single 'peak' (overload) LED, ideally with the signal level (in dBu) above which it lights, marked beside the LED.

And it's always possible to improve a mixer's metering for certain jobs by connecting 'outboard' meter units, such as up-market, large LED displays.

Using your ears as meters

It's worth listening to a single channel, through the PFL, on the cans you normally use, to learn what clipping on a given mixer will sound like.

Put various instruments (live rather than recorded will be best) up the channel, adjust for overload, and 'log' in your mind some of the sonic characteristics. Turn down the PFL volume as you turn up the channel gain, to keep a constant level, so you can hear the distortion without the loudness.

Then try the same thing with some boosted EQ. What does the EQ sound like when pushed too hard? If you learn/take-in these characteristic sounds, it could help you to reach for the right knobs quickly during a live show.

Monitor mixes from the FOH mixer

With very small and low-budget gigs/tours using mid-to-low-priced mixers, monitor mixing may be done from the FOH mixer using the 'Aux Send' knobs to control the signals for each monitor.

Your mixer might have two, three or four auxiliary controls, but they won't all be suitable for producing a monitor mix. That's because some (usually Aux 3 & 4) access the channel signal after the channel fader – these ones, called 'post' or 'post fader', are used for FX instead.

If the mixer has just two auxiliary knobs (Aux 1, 2), Aux 1 will be the one to use. This means there can be only one monitor mix. On more elaborate mixers with four auxiliaries, you can often switch the Aux 2 and 3 knobs between pre and post-fader duties. So, assuming Aux 2 and 3 aren't needed for FX, you would switch them to 'pre-fader', and then be able to make three separate monitor mixes.

Dedicated monitor mixers

Dedicated mixers for artist's monitoring are quite specialised bits of equipment – there are only a handful of pro-quality makers. They have different requirements from 'ordinary' (FOH) mixers, and so have some different controls and features (which we'll come to in a moment).

Historically, cheaper monitor mixers had no faders on their channels, only knobs, in order to create enough space for the extra controls on each channel strip. Today, thanks to miniaturisation and improved design, the better monitoring consoles do have faders. These come in handy to quickly 'swipe down' the level of a particular instrument that may suddenly come in too loud – for instance if the guitarist stamps on an FX pedal that dramatically raises the level.

The extra knobs on a monitor mixer are mostly auxiliary sends – one used for delivering each monitor mix to the relevant performer. There are no stereo outputs, just lots of groups: one per monitor mix – eight is the usual minimum, but some mixers have 24 or more. If the monitoring console has the capacity, you should be able to assign an output group to each individual musician, and similarly freely direct any input channel to any output group (mix).

Monitor mixers generally have the following facilities (some of which also appear on FOH mixers or multi-purpose mixers suitable for both jobs):

Phantom Power

As mentioned earlier, this refers to the power (usually 48V) that's required for capacitor ('condenser') microphones to work. It's provided on all professional mixers, but the job of supplying this power is nowadays mainly in the hands of the monitor console engineer.

Switching it in sends the required power up the mike cable 'invisibly', without the need for any other wiring, special connections, or external power sources. It shouldn't be used with unbalanced mikes (which you should preferably not be using at all anyway) – though even if phantom power is switched on when using a pro-grade, balanced moving-coil/dynamic mike, there may be a loud 'bang' from the PA (unless your PA has a front-end design that employs a 'soft-start' method to help avoid such problems). Otherwise there shouldn't be any long term ill effect, as the power stays 'invisible'.

But you still need to take care not to knock this switch on or off accidentally. Strictly, you should only ever switch it on once a capacitor/condenser mike is connected to the particular channel input. If it's on all the time, there's a waste of the mixer's phantom power supply (often of limited capacity). Far more importantly, plugging-in or unplugging any mike – even a capacitor type – when the phantom is already switched on, can damage some mixers' 'front-ends' – sometimes only adding noise or distortion, but occasionally zapping it.

Only switch on phantom power once a capacitor mike is connected to the particular channel

In general, to prevent unnecessary stress to the mixer, it's also a good idea to turn off all phantom power feeds as soon as a gig (or set) finishes – before the crew start 'tearing down' (disassembling the stage rig). Fortunately, some capacitor mikes you'll be using are likely be left pretty much alone throughout a set – for instance those used around drum-kits.

Random unplugging is more likely with vocal mikes, which can often be disconnected by artists, by interaction with an audience, or by stagecrew deftly swapping mikes. So special care is needed when using phantom powered capacitor mikes for vocals – unless you know the mixer is able to take the 'hammer-blow' of a mike being disconnected without warning.

Solo button

This enables any particular mix to be routed to an output. On a monitor mixer this can be fed through an identical amp and 'wedge' monitor speaker as the performers are using, placed next to the monitor desk so the mixing engineer can directly experience any of the mixes.

PFL/Off/AFL

A (typically) three-position toggle switch which allows rapid troubleshooting of any channels that are causing howls and screeches. Both the AFL (after-fade) and PFL (pre-fade) positions allow the channel to be heard on the cans (as with FOH mixers), or else over a wedge monitor, so the monitor engineer can directly experience what the musician is hearing.

The AFL is useful to gauge the relative loudness of the different sources or mixes. To do this, it has to be (and usually will be) 'post insert' – so taking into account the return from any inserted EQ units, which could well greatly alter the level.

EQ on inputs and outputs

This allows either overall EQ of the particular source, or else independent tweaking of the sends to particular wedges (usually in front of the artists, pointing up towards them), sidefills (at either edge of the stage, pointing in) and 'drum fills' (dedicated drummer monitors). (See Chapter 6)

On more sophisticated monitor mixers – for instance if the output EQ is a multi-parametric – the desk's built-in EQ may be good enough to avoid the expense and hassle of an outboard graphic EQ (see Chapter 3) for each output.

The monitoring set-up is more prone to feedback than the FOH PA, which explains why several types of EQ may be in use:

Ordinary mixer EQ – On multi-purpose and dedicated mixers, this will need to comprise the regular bass and treble shelving EQ, and at least two (and hopefully more) midrange 'bell' EQs,

with sweepable frequencies (as described in the FOH mixer section earlier in this chapter).

Graphic EQ (GEQ) – To be useful for precisely pinpointing monitor feedback, a graphic EQ needs to be at least a 28 to 31-band unit, with 1/3rd octave bands. For this reason it's usually a separate outboard unit, rather than built into the mixer – especially bearing in mind you may need one per channel. (There's more on how graphic EQs are laid-out and used in Chapters 3 & 4)

Parametric EQ (PEQ) – As already mentioned in the FOH mixer section (and see also Chapters 3 & 4), this is the most powerful form of EQ because it not only allows boost/cut of a sweepable frequency range, it also has the 'Q' control for adjusting the sharpness or spread of the effect.

With simple sweepable EQs only the exact frequency you want can be found. And of course adjustable levels on a graphic are at fixed frequencies – for instance if you need to tackle a particular feedback frequency but find it falls between two pre-set points on a graphic EQ, you might need to adjust two knobs, with the result that a much wider range of frequencies are affected unnecessarily.

The monitor set-up is more prone to feedback than the FOH PA, so several types of EQ may be in use

Unlike bulky multi-band graphics, pro-standard PEQs can be quite compact (because they use tunable knobs rather than individual sliders for adjustments). And because a PEQ can be steered to exactly where it's needed, you can do a lot with just a four-band unit. Either each band will sweep over most of the audio range, or the band ranges will be staggered so at least two or three of the bands can be deployed in any frequency area.

This is why multi-band PEQ is included in the

output channels of some monitor mixers. Alternatively you can insert an outboard PEQ unit into a channel, or place it in line with an output (see relevant sections in Chapters 3 & 4 for more on this).

FX sends

On a dedicated monitor console you have to use one of these (hopefully plentiful) outputs.

FX returns

A mixer's dedicated returns aren't often used for monitoring. Instead signals have to be returned to another input to allow full routing and EQ flexibility.

Output matrix

Found on 'high-end' mixers only. Part of the sub-groups section, it enables you to make a mix of channels and groups and outputs. It can be used to set up 'derivatives' of mixes: for instance you can use a matrix to set up mono sub-bass from stereo drums or keyboard mixes.

It can also be used to send a given mix to several monitoring locations at differing levels, also making it possible to mute those monitors on different cues. Practical examples might include upstage 'ramp-fill' monitors, drummers' headphones, or feeding different mixes to 'ego' ramp-fill monitors for solos (maybe the guitarist standing high up on a scaffolding or platform). Another use is to get lots of added monitor sends.

Stereo sends

Some musicians are treated to stereo monitor sends – for example keyboard players, drummers. Bear in mind this will double up the number of cabs, amps and EQs required.

Main (stereo) outputs

Simple – these aren't used for monitoring.

Mixer power supplies

All larger and many medium-sized mixers have an external power supply (also called a PSU – power supply unit) which changes the AC mains to various lower and DC voltages needed by the mixer. It does the same sort of job as a standard 'wall wart' power supply used for low-grade consumer goods, but is far more refined and rugged and provides more varied supply voltages and higher currents.

These PSUs may be rack-mounted, looking rather like power amps in shape, size and weight. Because such units are generally rather heavy, you need to be aware that the protruding 'panel parts' (switches, fuseholder caps, LED indicators) are fragile by comparison and may be easily damaged.

The PSU's front panel usually has some LEDs to show that the various different voltage supplies are working. Usually all the supplies must be present for the mixer to work. But one is the phantom power (+48V), and as long as there are no capacitor mikes the show will still go on if that one is dead.

Always carry a spare mixer power supply

The power supply is a critical part in a PA system. Nearly everything else in the PA is duplicated enough times that you can survive a failure here or there. But if the mixer PSU fails, it'll be hard to keep the show going unless you have a replacement supply to hand. Fortunately the internal electrical/electronic parts of mixer power supplies are generally pretty tough, but things can and do occasionally fail, so it's worth being well aware of the possibilities.

Over and above sometimes easily-damaged panel parts (such as plastic connector skirts, or flimsy plastic lamp covers), the main ways that the supply is most likely to fail are:

- Broken DC (multi-pin) power connector. Keep the unit well away from any position (such as passageways) where it could get kicked or tripped over, either by performers, crew or public. And carry a spare.
- Water getting inside. When using PSUs outdoors, keep them well 'sheeted' (set back under the mix position's tarpaulin covers). When indoors, be protective if there's liquid refreshment doing the rounds. Indeed, anywhere near the mixer's PSU is definitely a 'no liquids' zone – for electrical safety reasons

51

(more of which in Chapter 10) as much as for avoiding a failure.

- Overheating. Keep well ventilated and clean – regularly clear out dust from inside and from any/all cooling fins using brushes or a moisture-free airline (disconnecting the power first, of course). For more on gear maintenance, see Chapter 9.
- Fuse failure. This can happen randomly, from old-age. Make sure you carry two spares for all fuses. Make sure you know where all fuses are – internal as well as external types. If unsure, get a spec list from the maker.

On the more up-market PA mixers, power supply reliability is improved and worries reduced by the provision of dual PSUs (which may be in one larger or two separate units). Sometimes the idea is that one cuts in and takes over at the instant the other fails, or else both supplies share the work most of the time, and one takes over the whole job if the other fails.

The most important advice from experienced live engineers is to *always* bring a spare mixer power supply.

The DJ's mike is best plugged into the FOH PA mixer, if there is one

---

DJ mixers

Features & controls
The most important sockets are the deck inputs, which receive the signals from the turntables (those spinning devices used for playing vinyl discs). These sockets might also be labelled Phono (historically short for phonograph), Gram (short for gramophone), or RIAA (this was the name of an American audio standards body of the 1950s that established the best fixed EQ for vinyl records – all inputs for vinyl discs use, whether it's mentioned or not.)

The input sockets are usually phono/RCA types, as on hi-fi gear, but it's not hard (at least for

a technically-sussed user) to rewire a turntable and mixer to use more rugged pro-quality XLR sockets.

Some older or less mainstream mixer inputs may have some unusual switch settings on the back panel: 'CER/MAG' is a label found on low grade or very old gear, and was designed to make a tonal/EQ correction depending on whether your turntable used ceramic or magnetic cartridges.

Ceramic cartridges are now found only in the cheapest plastic 'hi-fi', but were the norm before Stanton and others introduced the DJ world to rugged and fairly affordable moving-magnet cartridges in the early 1970s. If you see this kind of switch, keep it set to 'Mag' – a wrong setting would be evident by distorted and tonally-peculiar sound.

Another switch label found on hi-fi oriented gear is MM/MC – where MM refers to the normal moving-magnet cartridge, and MC is its lower output relative, moving-coil. MC sonic quality is (somewhat subtly) higher, but signal output level is typically 10dB to 30dB (x3 to x30) lower, making electronics more expensive and possibly more prone to hiss and buzzes. If it's set to MC when an MM cartridge is used, the sound will be distorted due to overload; if mis-matched the other way, the sound will be too quiet.

Mike input
Essential for some DJs, unused by others. It's only necessary when the DJ's mixer is connected directly to a PA back-end (the amps and speakers – in other words where's there's no PA mixer). Otherwise the DJ's mike would be best plugged into the FOH PA mixer, as then the sound engineer has control, and the facilities are also likely to be better.

Where applicable, DJ mixer's mike inputs are similar to PA mixer inputs, but usually without a pad switch, phantom power, filters, or as good EQ. But they can be a lifesaver if all else fails.

Line socket
'Line' is anything else that plugs into the mixer – for instance CD, DAT, MiniDisc or MP3 players.

EQ
DJ mixers generally have less comprehensive EQs than PA consoles (there's far less space, for a start) – although the better types of DJ mixer do have powerful enough EQ to get a good vocal sound,

and/or tune away any howls/feedback. There may also be 'frequency cut' switches or knobs (used, for instance, to sweep the bassline levels up and down, or lose any particular sounds or vocals temporarily). Such EQ controls must be effortless to operate at speed.

PFL (pre-fader listen switch)

Much like the PFLs on a PA mix console, this switch permits you to listen to any of the signal sources for cueing-up and synchronising beats, etc.

BPM meter

Found on more up-market models, or you can buy it as an accessory – it helps the DJ check and match tempos between records.

Crossfader

Used for swapping quickly between alternate inputs (turntable to turntable, CD, or whatever). This needs to be very light, and easy to flick smoothly, but also rugged enough to stand up to constant use. For scratching, there may sometimes be a switch or knob that can alter the 'graduation rate' of the fader, which should go all the way from smooth segueing to on/off cutout.

EQ

The more up-market DJ mixers can include on-board FX, and tools such as a sampler, which is crucial for looping (playing snippets of riffs over and over) and time-stretching (slowing down or speeding up a section of a record without changing the pitch).

Daisy-chaining

Some mixers can be connected to others for multiple DJ events. With a PA system, this may be best achieved with the PA mixer – by connecting DJ mixers to separate PA mixer channels, then sub-mixing from there.

DJ Accessories

Outboard disc pre-amp/'headamp'

A standalone 'phono pre-amp' which handles magnetic cartridges and provides a line output which feeds to the mixer. This enables any PA-type mixer to be used by DJs, instead of being tied to DJ mixers, or ones with RIAA inputs.

It also allows more decks to be plugged into a given mixer which has all its existing 'phono' inputs in use. As a specialised unit, there's also a distinct possibility it'll be made to a higher standard than an integral unit. For example, some types may have balanced connections to help reduce hum and buzz, produce deeper bass, etc.

For best sounding results, outboard pre-amps made for high-end hi-fi may sometimes be used. Most outboard disc pre-amps have their own small outboard power supplies, ranging from wall warts to substantial metal boxes.

Never use ordinary phono (RCA) leads

Digital audio converter (DAC)

When CD players have S/PDIF outputs (the regular digital output, as seen on home players), improved sound can be achieved simply by making a good connection from here to an outboard D/A converter (DAC). There's a wide variety of both pro and domestic players, but even if the units are made for home audio, fitted with phono/RCA sockets, the S/PDIF's ground isolation means there should be less hum'n'buzz than if the CD player was connected directly to the mixer.

Professional quality DACs that cannot accept S/PDIF won't be any use, not unless the CD player is one of the rare (professional) units that has an AES/EBU output – the other regular sort of digital audio link for players.

Plugs & cables

Even if the S/PDIF connectors are phono/RCA sockets, you should never use ordinary audio phono-to-phono/RCA-to-RCA leads. If you want decent results, only use leads made with proper 75-ohm RF coaxial cable, which can cope evenly with the wide range of high radio frequencies in a digital signal – otherwise sound quality can be impaired through jitter.

With the air full of so many mobile phone signals (and other RF garbage), cable shielding is also crucial for clean digital connections and consequent good sound.

the fx rack

outboard EQ, compressors, reverb etc...

Quick view

If you want to use more than just the mixer's own internal EQ to enhance the sound, you'll need some additional 'outboard' tools and toys. 'Outboard' means not built-into the mixer, but joined to it. The most widely applied of these are 'signal processors' (commonly called 'effects' or FX) such as compressors, reverb, delay, plus perhaps pitch-shift and various 'special FX'. You may also need more powerful EQ units – graphic or parametric EQs. All these units are usually kept together, screwed into specially equipped racks positioned within easy reach of the mix engineer. They will be connected to the mixer via either the auxiliary send & receive circuits or the channel 'inserts'.

much of the rackmount outboard gear discussed in this chapter may be used for either front-of-house or monitoring purposes – each mixer will have their own dedicated racks. Those devices mostly concerned with sound control (such as precise EQ, compressors, limiters and gates) will be of greater relevance to monitoring than the more elaborate effects such as echo and pitch-shift.

EQ

Graphic EQ (GEQ) – mainly for monitors

With speakers aimed at, and so near to, so many mikes, monitoring requires quite powerful EQ to prevent feedback. Graphic EQ is one widely used approach. (It's called a 'graphic' because the position of the sliders gives an approximate view of the sort of frequency response curve that you're imposing on the signal.)

To be useful for monitoring, a graphic EQ needs to be a one-third-octave type (this means having narrow, quite highly selective frequency bands – each one-third of an octave apart). As the audio range comprises about nine or ten octaves, any third-octave GEQ will generally have between 28 to 31 bands (and thus sliders).

They also have to be big enough to give fast-moving professional fingers proper access, which explains why GEQs are almost always outboard, rather than being found built into the mixer – especially considering that you may need one graphic per output channel.

This creates a pressure to use compact devices – single-channel graphics are typically 2U (two units high), and two-channel units are in 3U high cases. But if you can afford the rack space, the more 'classic' 3U single-channel, with the longest 60mm 'long-throw' size of sliders, can offer finer control. Case height alone is not the key, as some older 3U-high GEQs have shorter sliders. Dinky 1U units are to be avoided for this job.

Also note that for working on monitors, where the graphic is being used solely to suppress howls, boost settings won't be used to the same extent (dBs of knob travel). Some specialised PA units have a 'cut only' facility where all the faders are 'in neutral' (set at 0dB = no effect) at the top position.

This is less useful when there are genuine reasons for boosting – such as correcting a mid-range dip (caused by whatever reason) that's affecting the 'intelligibility' of what the players are hearing.

Beginners often make the mistake of thinking that the more boost a graphic offers the better it is. A good GEQ will have more cut than boost (say +6dB/-14dB) – so-called 'asymmetric' EQ.

Occasionally a GEQ is used on an individual FOH channel (often vocals) to achieve effects that the desk's channel EQ can't. It's connected through the channel insert point, and is located in the FOH 'inserts rack'.

(There's more about using GEQs in the 'House Graphic' section in Chapter 4.)

Parametric EQ / PEQ – for FOH & monitors

As already mentioned (in Chapter 2), this is the most powerful form of EQ because it not only allows boost/cut of a sweepable frequency range, but also has the 'Q' control for refining the spread of the effect. Its greatest value is for suppressing feedback by reaching the precise frequency area causing the problem.

Compared with graphic EQ (where the frequencies at which boost/cut takes place may be plentiful but are also fixed) and sweep EQ (where the width of the affected frequency area is fixed), parametric EQ allows you to find both the exact frequency and narrowest area you need to suppress/cut.

For instance, if you need to tackle a particular feedback frequency but find it falls between two pre-set points on a graphic EQ, you'd need to adjust two knobs. This not only takes longer, but means the end result would be that a much wider range of frequencies are also affected. This gets serious when there are maybe ten howling frequencies – then maybe two-thirds of the knobs on the graphic would be set to cut, leaving a rather 'bumpy' frequency response.

Unlike multi-band graphics that are big enough to have solid and accessible sliders, outboard PEQs can be quite compact. This is mainly because you need far fewer channels, since each can be steered to exactly where it's needed. In fact, you can do a lot with just a four-band PEQ. Either all bands will sweep over most of the audio range, or the band ranges will be staggered so at least two or three bands can be deployed in any frequency area.

Digital EQ insertion – for FOH & monitors

Benefits of digital EQ include theoretically perfect matching between any number of channels, and tracking of that whenever knobs are tweaked. Also, boost should be quieter (as there is no EQ circuit to create noise).

With a modern digital processor (such as XTA's SIDD – Seriously Intelligent Digital Dynamics), it's possible to see a graphic-type display of the frequency response that the graphic is creating on a hooked-up laptop computer. This can help make PEQ easier to use in the high-pressure environment of a live show.

Reverb & echo

Low-budget units often provide both of these together. If you're on a tight budget, reverb is the one essential effect. It's the first FX unit that a band's rider (their contract with the PA supplier) is likely to specify.

Reverb

This is useful on most instruments, but

Rack gear

Rack-mounted gear either comes in a universal width of 19 inches/482.6mm (that's the length of the front panel, from ear to ear), or if smaller may be adapted to fit using a 19in housing. The height of rack gear is measured in 'units', abbreviated to U. 1U is equal to 1.75in/44.45mm, 2U is 3.5in/88.9mm, and so on, up to 6U, which is 10.5in/266.70. Depth (and weight) of units can vary considerably.

particularly on vocals. It adds various kinds of 'space' (selectable via menus with everyday names like 'large room' on digital units), recreating various natural or deliberately unnatural ambiences, or it can just generally 'sweeten' the sound.

For outdoor sets, where there is nothing solid to bounce against, the PA would sound very flat (dry) without reverb

If the aux send used for the reverb is suitably EQ'd, then (for example) a venue with a naturally 'dark' (bassy) reverb can be 'lightened'. In this case you'd boost the reverb's HF and also cut the LF, so the PA adds reverb only at higher frequencies.

For outdoor sets, where there is often nothing solid for any of the stage sounds to bounce against, the PA's overall sound would be very flat (dry) without any reverb to add some lushness. (Reverberated sound is often called 'wet' sound.)

Echo
This is more of a 'fun' FX – although it also creates its own kind of impression of space, and a certain 'mood'. A short type of echo (sometimes called slapback) was a familiar part of the original mid/late 1950s rock'n'roll sound (Presley, Little Richard, et al), and John Lennon also used it a lot in the 1970s. This was originally achieved using short loops of tape (as in the Watkins Copicat Echo), but has been largely superseded since the mid-1970s by ADT (automatic double tracking) or DDLs (digital delay lines).

Echo is commonly 'spun in' (manually added)

by turning up the appropriate aux send knob, just to catch the end of a phrase or last words of the vocalist – usually to effect a rhythmic repetition for emphasis, or simply to fill a gap.

The echo can also be made to (creatively) feedback on itself, for greater effect – the echo is returned into a separate channel, and fed back via the same aux bus as the original input, using the channel's fader to control the amount of 'regeneration' (echo feedback). Indeed, modern reverbs also have a knob that does the same thing, but it's not half as easy to use on the fly as the mixer's fader – especially if the echo's control panel is of the ergonomic nightmare type (layers of annoying menus, penny-pinching plastic controls et al).

Be careful, though. If you increase the return fader level beyond the point where the gain exceeds losses, electronic feedback can take control. This might not be the effect you want, particularly if it damages the PA drive units, which it might (depending on the feedback frequency). Ease the fader up slowly, and note where the sound starts to change and 'lose control', and maybe leaps into a squeal. A piece of tape (or Blu-tack, Buddies or similar) can be placed to stop the aux return fader going beyond a safe point – say 3dB below where electronic feedback occurs.

(Noise) gates – FOH or monitors

Gates are basically channel-muting devices which can be set so they automatically switch off the signal when the levels are too low to be contributing anything useful. In effect, they automatically cut off a mike deftly when the sound its receiving (and broadcasting over the PA) is deemed inapplicable, then automatically reconnect it whenever a 'relevant' sound is received.

For example, a gate could switch off the signal from a snare drum mike in-between snare beats (allowing for the strike's decay/sustain to finish), or

Open & closed confusions
The terminology used to describe the working of a signal gate is similar to that of a tap or garden gate: in other words when the gate is 'open' the music passes through; when it's 'closed' the signal is impeded, either slightly or completely. For any readers who are trained in electrical engineering or electronics – and everyone else who may move between reading about gating and the rest of audio switching – it's important to note that this is the total opposite of the terminology used with other forms of switches and switching, where 'open' means that the passageway is stopped (like when a swing-bridge opens and traffic can no longer pass along the road).

while a trumpet player takes the instrument down from his/her mouth and is waiting to play again.

Gates were originally designed to be used in studios to help reduce noise (such as hiss) from a mixer channel that was temporarily empty. In PA work, mikes aren't manually turned on and off during a set, and temporarily unused channels' faders can't always be dragged down, in case the mixer person misses a cue or the player unexpectedly improvises. So each open mike ends up relaying random 'spill'. This is much worse (for live sound) than a spot too much background hiss.

Also, unlike with recording, an open, ungated PA mike can cause feedback, or at least it lowers the threshold level.

Gating can clean up the FOH sound – stopping unnecessary random spill entering the mix

So in its basic mode, gating can be used to clean up the FOH sound – stopping unnecessary random sound spill entering the mix, as well as helping prevent feedback from unwanted mikes. But gating can also have sound-enhancing effects. To extend this artistic side, some gates are 'expander' types that don't turn sharply on or off; instead the signal can be faded up or down – for instance the 'closed' gate's signal isn't fully turned off, just 'gain reduced'. In this way, gates become creative dynamic processors, and the line between gates and the compressor family (which we'll come to later in this chapter) gets rather blurred.

It can be used, for instance, to tighten-up a bass guitar by removing messy between-note fingering or unwanted noises. It can achieve a cleaner sound on a kick drum, snare or tom by removing the sound of the drum resonating sympathetically with the other drums. A gate can also be used to make one sound trigger another – for example replacing a poor-sounding (but in-time) kick drum by using it as a 'key' to trigger a better (perhaps sampled) kick drum sound.

Gate controls (in order of functional importance)

Threshold – a knob that sets the signal level below which the gate closes (stops signal) and above which it opens (passes signal). A typical practical range is wide, from +20dBu down to -50dBu.

Key Selection – a switch which determines whether the gate triggering is from its own signal ('int', meaning internally derived), or some other signal ('ext', from an external source)

Key Input – the input for the external source. Only used if 'external key' (above) is selected. The signal employed will have some sibling, sequential or artistic relationship with the signal being gated. (An example of 'sibling' type is that the top snare mike could be used to key the bottom snare mike's channel.) Connection is usually via a quarter-inch (jack) socket. This may be unbalanced – if so, see 'Sidechain Insert' remarks in a moment.

Key Filters – usually low and high-pass (LP & HP) sweep filter – much as covered in Mixers/Chapter 2. In this case, the two knobs are used differently – to home-in on any specific frequency range, be it wide or narrow, that contains the sounds that are required to perform the keying. The filters can be used whether the key is internal or external.

To be useful, the filters will sweep over most of the band, with an overlap between their respective 'ends' – for instance LPF = 20Hz to 4kHz; HPF = 200Hz to 20kHz. With the overlap (in this example between 200Hz and 4kHz) care needs to be taken not to set the two filters so they remove all the signal. This won't happen if the LP filter's frequency setting is always kept below the HP filters' setting. In higher quality, more modern units, these filters will have -24dB/octave slopes for tighter selectivity.

It's important to realise that the steep filters used in gates will delay parts of the signal that are at frequencies near the filter's cut-off frequency (as marked on the panel). This means that the gating will be nearer the speed you expect if you back off the filters, so they don't start to act at frequencies only just below those occupied by the required key sound.

For example, say the key sound is affected if the HP filter is set at 280Hz – then back off the filter to, say, 140Hz, and accept the small extra amount of unwanted signal 'stuff' that this lets slip into the key drive.

Key Output – this allows the key signal to be passed to (and trigger) other units (or channels in the fx rack

one multi-unit). This enables the first unit to act as master for the others, avoiding the problem of matching and keeping matched more than one lot of control settings, not to mention the problem of calibration differences across various units.

In other words, seemingly identical units may not give exactly the same result even if the knob is set in the identical position. Ideally the signal passed on will include the EQ, otherwise the EQ knobs on the next unit will need to be set and carefully matched.

Key Listen – a switch that on some units enables you to hear the key signal. This is ideally after the key filtering, since this helps you set up the key – because you can be sure, for example, that you do really have the feed you think you have, and aren't going to filter the wrong sound, or that the knob settings aren't just seeming to work right by some coincidence.

Sidechain Insert – a socket (or sockets) that can be used to insert (patch) a PEQ or other EQ (or other FX) into the key signal's processing (also called 'sidechain', this is explained under 'compressors' below). A PEQ is more able to latch onto the required signals' frequency. More creatively, try latching onto the signals' harmonics (at frequencies x2, x3, x4, etc above the lowest frequency component)

As on a low-budget mixer, a single quarter-inch, three-pole ('stereo') jack socket is usually used for the insertion send and return connections. This signifies unbalanced connections (a balanced connection would require two such jack connections) – but fortunately these need not be troublesome, as the key signal's functioning shouldn't be affected by the amount of hum or buzz that the unbalanced connections can sometimes cause, and which would be upsetting as the audible part of the PA's signal.

(There's more on connectors and connections in Chapter 7.)

Gate metering – gives an 'at-a-glance' feel for the gate's status and whether it's operating correctly. Gate metering varies between basic two or three-LED traffic lights (open, shut, or hold) on standalone gates; and bar-LED meters showing the gate's expansion or gain reduction on expander-type gates.

MIDI interface – available on some gates, and useful for triggering from note on/off information, sequencers, keyboards, etc. Example: for a techno outfit, the keyboards may be gated as follows – the keyboard is set to play a 'straight pad' (a continuous chord); a sequencer then controls the gating of this, chopping up the keyboard sound and reacting a percussive effect, at a rate that would be physically unplayable.

A gate can also be used to make one sound trigger another

The following controls are more for artistic than operational effect. These features are mainly shared with the compressor family (more on which follows over the next couple of pages).

Range (Depth) – only found on expander-gates, this knob sets attenuation levels (other than totally muting the signal, as on plain gates). A lesser attenuation depth (of say only -30dB) gives most of the benefits of plain gating, while making the transition between open and closed speedier and more subtle. The latter could seem worthwhile for some quieter folk or solo artists, though whether

***Tip from the top* – using gates**

A typical mistake made by beginners is to set too low a hold time and a fast attack. Then the gate keeps attacking, staccato-like. The solution is to set a longer hold time, and with a higher threshold setting.

To deal with 'ringing' toms, first set the attenuation to about -30dB, refine the other settings, then readjust the attenuation later. Expander-type gates with a ducking facility can be very useful. For example, it's normally hard to achieve the correct threshold on toms, since the snare drum is inevitably the loudest local sound, and will often trigger the tom gates to open – unless the threshold is set so low that even the tom itself won't open its own gate.

One way around this is to apply a stereo ducker across the tom group (a group on the mixer just for toms) and externally trigger it with the snare drum. Then each time the snare is hit, the ducker ducks the tom group, allowing the engineer to set the tom gates less hard. They still open, but the ducker is ducking any snare spill at the group end.

or not the subtleties of expander-gates are worthwhile may also depend on the background noise levels of the audience and the venue.

Attack, Hold, Release & Decay controls – alter the sound of the gating. With some sounds and threshold settings, they may also be used to prevent stuttering. This occurs when a signal hovers near threshold, causing the gate to keep turning on and off – which creates a staccato sound. The attack, hold, release and decay controls can prevent this by slowing down the opening and closing of the gate.

Be aware that despite the similarities in the available knob settings, the actual performance of the attack and release varies between different makers and models, and has quite an effect on a gate's sound quality.

For example, some gates have a linear decay – like a bell. On others, the gated signal decays faster (accelerates) as it fades away. These ones can sound better for percussion. The four sorts of setting are all pretty interactive and have a main part to play in the sonic quality of the gating. Hold provides an adjustable pause period between the gate being open then starting to close.

1U-high multi-channel gates tend to have fewer features

Auto-dynamics (shortened to various fancy three-letter abbreviations like 'ADP') – these generally provide preset settings. They are not much used for professional live sound.

Ratio – a control knob on the expander-type of gates only. Useful when trying to gate the tail-end of (dying away) sound. Also helpful for softer sounds, such as vocals.

A Ratio control is rare, but helps you refine the triggering by tightening or loosening the trigger source's dynamics. It helps to stop the stuttering or 'combing' effect when the trigger source is close to the threshold point, as in long reverbs, or a drum roll on the snare drum. A tight ratio may be good for independent, well-defined snare beats, but a looser ratio will be good for a press roll –

where the quieter parts might accidentally be gated out.

Mask – a knob that prevents the gate from retriggering for a time, variable from typically a few milliseconds (thousandths of a second) to perhaps two seconds. So you could trigger on the first sound of a rapid series from a percussion instrument, then avoid retriggering until the next sequence starts.

Invert – a panel switch that changes a gate into a 'ducker', as used by DJs and announcers to quickly and smoothly lower the music's signal level when they're talking. The switch 'flips' the gate's action, so that the voice signal sent to the key input closes rather than opens the gate. The gate is usually the expander type, so the music can be made to fade but not disappear.

Auto-bypass – if the power lead falls out, or the gate dies in some way, the signal will continue to be routed through. You won't get any gating, of course, but at least you don't lose the instruments altogether. Manual bypass may be provided – it's useful for comparing the ungated sound.

Stereo link – useful in two-channel applications, enabling the threshold and possibly dynamics settings to be shared, so the gating on two channels is precisely matched.

General notes on gates

Some gates don't have a very flat ('linear') frequency response at the extremes. So instruments which cover the extremes – such as organ (low end) or cymbals (top end) may benefit from some LF or HF shelving EQ (respectively) from the mixer to compensate.

If using the bypass function, beware of a slight gain increase, say +1dB, exhibited by some units. This could make the ungated sound appear better in some comparisons.

With some gates, it's possible to 'lose the key' – meaning the filtering has set too narrowly or overlapping, and possibly not able to follow wandering key sounds. Then the gate won't open.

Most gates come in 1U packages and are at least stereo (two-channel) units; four and six-channel versions are available, though each gate channel has less room for its controls, so 1U multi-channel gates tend to have fewer features.

That's why some multi-channel gates are in 2U

or 3U packages. Some of these are modular, comprising an array of vertical, frame-mounting plug-ins. In such cases, gates are often just one of several different FX or processing modules that can be plugged in the overall 'main' frame, enabling compact, powerful bespoke processing to be assembled 'on the fly' by any sound engineer.

The compressor family

Compressors and their relatives are used to control or creatively alter signal dynamics (ie changes in level). Limiting is basically a more abrupt form of compression. Together they help keep the unpredictably variable levels of instruments and artists in check – partly so the PA isn't blown up, and also so average levels can be kept as high as possible (and cut across audience noise) without excessive peaks.

There are several related devices in the compressor family, as follows:

Compressor (FOH & monitors)
Used on individual instruments or voices, a compressor can have the effect of 'tightening' the sound – and making performances seem punchier, as well as louder.

Most vocalists (but particularly inexperienced ones, or those with poor mike technique) have unpredictable level jumps in their singing – some notes in some registers just come out louder than others. The same goes for players of instruments like horns and basses. Compression can help compensate for this – if used judiciously – and give a more professional, tight result.

It can also lift an instrument, or part of a mix, to a higher level (of apparent loudness) without clipping the amps, or indeed raising peak levels too much. A fair amount of compression generally makes sounds fatter, and sometimes bassier.

Used across the whole PA, compression can increase the apparent overall volume by smoothing out the most dramatic leaps – though if this is overdone there's a danger the music will be deprived of much of its interest and excitement, which is partly provided by variations in level (otherwise known as dynamics).

Compressor-limiter/comp-lim (FOH & monitors)
A compressor is safe to use alone, as long as either a (peak) limiter is supplied elsewhere in the path or is not required on a given instrument (perhaps because the maximum level has a definite limit – as with, for example, a DI'd organ.) But it's more usual, and useful, to have the two functions (compressor & limiter) available in one unit, though with the two parts ideally able to be set up separately. The limiter part is used to put a 'hard stop' on levels exceeding a given threshold, mainly to protect equipment and hearing.

Leveller & automatic gain control (AGC)
A leveller is a compressor which works to smooth out level changes over longer periods (tens of seconds and minutes). Levelling makes quiet

Peak, RMS & crest factor – some facts
Crest factor is a number that describes how much the average levels (RMS) of a sound signal differ from the instantaneous peak levels. When compressors and limiters are used to protect a PA system against damage, it's helpful to know that RMS levels cause heating, while the peak levels cause mechanical damage. Limiting reduces peak levels, whereas compression, which is mainly controlled by RMS levels, also reduces the peak levels. At the same time, both compressors and limiters can reduce

the average 'peak-to-mean ratio' (PMR – alias 'crest factor') of the signal. The result sounds louder, because it has higher power density, while not actually exceeding the PA's headroom limits. The outcome of creating too low a crest factor is dead speaker drive-units, just as with the clipping that the compressor and limiter are often used to prevent.

One solution, automatic in some specialised PA compressor-limiters, and also performed manually by skilled engineers, is for the input level to be reduced by a few dB (a few per cent), each time the gain reduction meter

shows compression is causing more than 6dB to 10dB of gain reduction. The idea is to keep the crest factor above 6dB. This dB figure is not directly related to the gain reduction value, and compressors don't have crest factor meters, but gain reduction still gives an idea.

Effectively, if you have to drive the PA so hard that even the compressors and limiters cannot protect the speakers from damage, then all you can do is reduce the level. As ample compression will have already made the sound seem louder, this is not quite such a backward step as it sounds.

sounds louder and louder ones quieter – this can be useful in small amounts, but taken to excess it creates bland 'muzak'. A similar effect can be gained to some extent by manual 'gain riding' of the main faders – if you know the music well enough. AGC is similar, but with a longer release period (which we'll explain in a moment).

De-esser

By making compression 'frequency-conscious' (meaning specific to a certain frequency area), it can help with problems like over-sibilant 'tssy' vocals or speech. Beware that there are narrow and wide-band types of de-esser that are suited to different jobs. You could try either with solo vocals, but must use a narrow-band type for mass vocals.

Whereas compressors turn down loud sounds, expanders turn down quiet sounds

---

Downwards expander

This applies a sort of 'backwards' compression – quiet sounds are made proportionately quieter, while loud sounds remain loud. In other words, whereas compressors turn down loud sounds, expanders turn down quiet ones. At a subtle level, this seems the opposite of what's usually required for PA, but it has its creative uses, such as creating dynamics where none existed before – perhaps aiding a monotone vocalist. Or to recover sounds that have already been compressed – which includes most recordings and radio broadcasts. At

a more pronounced level, with lots of dBs of expansion downwards, it becomes an expander-gate (mentioned earlier).

Compressor/limiter controls & features

Threshold – this control sets the levels above which the compression (or limiting) begins to take effect. At levels below the threshold, there will be no effect. For PA work, the threshold range needn't go really low (-30dB would be enough), but should go high enough, say to +10dB. That's because you may end up with a high signal in a real situation.

Ratio – this knob dictates how abruptly the compressor works. A high ratio (say 10:1 and above) corresponds to an abrupt setting, approaching a limiter in effect. This will be more audible than a low ratio (say 1.6:1), which has a more subtle effect.

Limiter sections (in comp-lims) have no ratio control, as the ratio is preset to be (theoretically) infinitely large – at least 20:1.

Some compressors offer 'soft-knee' compression (called 'over-easy' by dbx, the US company who largely pioneered the effect). With this facility the ratio is applied at all levels, very shallowly at first, but it gradually increases with higher levels so there is a transition from low compression to higher and higher compression, almost becoming a limiter. The difference is that a dedicated limiter senses peak levels, while a compressor with a high ratio still senses only the average levels (see the sidepanel opposite on 'peak, RMS and crest factor').

In theory, the result of using a soft-knee compressor is that the average level is better controlled over a range of conditions, and the compression is less abrupt. With normal compression ('hard-knee') there can be an abrupt change in the sound's quality as the signal passes

---

Tip from the top
Making compressor settings 'line-up':
If alignment of settings between two or more channels of compression (or comp-limiting) is important (on a stereo feed for example), and you can't use or don't have a gangable stereo unit, or else need more than two channels to line up, the setting accuracy can be greatly helped

by resetting all the knobs mechanically, so their pointers are all indicating at the same place.
This can be checked when placed, say, hard over to the left (counter-clockwise). The pointers are not in line usually because the knobs are 'collet' types that can be set at any position and then tightened down. To undo, remove the

cap and slightly slacken the central screw or nut. To help line up the pointers, look at the panel 'square on' (not at an oblique angle), and use a horizontal line or similar 'azimuth' as a reference to see which pointers are out of line. Then tighten back up, taking care not to move the knob against the shaft while doing so.

above the threshold. Soft ratio is just a bit like having a compressor-limiter.

At the same time, be aware that there is no free lunch: the signal's density is rising as the peak levels are squashed down (again, see the 'peak, RMS, crest factor' panel, p60).

Soft-knee units may also be harder to set up to get the results you want, and are felt to damage sound quality, since some degree of compression is occurring at all lower signal levels – not just with excessive peaks, as with ordinary compressors.

Applying too much compression (a high ratio coupled with high gain reduction) takes the life and 'balls' out of a live sound. But (perhaps surprisingly) it also increases stress on thermally-vulnerable PA drive-units by increasing the long-term average level to which they're exposed (see sidepanel, p60).

Applying too much compression takes the life out of a live sound

With vocals, setting the ratio to give about 6dB of gain reduction at the average vocal level – by watching the gain reduction meter – will protect the PA from wild variations caused by varying vocal intensity and movement towards and away from the microphone.

Attack – a knob that varies the time taken for the compression to start to act, down to about 10mS (ten thousandths of a second, where 'm' stands for a thousandths) and up to about 100mS (100 thousandths, or one tenth, of a second). Longer times allow momentary overdrive, usually of a few thousandths or tens of thousandths of a second – not enough to harm the PA, but enough to let the instrument's natural leading edge be heard, so it sounds more natural without losing the benefits of compression. But taken too far (if the attack time is lengthened towards or beyond say 100mS = one-tenth sec), the compression is defeated long enough to risk damage (to HF drivers), and also let through audible distortion.

Release – this is intended to vary the time taken for the compression to stop acting, after the (peak) signal levels that triggered it are over. It may vary

from 10mS up to two seconds or so. Too short a release can cause 'pumping'. An example is compression on the drum sub-group causing the background level (other drum sounds) to go down with each hit, then return between hits. With longer release (but not too much) the backing would be reduced at first, then stay down, only fading back after the drum hits were suspended or more spread in time.

A long release time prolongs the period of gain reduction, which also prolongs the period in which the audio quality is changed and/or degraded by the compression action. It could also be a problem with the kick drum again (as above), if set high enough so the compressor doesn't recover in time for the next beat.

Using ratio, attack & release:

- For a DI'd electric bass guitar, you might have a ratio of about 4:1, with a fast attack – say 1mS – and a slow release of say 500mS. Or, for a punchier sound, increase the attack's setting up to around 20-50mS.
- For a 'medium' vocal, try a ratio of 3:1, attack 10mS, release 80mS.
- For a 'hard' vocal, try a ratio of 4:1, attack 1mS, release 50mS.

Peak Limiter (threshold) – this knob appears only on compressor-limiters (comp-lims), and sets the threshold of the integrated limiter. This threshold must be set at least 6dB, if not 10dB above the threshold of the compressor's section (the sidepanel below outlines why). The limiter section often has its own preset, fast attack and auto-release 'time constants', meaning there are no other knobs to concern us.

Master/slave – a switch (generally on stereo/two-channel units) that allows one channel's settings to act on (usually) two channels. This is very useful for stereo signals where, if not linked together, a central panned signal might dodge (leap about) between left and right, when one side has more gain reduction than the other.

Metering – Often three modes, switchable between input, output, then showing the difference as gain reduction. This shows the amount of signal level in dB that has been subtracted ('squashed') by the compressor action. The bar display sits at 0dB (on the right) and dips to the left as the compressor works harder.

Some meters claim to show RMS signal (rather than average) levels, which tend to be most accurate for assessing apparent loudness. More normally the response is 'average', which is similar to RMS but not quite so accurate. Some compressors have reverted to using VU-like mechanical 'swing-needle' meters (but scaled in gain reduction units), the average display of which again shows apparent loudness in a way many engineers prefer.

It would also be useful (but is almost unknown) to have simultaneous peak level metering, so that crest factor can be gauged.

Sidechain output & input – these send and return sockets enable the controlling 'sidechain' signal (which is derived from the input and/or output signals), to be sent out for EQ or other processing, then returned into the unit. Enables EQ (usually, but it could be other processes) to be patched into the sidechain.

A limiter restricts loud signals... a gate restricts quiet signals

---

Compressor family relations
In perspective, it can now be seen that the most useful members of the compressor family are all symmetrically related:

- A compressor turns down loud signals.
- An expander turns down quiet signals.
- A limiter restricts excessively loud signals.
- A gate restricts excessively quiet signals.

And, in practice, this means:
- A compressor with a high ratio becomes a limiter.
- An expander with a high ratio becomes a gate.

Delay (FOH & monitors)

Although mainly thought of as being in the PA's drive rack (see Chapter 4), delay – which we encountered earlier used as a kind of echo – may be employed to enhance DI'd bass. The idea is to delay the direct send to be in sync with the miked-up signal.

Finding the optimum delay setting – two milliseconds (2mS) at most – is made easier by temporarily reversing the polarity on one feed (using the 'phase reverse' switch on the mixer channel) and adjusting for the most cancellation (level drop).

Pitch shift

Creates a higher or lower-pitched version of a signal, which can replace or be added to the original. Its effect can be extreme and unnatural, but used subtly it can be good for vocal or instrument doubling, or to give body to a sound – for example making a six-string guitar sound like a 12-string one. It can also 'retune' a persistently flat or sharp singer, or certain instruments. (And see panel below.)

Exciter (psycho-acoustic enhancer)

Used mainly on vocals. Pioneered by Aphex (in the US) back in the 1970s, these add modest amounts of the 2nd harmonic at high frequencies. This addition of what is usually regarded as 'distortion' actually (and seemingly paradoxically) makes the sound seem clearer. The added harmonics make the HF more dense, which makes for a loud sound (lots of RMS) without overdriving the PA system too readily at HF.

***Tips from the top* – pitch shift**
When using pitch shift, first try panning 'dry' vocals or guitar into the centre. Then send the same vocals with slight pitch shift upwards to the right side. Finally, send the vocals to a second pitch shift unit, set for a slightly lower pitch, to the left side. This gives a great sound with a wide image, and makes any basic band sound a bit 'fatter'.

Pitch shift can also be used to add extra octaves on bass. If you're feeling adventurous you can even mix in intervals like major 3rds, 4ths & 5ths (get musical help to do this if your unsure of harmonic theory) – if carefully done this can create complex overtones and richness.

The added harmonics can also make vocals sound as if they are right 'in your ear'. Some people like this, others find it intensely annoying. It's widely heard on modern, R&B/pop, providing exaggeratedly breathy, whispering intimacy to vocals.

Compatibility of outboard gear

You would think there are enough complexities in a PA system without introducing outboard gear made by 101 different makers, produced to meet individual, and often parochial standards.

Sadly there is no 'joined-up design' in the world of PA. The best units (or at least highest quality for the price) in each category of gear will very likely come from different companies. So care must be taken to check that each different make and type of processor, FX or EQ unit won't cause grief when plugged into the system. Never assume anything.

Some FX equipment isn't fit for the road – it doesn't meet professional requirements – even though it may be packaged and promoted as such. Certain makers continue to produce and sell 'consumer-grade' equipment to PA users, sometimes with out-dated technology and wiring methods, dressed-up to look pro, complete with 'rack ears' for slotting into your racks. And people continue to buy it through technical ignorance (and perhaps blind faith in a name that's domestically known).

This sort of non-pro equipment should be avoided for serious live work, unless you're really desperate – it's very unlikely to last the pace. But how do you spot it? Look out for any of the following (these points in fact apply to most PA gear in general):

- Phono sockets (RCA jacks) or quarter-inch (phone) connections only – with no sign of genuine professional XLR in/out connectors.
- Unremovable power cable – not simply fixed, but no removable gland, just a sealed plastic grip.
- Hardboard or plastic rear panel – pro audio gear requires proper shielding, unless you want to broadcast passing taxi messages through the PA.
- Plastic casing – some plastic is fairly sturdy nowadays, but most will break, crack or snap

when used on the road, where metal cases wouldn't; not to mention the fact that plastic often gives off potentially deadly toxic fumes if an electrical fault starts a fire.

Some FX equipment isn't fit for the road – even though it may be promoted as such

- Two-core power cable – comes hand-in-hand with the plastic enclosure just described, or a 'double-insulated' ('class II insulation') metal case. The latter combination can be acceptable, as the user gets most of the benefits of metal casing without the hum risk that the third earth/ground wire would cause.
- Low-budget 'paxolin' circuit boards – these tend to be either golden brown or beige colour, and equipment using them should be avoided. This type of board absorbs moisture, and can snap easily (and sometimes invisibly) when stressed. It should only be used for home hi-fi (and stationary set-ups) where moisture is driven off by constant warmth – not out on the road.
- Unserviceable construction – where the unit has wires or looms inside that have to be hand de-soldered. In much low-budget, high-profit equipment, fake connectors are used which disguise soldered joints. The easy way to tell is to gently tug and try unplugging any wires, cabling or looms that appear to be connected to plugs or sockets.

If 55 wires have to be desoldered to take something apart, it will be costly to mend. Another bad sign is where the circuit board is held in place by bent metal or plastic clips rather than proper re-useable metal screws.

There can be exceptions to the rules above – there are occasionally bits of gear which transcend ideas of build-quality and, over years of widespread use and approval, become 'classics'. These might be worth the extra hassle their use involves.

Then again, you might just find a newer, better-made unit that can out-perform it. Don't blindly idolise old technology. Some of it is good, some is over-valued.

A lot of low-budget or older oriental-made electronic gear may well be unsuitable for pro-sound purposes – although it's undeniable that the reliability and longevity of modern Japanese processors (for instance) is very good.

It's also fair to point out that some apparently solidly professional British and US gear can have reliability problems. In most cases, this is because it was designed for studio use and the designers had no knowledge of the requirements for road use, where prolonged vibration and occasional shocks will need to be sustained.

Finally, beware of look-alike pirate copies of some reputable models – unusual finishes and colours, and plugs that jam in place are signs to watch out for.

Input & outputs

The inputs of professional processors, EQs and FX units should always be balanced, and connected as such. The outputs may or may not be balanced. If a balanced output is provided it's best to use it, but that's only feasible if the return destination (usually at the mixer) is a balanced input.

The inputs of professional processors, EQ and FX units should always be balanced, and connected as such

---

Always use more reliable XLR sockets, even if quarter-inch (phone/jack) connections are offered as an alternative.

Gain structure when using FX

In an ideal world, gain (the difference in level between the input and output) should be 'unity' – a term used to refer to 0dB, or no change. But with EQ and some FX, what unity gain actually

means is less clear-cut. For example, if an EQ is set to cut at some main frequencies, the sound will seem quieter on average. But signal peaks (momentary loudest parts) at some of the other, un-EQ'd frequencies will require just the same headroom as they did before the EQ was set.

This means:
- add gain with caution, listening for overload on peaks as you do it.
- add as little gain as you can get away with: +6dB is a suggested limit.

If levels have to be cut or boosted, there may be a benefit in boosting or cutting level before or after the FX unit. For example, if you cut after the EQ, overload could be a problem, but noise will be reduced. So that would suit lower level signals. Well-designed EQ and FX units allow gain structure to be optimised by having modest gain cut (-dB) and boost (+dB) controls at both the inputs and outputs.

At the same time, such controls should have a definite unity (0dB) gain setting, such as a default switch setting, rather than some consumer-grade detent that offers 'fake' certainty – because the mechanical position may not exactly correspond with the position where the control actually gives 0dB.

If the FX or EQ unit has no overall input or output gain control, then various of the mixer's gain controls can be used instead – either Aux Send level, and/or Aux Return or channel input gain. These (as available) can be used to compensate for gain loss or excess.

the drive rack
house eq, crossovers, delays

Quick view

The stereo outputs ('main feeds') from the mixer connect to the drive rack – a flightcased box of rackmounted gear that organises and 'drives' the signals on to the relevant power amps and speakers. It's usually near the FOH mixer, with a separate one if needed at the monitor desk. The drive rack mainly includes 'house' EQ for adapting the sound to the venue, crossovers to split up the signal into various frequency bands, and delays to synchronise different sets of speakers.

the drive rack normally comprises just three units – a multi-band or 'graphic' equaliser (EQ), an active crossover, and a signal delaying unit – though it may also include some 'system control' devices.

The EQ here is not for adjusting particular instruments or signals, but for final fine-tuning of the overall sound, ensuring the music will sound OK in a particular room, hall, park or stadium.

A crossover divides the music into frequency bands (at least two, but perhaps up to six or so), sending each of these bands to a separate power amp, which then feeds suitable speakers (bass sounds go to large, bass-friendly speakers etc).

A delay unit, in this context, has the delicate job of holding back particular feeds to certain speakers (which may be spread around a venue) in order to synchronise their sound together, avoiding echoes and mis-matching.

Sometimes all these units are brought together in multi-processor devices, and/or operated by digital 'system controllers'.

The drive rack is a vital but easy-to-overlook stage in the live audio chain. Apart from anything else, the FOH PA signal – which is what most of the audience have paid to hear – is particularly vulnerable at this point (just before and inside the drive rack).

For a start, the signal has been reduced to just two circuits – left and right (plus perhaps an extra centre signal, if it has been created before this stage – see side-panel on the opposite page). This means that if just one connector falls out, or one graphic equaliser fails, either half or all of the entire FOH show stops happening.

We're also dealing with a complex, composite music signal, covering the full frequency range – it contains all the instruments and voices mixed together – rather than the simpler, lower-level signals that the mixer receives from the individual sources. This can be more arduous for all electronics – analog and digital – to handle perfectly. Even after this point – when the signals re-emerge from the drive rack, split into different frequency bands by the crossover – there's less hard work for the electronics to do.

In other words, even more than elsewhere in the system, this is an area where high quality audio equipment is vital – it's one place not to take chances with the sturdiness of the gear, quality of

House graphic EQ

'House' here refers to the venue (as used in 'front-of-house'), rather than any particular style of music that shares the name…

A graphic equaliser (called 'graphic' because the position of the sliders suggests the graph of the frequency response the settings are creating) is usually required in line with the main feed. This means it stands between the L&R outputs of the mixer and the rest of the drive rack – all FOH signals pass through it.

The purpose of this particular EQ is not to supplement the desk EQ (though you could use it that way if desperate). Instead it's to have some control over the sum effect of the following speaker cabs and the house (meaning FOH) acoustics – for instance if the PA cabs had to be used in such a way that reduces their bass output (due to a cramped venue, say), or if the venue happened to have a very bright acoustic sound.

Graphic EQs suited to serious PA use have 'one-third-octave-wide' bands – which means each slider operates on a 'finger' of frequencies that's a third of an octave away from the one next to it. The centre frequencies (those in the middle of the imaginary finger) conform to certain world (ISO) standards. Third-octave graphics have been the main choice for house EQ since they were introduced in the early 1970s (before then there were really no useful tailor-made tools).

'Third-octave' EQs vary in the number of slider knobs they have, but typically there are between 28 and 32 bands (hence that number of sliders), which divides the audio spectrum into enough segments for reasonable selectivity. As the centre frequencies are fixed, it's clear that models with more

sliders/knobs reach a wider range of frequencies. The extra ones are mainly at the low end, though occasionally at the top – for example on many units the highest knob is marked 16kHz, but a few have a 20kHz knob. Similarly some graphics stop down at the 25Hz knob, while others have the next two frequencies, 20Hz and 16Hz.

A good graphic should be fairly quick and easy to set up – this can make a big difference to the time it takes to get the PA ready. Here's a list of other features to watch for on all graphic EQs used in PA (note that graphics are also used on monitors, and much of what's said in this section applies there as well – but there are differences, which are covered in Chapters 3 & 8).

Faders/sliders

The better and more modern 3U graphic EQs have long 60mm (2.4in) travel sliders. Obviously if your graphic is of restricted height (1U or 2U) it will have shorter slider controls, and these may make fine adjustment harder. If short controls must be used, their movement must be very smooth – gear with jerky controls should be avoided.

The controls should also stay firmly put – not move too easily, or slip back in any way when you let go.

Flashier graphic EQs may have a facility for auto-resetting of stored settings – motorised sliders can be returned to memorised positions at the push of a button. The same result can be obtained in purely analog, manual mode, using a stiff cardboard 'stencil' as a guide – initially marked out with a pencil, cut-out to profile the knob positions, then placed up against (and used to reset) the sliders when you want to move them back to the desired settings.

Boost/cut switching

A few units are supplied (or can be set to work) in cut mode only. Others, usually older designs with 45mm (1.77in) sliders, have a range switch which

Centre signals

The signals from most sound sources (mikes, DI boxes) are routed roughly equally to the left and right sides. A centre signal can be created (in the master section of some mixers – see Chapter 2) simply from a summation, an equal blending back together of these two. The resulting mono signal is then fed to central speaker clusters. In some equipment in the drive rack, centre signals can also be created ('derived' in PA-speak) in the same way – the 50/50 blending of left and right. This is dealt with later on in this chapter – see 'Mono for sub-bass cabinets'.

On some equipment back panels, centre signals can also be called 'mono' or 'summed outputs' or 'L+R' (signifying 'left channel plus right channel') or 'LR sum' or variations on these themes.

changes the boost and cut ranges from ±12 down to ±6dB. This is done to give more slider travel for making fine adjustments of below 6dB – helpful in both a visual and tactile sense when trying to match up the slider settings exactly on the left and right channels.

Peak indicator LEDs

On a few graphics (usually models aimed more at installation and semi-pro users) a red warning 'peak' LED lights up to show you the signal level is approaching clipping (harsh overload). The signal level being shown may be that at either the graphic's input or its output. As the signal level may be largest at either point, depending on whether the graphic's sliders are largely boosted or cut, levels at both input and output need to be indicated if excessive, but few graphics are designed to do that.

So it is possible to have clipping, or the gradual increase in distortion that occurs beforehand, without this sort of LED indicating – which is not what you'd expect. This could occur if the gain structure is set up wrongly – for instance high input, boosted controls, but output turned down. The lesson is to believe your ears over what LEDs tell you.

High & low-pass filters

See Chapter 2 for a full explanation of what these kinds of filters do. Here they may be simply switched in/out at fixed frequencies, or else they may offer tunability, with knob-sweepable high and/or low-pass 'cutoff', 'corner' or 'turnover' ('beginning to filter') frequencies. They're not normally used or needed, but can solve some tough problems when called upon.

For example: the venue/hall is found to have a terrible rumble (low bass reverberation) when the band do their soundcheck. Rather than pull down the last three faders of the graphic (which can only give a narrow range of reduction, and maybe won't even cover the very low frequencies involved), or else try to engage every single channel's high-pass filter on the mixer, a flick of the high-pass filter (HPF) switch on the house graphic – and maybe some turnover frequency tuning – sorts it out.

The graphic's HPF can be helpful in a related but different way – if environmental noise is a problem (see Chapters 8 and 11). Other uses for the HPF are removing sub-harmonics' (signals artificially generated by synths at frequencies that are lower fractions of the fundamental note), and removing rumble from DJ's turntables. In all such cases, the benefit is that you're not wasting the PA's power on stuff you can't hear or appreciate. Suddenly, the PA will have more headroom and 'space' to deal with useful parts of the music.

As mentioned in Chapter 2, one way to set the HPF is to tune it upwards in frequency while the lowest bass parts of the set are happening, until you can hear or feel something is missing. Then just back it off slightly – to half the frequency maybe – to be sure of not losing any wanted, musical parts.

The low-pass filter (LPF) is useful if you suspect that some signals above audibility are using up headroom. For example – some samples or synth sounds output high-level signals above 18kHz which will be inaudible to most of the audience and contribute little of practical benefit to the performance. Such high levels may be observed either on the mixer's metering, or because the limiters on the crossover are hitting their 'stops' (see later in this chapter).

Suddenly the PA will have more headroom and 'space' to deal with useful parts of the music

A related use is emergency RF (radio frequency) filtering. In other words, if the PA was picking up strange sounds that appear to be radio or TV broadcasts, or other RF such as taxi cab messages, you could try using the LPF control to reduce or prevent these. Whether this works would depend on a variety of factors – such as where in the PA the radio signal was getting in (if just beforehand, it might help; if afterwards, it cannot), and where it was being 'detected' (meaning 'turned into an audio signal').

Some graphics offer a choice of 'slopes' (the rate of reduction), from the basic -12dB per octave (this is the commonest, which quarters the signal voltage for every doubling of frequency beyond the turnover frequency) to the next standard, steeper rates of -18 or -24dB per octave. The steeper slopes are useful because they are more selective. For example, you could cut rumble quite deeply even if it was only one octave beyond a major sound, like the bass guitar, which obviously mustn't itself be reduced.

But steeper slopes can come at the expense of sonic effects (as we've said before, there's nothing for free). With high-pass filters, the main drawback is caused by the steeper slope's extra phase shift – rather than this causing random sound cancellations, as in the mixer channels (see Chapter 2), the more audibly noticeable effect is that it slightly delays higher frequency bass sounds (ones not directly affected by the filtering), compared to mid and higher frequencies.

This may be subtle, but an accumulation of such effects make the difference between a superb live sound and a nondescript one.

A similarly subtle yet sometimes audible effect with both high and low-pass filters is 'ringing' – sudden transient sounds get stretched out a bit in time, like a bell's sound (or the springy-sounds in old cartoons).

Gain

Many graphic EQ units have gain controls or preset gain switches/knobs – these must be deployed almost as carefully as boost (which we'll come to shortly). The gain on offer may be required when a lot of cutting causes perceived sound level to drop, but bear in mind that signals at the un-cut frequencies will be pushed nearer to overload by the amount of gain used. Use it gingerly.

Operating level

Some not-fully-professional units (made really for small bars or home studios) have slide switches or other settings hidden around the back, or inside, that set the 'zero level' or operating level (the average signal level that's going into and out of the graphic or other gear, were the PA to be driven flat out), usually with a choice of +4dBu or -10dBu. For PA, always use/set +4dBu.

Automatic & manual bypass

The whole gig can depend on the FOH graphic and other gear in the drive rack, in the sense that

if the graphic's internal signal path fails – perhaps just because a ten-penny part in the power supply dies – it's essential that a gig is not held to ransom. Better to lose the graphic than the gig. To deal with this, all professional standard graphics have automatic bypass, so if the graphic's power fails, or is turned off, or signals aren't passing through due to a fault, the input is connected directly to the output, and both input and output are disconnected from the graphic's circuitry.

If power is restored (for example if the power lead that someone tripped over is then plugged back in), the graphic is automatically reconnected, as it was.

The whole gig can depend on the gear in the drive rack

In a good unit this changing over should be accomplished without rude noises, like bangs or plops, which can occur with some less well-made models. Such behaviour may go un-noticed in emergency circumstances, but not if it damages the speakers. To avoid any risk of this, use the crossover mute buttons first to protect the PA, then turn the graphic back on, then unmute the crossover (see 'mute switch' in the crossover section later in this chapter – p76).

Manual bypass is a switched version of the auto arrangement, which additionally allows the engineer's ears to quickly make a sanity check of EQ'ing – namely, "Are these graphic settings actually making any difference?" It's a rather vital tool for setting up.

Also, as a measure of any GEQ, when switching between the 'bypass' (graphic off) position and the 'on-but-flat' setting (all sliders at centre detent), there should be no greatly discernible difference in the overall signal, other than a slight noise (hiss, hum) increase. Level and sound balance should remain the same.

Graphic 'headroom'

A higher-than-average maximum operating level (MOL) spec is helpful – say of above +20dBu – to give more room for boosting while keeping some decent headroom. For example, if the signal going into the graphic is peaking at +8dBu and exceptionally requires +8dB of boost in one frequency area, signals in that area may then peak at 8+8 = +16dBu. This level is too high for the amps, so it's reduced by adjusting the graphic's output gain (level setting) control back, so the level received by the next unit in the line is set to peak at 4dBu.

This suits the later gear, but inside the graphic the signal (still peaking at +16dBu before the output fader) may be dancing within just 2dB of clipping – assuming (say) the graphic's MOL is +18dBu. If you had a more professional model, that offers a MOL of +24dBu, you'd have a breathing space of 8dB, which is much more comfortable. With just 2dB, each time the instrument or vocalist operating in the relevant frequency range surged in level there could well be some bad distortion, which you might not have time to track down.

Constant Q

This is a feature you'll see touted by some makers, but the jury is still out on its usefulness. It's supposed to ensure that the 'width/sharpness' (Q) of controlled frequencies for each slider stays more constant in a way that should reduce interaction as bands are adjusted. With ordinary graphics, the Q

of each band does vary slightly with more or less boost or cut, but in practice there is no big problem with this. With so-called 'constant Q' graphics, professional PA users can't detect much, if any, advantage, and in many real conditions they've been found to actually give a less smooth response.

Hints on house EQing

Using the house graphic differs from using a similarly-placed graphic on the monitoring. The former is mainly used for tweaking to get a good overall sound, rather than for firefighting, as in dealing with acoustic feedback on the monitors (for advice on curing feedback, see Chapters 2 & 9 – and for specific mix EQing, see Chapter 8).

● Bear in mind that all EQing introduces its own relatively subtle distortions – of signal timing (aka 'phase shifts') as well as coloration (tonal aberrations) and damping (signal level dynamics). Always avoid using EQ as far as you can – the sound should be better for it. Can you move a microphone to a better position (see Chapter 1), or trace a wrong connection or misused switch, or alter the PA cab positioning or angling (see Chapters 6 & 7) to avoid the need for EQ altogether?

EQ should be applied gently and minimally –

Graphics and noise – hiss & hum

Hiss (the steady 'tssss' noise – like escaping steam – made by the graphic itself) is not much of a problem on the more professional units. It's most likely to be an audible problem on bad or low-budget designs, and older or damaged units. With a decent unit you shouldn't hear any hiss at the FOH mix position at all, over the FOH PA, when the bypass switch is flicked in and out, and while the sliders are all at 0dB.

Of course the prevailing hiss level will be made more audible if you have to boost frequencies above 5kHz, or if you're in a small venue where some of the audience are unavoidably seated close to the FOH PA cabs. Hopefully, they won't notice it over the music.

Lower-frequency noise (buzz and hum) may be induced if graphic EQ's are placed (for any reason) near to 'leaky' AC transformers – meaning their magnetic fields leak outside their casing and affect other equipment. Graphics are more likely to pick up hum than other units in the drive rack, so before finalising a drive rack set-up, it's a good idea to spend a little time experimenting with the relative locations and spacing of the units in the rack, with the graphic placed furthest (if feasible) from any other units that induce hum.

To make good and bad positions easier to locate, you can temporarily exaggerate the problem by fully boosting the sliders at 50, 150, 250 and

350Hz (or in the USA and wherever the local AC power frequency is 60Hz, use the sliders nearest to 60Hz and its multiples). For this test, nothing is plugged into the graphic – though any gear around it must be powered-up and switched on – and it's simply plugged straight into a PA amp and a bass bin, or any bass speaker made to reproduce frequencies between 50Hz and 400Hz.

Note: Although you can use the graphic to exacerbate hums in this way to help positioning, never use a graphic to eliminate hums. You'll reduce the hum, but also cut an unnecessarily wide range of bass frequencies. For a suitably selective cutting you must use a parametric EQ (PEQ), as described back in Chapter 2.

if it's used solely to create an impressive sound, it will usually tire the ear – it's rather like the over-use of powerful spices in cooking.

The bypass switch enables the effect of any EQ you apply to be tested at any time against the original. Keep making comparisons until you're satisfied.

- In general, far more cut can be used than boost. Most EQ's have more dBs of boost on offer than you should readily bring into play (see side-panel on 'Boost & headroom' below). If you boost a frequency range by a high number of dB, bear in mind that the drive rack's limiters (which we'll come to in a moment) will 'dynamically defeat' some of the intended effect at higher levels. In other words, the loudest parts of the signal at the 'over-boosted' frequencies will be reduced by the limiter. The effect is an extreme sort of 'frequency-conscious' compression (as described under 'de-essers' in Chapter 3).

In general, far more cut can be used than boost

If you find you're boosting a lot (more than half) of the graphic's sliders, even by a small amount, then try an alternative approach: reset all the boosted frequencies to 'flat' (0dB – the centre detent position), then set the remaining ones to cut, until you achieve the sound you want. By using less EQ to get the desired result, the FOH sound will gain from lower accumulated distortion, and likely a smoother overall response curve, while valuable headroom is saved. This may seem subtle on its own, but good PA sound is an accumulation of these little points.

- When the knobs are all in a straight line anywhere other than the centre detent position (0dB) – ie at any boost or cut setting – that's wrong. The response will not be ruler flat as the sliders might suggest. At all positions other than the central detent, the knobs only give a fair approximation of the actual response curve. That's partly because the real response curve of the frequency band controlled by each slider is 'rounded off' at the edges, just like a row of equal length fingers. So if you set all the sliders to any setting other than 0dB, the response curve will be flat on average but also uniformly wavy – like corrugated roofing seen in profile.

- A single knob/slider can be used to remove a particular frequency. But if boosting above a few dB, or if more knobs for one or more adjacent frequencies are involved, the response change must be gradual. In other words, keep it 'curvy'. Zig-zagged (+6 and -6dB on adjacent sliders) or 'rectangular' knob arrangements (+10dB, +1dB on adjacent sliders) will not sound good and shouldn't be required – except when the boost and cut are under 1dB or so.

To sum up, you don't need to move every slider or use wild settings. The better you are at using PA, the less you'll move any sliders. Unless there are big problems, most will be left resting in the centre – it's all about learning which ones to move, and by how much.

Making it sound right

EQing away a PA's acoustic problems is not an exact science. All EQ – it doesn't matter what sort or how smart – can only make approximate fixes at the best of times. For one thing, sound in venues (as in any large rooms) varies all over the space – so we may have 101 or 20,001 different listening positions. The common compromise is to EQ so the sound is 'about right' at the FOH mix position.

Alternatively, you could check the response at some randomly chosen audience locations and tune the graphics to help these areas a little. But this attempt at being perfect could make the FOH mixing difficult, particularly if the corrections for other room positions are taken too far, as the

---

Boost and headroom

Boosting any frequency range uses up headroom (or 'space' before overdrive/ clip) by a proportional amount – meaning +8dB boost reduces headroom by 8dB. But this is only the case if the incoming signals in that frequency range don't really need boosting to make them the same as the overall average level. In other words, if you're not really boosting the level, but just correcting it, then there need be no loss of headroom – as long as this sort of EQ adjustment is properly made on the relevant mixer channel(s).

the drive rack

engineer will be struggling to work out what the 'real' sound should be.

More often, signals in the frequency area that needs to be boosted are reduced later by the speakers and/or hall acoustics. So when these frequencies are boosted, headroom is used up, even if you don't hear an excess.

Signals in the boosted frequency area are often reduced later by the speakers or hall acoustics

---

It's important to keep in mind that a graphic (or any other EQ) can't increase the maximum power delivery of a PA. When boost is used, if the signals in that area are larger than the average signal, and are so subject to being clipped (or limited) before the main parts of the signal, then either you allow these boosted parts of the music to be reduced rather brutally by the PA's protective limiting (see later in this chapter), or else the level of the PA has to be turned down to accommodate the boosted parts of the signal. So the PA is quieter overall, except in the boosted range.

As a practical example, let's say a 10kW PA needs to be pushed nearly 'flat out', operated within 8dB of clip, and we want an averagely loud frequency to be boosted by +6dB. We certainly can't spare any of the 2dB of remaining headroom. To regain the +8dB of headroom (without using a bigger PA), the overall sound level would have to be dropped back by 6dB (that's cut in half, in terms of signal voltage). This makes the PA only as loud as a comparable 2.5kW one – in other words, only a quarter of the power.

Looking at EQ boosting in this way, a massive three quarters of the 10kW is 'used up' by a mere +6dB of EQ boost in just one frequency area. Alternatively, to achieve the +6dB without reducing the overall level, we'd need to notionally uprate the PA from 10kW to 40kW – but this four-fold increase in power means a lot of extra gear and cost.

The lesson is that it's easy for EQ manufacturers to create and sell you boxes that boost signal levels, but the real-world ramifications are something they don't very often mention.

Bad graphic EQs

Not all graphics work or sound the same. Beware of low-quality, older or 'fake' graphics (copies perhaps with a mis-spelt famous name, and/or 'consumer-grade' circuitry inside).

Over the years there have been quite a few over-adventurous or naive designs (even made by well-known makers) that don't 'cut the mustard'. If you've never seen an old 'XYZ' make of graphic in use in another PA system, maybe there's a good reason. If there are any good graphics from new or relatively unknown makers, other PA users will discover them. Word gets about.

Bad graphics have the same 27 knobs as good ones, but it's the effect they have that's the problem. As an example, say there are two adjacent faders, each set at +6dB (with others either side sloping back down to 0dB). You'd expect the signal to be raised by +6dB at both the centre frequencies. But on some lesser graphics the controls interact to give a peak of +12dB (that's twice as big as +6dB) in the frequency area between the two centre frequencies, and also possibly much less than anticipated on either side.

A likely outcome at some stage is that signals in this frequency range are heavily limited or that the PA is overdriven/clipped in this area. This could be quite puzzling to any sound engineer, and they may not have time to work out what's causing it. They'd also have to be confident enough to be able to believe their ears more than what the knobs were saying.

A similar misbehaviour in low-budget and fake graphics is demonstrated where one slider knob is set to boost, the one next to it set to cut, and the next one to boost (these settings are allowable if the boost/cut is below 1 or 2dB). But instead of a 'wiggle' curve that the sliders suggest, the two outer boosted sliders override the cut one, leaving just an area of only slightly lowered boost.

'Smart' digital EQs

Some modern 'advanced' DSP (digital signal processor)-based EQ systems are claimed to 'read and learn' the responses and relationship of the speakers and the room they're in, and then make a near perfect correction from imperfect acoustic conditions.

the drive rack

Not all of them satisfactorily deliver what they promise, though. The more basic ones correct the sound for only one location (no better than a FOH sound engineer already achieves) and in fact can be easily 'fooled' by everyday situations perhaps not anticipated by the programmer. Even some good ones, which deal quite well at suppressing low-end room resonances, are less good at correcting higher frequencies, and can sometimes 'kill' the venue's sound.

The truth is that the enormously complex reflections created by loud music in large spaces is something that even the most powerful DSP can't currently handle very comprehensively. An astute FOH engineer will regard this sort of system as just another tool – a second opinion – not something that can replace an experienced pair of ears.

The crossover

Crossovers are necessary because no single loudspeaker transducer (drive-unit) can handle the entire range of audio frequencies – at least not while it's delivering high power and a uniform frequency response. To achieve this the music has to be split up into frequency ranges, or 'bands'. This is the job of the crossover.

There are two types of crossover, active and passive. The latter sort doesn't really concern us at this stage, as they're only found inside PA cabs, in so far as they are used at all (there's more about passive crossovers in Chapter 6) – though they still need to be mentioned here as a point of comparison. The crossovers you find in the drive rack, however, are the active sort.

Some companies use a more descriptive term for their active crossover units, namely 'frequency dividing systems'. It's apt because any crossover splits the composite (full mix, full frequency range) signal coming from the mixing desk, and divides it into two or more frequency ranges that are best handled by dedicated speakers (bass bins, top-end horns, etc).

Some companies aptly call their active crossover units 'frequency dividing systems'

Why active crossovers?

Active crossovers are known in some quarters as 'electronic crossovers'. This is a misnomer, as passive crossovers are no less 'electronic'. What's really meant is that in active crossovers the filters are built around powered electronic circuitry, the benefit of which is that the filtering can be more refined, and quite sophisticated filters can be made cheaply and without the signal power wastage to which passive crossovers are prone.

Active crossovers are essential in all but the most basic PA systems. With an active crossover the splitting of the frequency bands is done before the amplifier – which essentially means you will have to use (at least) as many amplifiers as there are bands, in order to provide a separate amplified signal for

Crossover control

The 'frequency range division' carried out by any crossover involves not just filtering to create each band, but also using complementary pairs of high and low-pass filters. In other words, in the changeover frequency areas, where one signal is gradually filtered away in one band and coming up on the other, the two signals have to be matching opposites, capable of re-joining seamlessly to give the original frequency response – without signal level reductions (response dips) or increases (response peaks) in the region where the two ranges are sharing the work.

This is delicate enough while the signals remain electrical. But in a PA system the bands will be recombined acoustically, not electrically, after the signals have been changed into sound and emerged from the drivers and horns. This introduces extra complexities of distance and timing. Because speakers are bulky physical objects that cannot all be in exactly the same place, nor always just the small size or convenient shape you'd like, signals in different bands can end up slightly out-of-time with each other. Although the difference may be just a few thousandths of a second, musical quality can be quite audibly affected. For the different parts (frequency ranges) of the music to be 're-synchronised', active crossovers commonly include adjustments to enable phasing or delay (something which wouldn't be feasible with passive crossovers).

each speaker (or speaker group). At first sight this appears to add complexity – more amps, cables, more things to get set up right. But active crossovers do have some crucial advantages.

One major payback for using separate amps for each band is that it allows finer control of limiting – needed as protection for the speakers. By contrast, in a basic PA system, it's not readily possible to limit the power to the more delicate HF speakers without limiting the bass part as well.

There is one possible disadvantage: limiting of one band before others causes 'spectral shift', which is heard as a tonally unpleasant sound (perhaps hard or over-bright). Then again, this is the PA's way of telling you that you don't have enough bass bins (or whatever) in your system for the sound level required.

Making the most of the amps & cabs you have

Because an active crossover requires at least one power amp for each frequency band output (low, mid, high etc), there's less strain put on each

---

Crossovers – how many 'ways' ?

Whether the crossover is set up for two, three, four or more bands will be largely pre-determined by the PA's speaker system. Here's an outline of what the different numbers of 'ways' generally accomplish, and how they're placed.

Two-way

Two-way crossover systems require a minimum of two amps, so they're referred to as bi-amped. The split signal feeds just low frequency (LF) and high frequency (HF) drivers. Such a simple set-up is typically restricted to monitors or mini PAs, because for most kinds of music it's not possible to cover the audio range fully enough (particularly with low-enough bass) with only two types of drive units. (See Chapter 6 for more on speaker abilities.)

Three-way

Three-way systems (sometimes called 'tri-amped') feed low, mid and high frequency drive units, and are the most widely used configuration for ordinary FOH PAs and more elaborate stage monitor cabs.

If you require a three-way PA system on a tight budget, you can use set-ups such as 'bi-amped + passive split' or 'two-way active with passive split'. All that's required is a single two-way stereo crossover, a minimum of one stereo power amp per side, and three-way cabs wired for this approach. The system is still bi-amped, but is also three-way. It's achieved by 'splitting'

the HF band's signal (really mid+high) using an additional passive crossover in each mid/high speaker cab. Hence 'passive split'. The active crossover splits the bass from everything else. The passive then splits the HF from the midrange – like a sequence of two forks in a road network.

Despite the limitations of passive crossovers, passive splitting of HF (only) can work quite well (there's more on passive crossovers in Chapter 6).

One restriction of the 'passive split' configuration is that the top-end's level can't be limited with any discrimination – any protective limiting is 'lumped in' with the midrange (we'll cover limiting in more detail shortly). But this only affects the ultimate sound level capability. The excess levels in one or other will turn down both. Still, the limit has to be set low enough to protect the tweeter, rather than the mid, which would handle more power.

If higher sound levels are really needed, more active 'ways' are simply going to be required. But you can still add a passive split to these.

Typically this is done for occasions when super-tweeters (working at frequencies above about 14-16kHz) might be used – for example providing a PA in a small venue where high frequencies won't be largely absorbed before they've reached the audience.

Four-way

Four-way crossovers (quad-amped) will either send signals to a sub-bass driver,

plus low, mid and high units; or else to low, low-mid, high-mid and high frequency speakers. It's mainly used for more up-market FOH PA.

Five, six, and seven-way systems

These were more common in the past, when PA developers experimented with different schemes – and before accountants worked to prune tour costs to the bone. The frequency ranges became further sub-divided and the speakers used were increasingly specialised. These higher-way systems are more complex, but offer potentially higher sound quality and 'maximised' power handling.

But the 'law of diminishing returns' sets in fairly steeply once the crossover has split-up the audio range into four fairly equally-sized bands. For one thing, the weight of amps and cabling, as well as the wiring and rigging complexity, is bound to start increasing substantially, but without much worthwhile increase in sound level or quality. There's also an acoustic trade-off, in that it can be increasingly hard to synchronise larger numbers of separate sound sources.

Despite this, you will occasionally still meet five, six and seven-way systems. There are commercial analog (active) crossovers with five and more bands, some of them flexible modular types. Or else they're bespoke (custom-made), combining two ordinary three or four-way crossovers. The more upmarket digital crossovers also typically offer up to six bands.

individual amp, as the frequency range it's being asked to handle is narrower than it otherwise would be, so the signal is less 'dense' (there's not as much happening at each instant).

This means less-than-perfect gear can be helped to sound reasonably good, and good gear even better, because there is less of the 'friction' or 'noisy argument' – technically called intermodulation distortion – that usually results when many signals at varying frequencies interact. Equally, particular makes and models of amplifiers known or felt to sound good in certain frequency ranges (possibly along with particular PA cabs) can be used where they work best.

Active crossovers also help to get best performance from any given set of speakers. Passive crossovers are commonly placed between the amp and the speakers but this is not ideal, as a direct, low resistance connection between the amplifier and speakers is needed to provide 'good damping' (electrical control). This is particularly true with bass speakers and drive-units. Here, the passive crossover will necessarily involve a low-pass filter with a series inductor – which places a moderate resistance in the way of the amplifier-speaker connection, reducing damping and upsetting the speaker's sound – while the active sort doesn't.

Passive crossovers can also present difficult loads to the amplifier at certain signal frequencies (on top of the complicated and difficult loadings all speakers already present), and this isn't very welcome because, again, sound quality is affected.

It could be argued that passive crossovers, as they require fewer power amps, are less complex and costly. But in reality as many or even more amps would be needed for a solely passive system to sound as loud and keep the sound sweet – and without any of the benefits mentioned above. The wiring might be simpler, but the sound quality far less robust when pushed – and the truck would be heavier, with the extra amps and cabs.

For all these reasons, and more, PA systems mainly use active crossovers – hereafter just called 'crossovers'.

Analog versus digital crossovers

The crossover was one of the first areas in PA where digital processing became widely accepted. Any early sound issues have now been resolved, and digital crossovers can boast some powerful capabilities – such as signal delay (which we'll cover in a moment) – to help make the acoustic integration 'seamless'.

Also, while already in the digital domain, it makes sense (and it is now straightforward) to offer other digital features, such as EQ, limiting, and general delay, which can be provided at relatively little added cost – both in money, and in terms of sound quality. This is partly because only one set of analog-to-digital (A-D) and digital-to-analog (D-A) converters are required: good D-As and A-Ds are costly and all such conversions must be done correctly to the nth degree to preserve audio quality. The fewer the conversions a signal has to go through, the better.

Passive crossovers can present difficult loads to the amplifier at certain frequencies

But analog crossovers remain in widespread use (especially in PAs using older gear, which beginners may well experience first). And since they also remain the baseline of comparison, even for engineers now using digital units, we'll have a look at the features of the best 'traditional' analog units and compare them with the best front-line digital crossovers, where relevant.

Band selector switch (two/three/four-way)

Some crossovers have just two or three fixed bands (and so are known as 'two-way' or 'three-way' crossovers, where 'way' means the same as 'band' – see panel on the opposite page). More flexible models have a switch (or switches) that commonly select two, three or four-way operation – according to what the PA cabs require.

Usually the basic design is stereo two-way, with the crossover reverting to one channel or mono (sometimes called 'monaural in the US) in three or four-way mode. That's because the crossover you've paid for has four sets of crossover filters, and either you share two sets between two channels (usually L&R), or opt for one channel with four sets of filters – or three, if you need that number.

Be aware that two/three/four-way switching can be dangerous to the health of drive units if changed-about while in use. For example, if high-frequency drivers receive signals well below their ideal frequency range, they could be damaged. Considering the cost and potential show-damaging implications of this, crossovers where the mode switching is recessed and not easy to reset or switch accidentally are going to be preferred for PA. With the reduction of accidents in mind, some units feature indicator LEDs that confirm the 'number of ways' setting.

On a typical digital crossover this situation alters only slightly. The two/three/four-way switching is nested down in the menus, though it could theoretically still be accidentally changed with the light touch of a button – or a different stored crossover setting might be called up by mistake. On the better digital crossovers, though, settings can be locked with a password.

Crossover frequency selection ('crossover point')

The crossover frequencies ('points') mainly depend on the speaker system in use, with some room for experimentation.

In pro-grade analog units the crossover point may be pre-set (and re-set) by plug-in 'cards' (small fibreglass circuit boards) or plug-in modules (tiny plastic boxes) to suit the speaker system in use. Alterations to these points usually require you to take the crossover's lid off, which to some extent prevents damage to speakers or wrong operation caused by casual tampering. (Spare modules need storage in a safe place, as they generally contain exposed, delicate electronic parts and connector pins/prongs.)

Some less professional units allow the crossover frequency to be selected variably with a knob. This is better kept in the PA user's R&D (research & development) lab, where it's genuinely useful – in a live situation just knocking the knob could sweep the crossover point down into damagingly low frequencies for the driver concerned.

With digital crossovers, the crossover frequency can be varied continuously (meaning in fine steps). The big difference is that the adjustability comes with almost perfect tracking. The chosen frequency can also be 'locked' by simply not entering the digital crossover's frequency-setting menu. The ease of accurately altering crossover points, together with that of storing settings, would be handy if you had to regularly change crossover points, between using different PA cabs.

Slope & alignment

'Slope' refers to the rate that the signals in each band roll-off or fade away at the frequency band's edges. Having a suitable slope is important for almost any PA speaker system.

The alignment describes the initial slope, where the roll-off starts

All modern crossovers for PA should have slopes of -24dB per octave. That's quite a steep slope, but it's no arbitrary choice. All normal crossover slopes come in increments of 6dB/octave. In the past, shallower slopes of -12 and -18dB were felt to be satisfactory for speaker crossovers. In the 1970s electronics engineer Siegfried Linkwitz calculated that a higher slope of -24dB/octave would ensure smoother recombination of the sound between two speakers. It meant the frequency response of the sound in various positions in front of the speakers wouldn't be subject to dips or peaks, which can sound nasty and may well prove fatiguing to the ear.

In America, crossovers with -24dB/octave filter slopes became known as 'Linkwitz-Riley' crossovers. Although -24dB/oct is now the standard, some flexible crossovers can be switched lower (slopes of -18 or -12dB per octave can be chosen). These should only be used when instructed, or when it's known to work OK – the result of tentative experiments. That's because use of lesser slopes means lower frequency signals can reach the drive-units, and could harm some drivers when pushed hard (to maximum levels). Passive crossovers commonly have a similar limitation.

Related to a crossovers' slope is its 'alignment'. This describes the initial slope, where the roll-off starts. Alignments are adjustable in some analog crossovers, sometimes switchably but usually by altering the frequency setting cards or else with dedicated alignment cards.

Pro-standard digital crossovers can readily provide a wider range of slopes than is justifiable or affordable in analog units. Few users will need most of the exotic variations, and they can be safely ignored most of the time. More important is the fact that, if required, the slopes will be highly accurate – to within less than 0.1dB. By contrast, with analog crossovers, makers are tempted to use parts with wider-than-ideal tolerances to save a few cents – but these can cause disproportionate inaccuracies in the slope and alignment.

Level (gain)

Levels are dependent on the sensitivity and arrangement of the gear that follows the crossover – the amps, speakers, etc. The variables include the room acoustics, so these knobs are re-set for each gig.

The different band levels shouldn't need to differ much from each other. For example, setting the bass at 0dB but the treble at -25dB would be un-usual. If this is required to get the PA to sound right, something is amiss – the amplifier power ratings, sensitivities, or gain settings, need to be looked at (see Chapters 5 & 9 for more on these settings).

Levels are dependent on the sensitivity and arrangement of the gear after the crossover

Pro-standard crossovers provide a limited range of level adjustment, typically ±6dB (in other words a total of 12dB). Making fine and visually-matched knob settings is easier over this limited but adequate range. Less competently-designed units provide a much wider range, more like a volume control's 60dB or so. This makes band level adjustment far more fiddly, making it impossible to match closely between L & R channels.

On the better digital crossovers, the band output levels are set by knobs, or a single knob which sets (then saves the setting of) any selected band. The range can be quite wide (maybe 60dB) provided the knob can work in fine increments, making digital crossovers particularly flexible.

Mute switch

A button you push to deaden one, some or all of the band outputs. Used for testing and set-up (more of which in a moment) so you can, for instance, hear just the bass section, then the mid, and so on. It also helps speed-up diagnosis if a wrong crossover point card has been inserted, or band outputs are wrongly wired – for example, maybe the mid has got crossed with the bass, and no one has noticed.

On digital crossovers the menu may have a password which can be used to lock out all the mutes – handy when you leave a PA set-up to take a break.

Polarity switch

Also called 'phase invert/reverse', this is a button you can push to 'flip over' the polarity of one or more of the band outputs (see Chapter 7 for more on polarity). Mainly used to solve system problems while setting up. Should be used specifically and ideally only temporarily to fix 'inter-band' polarity problems. Improves the sound when used correctly. The button does the same job as inserting a polarity changeover lead (see Chapter 7) in line with the affected output cabling.

In a well-organised PA system, reasons for using this button should only be teething problems caused by wiring errors. Examples all occur in the gear connected after the crossover outputs – mainly amps and speakers – and cables. Polarity 'flips' within these might well arise when you hire-in other gear or speakers, or when you first use some newly made-up cabling.

On some units, the 'polarity invert' button is available on all outputs, as you might expect. But since polarity is a relative matter between the bands, on other units makers save a few pennies by providing the polarity changeover button for all but one of the bands. This band (often the mid-range) is arbitrarily assumed to be correct, and polarity changeover buttons on the others mean they can be made to comply.

That way works fine, but it does mean pressing more buttons all at once to do a check (an 'A/B check') on whether the polarity changeover is helping the sound quality or not.

Polarity reversal switching is just the same in digital crossovers, except it may be buried in menus, and less easy (or not even possible) to change several bands at once. On the bonus side, *the drive rack*

there's unlikely to be any switching omitted on one of the bands, since it adds little to manufacturing costs.

Mono (for sub-bass cabinets)

'Mono' can refer to one side of a stereo signal (the left or right), but here, in the crossover, it means mixing the left and right sides to recreate a real mono (aural) signal (see also earlier in this chapter for a different example of this, the centre channel).

'Mixdown' to mono is usually only performed on the lowest band output, to the sub-woofer. It's done on the basis that the ear can't hear direction very well at low frequencies, below 100Hz, so stereo isn't very useful down there. It also happens that sub cabs (low bass cabs) are operationally best placed together anyway for most mutual reinforcement.

In two/three/four-way crossovers that are set to mono (here meaning one input channel, for usually three or four-way operation), the mixdown of the sub signal usually requires a special cable link between two units, so that one's sub-bass outlet can provide the 'mono' feed. Methods differ, so consult the maker's manual.

Phase adjust

This control is used to tune the PA speaker system for the best sound when set-up. The knob adjusts the phase (signal timing) between bands – sometimes just at the band edges. Unlike the polarity switch (described earlier), which adjusts 'phase' in chunks of 180 degrees, a phase knob provides a fine incremental adjustment, typically say 0 to +180 degrees, degree by degree.

The idea is to help make sure that the sound from different drivers mixes smoothly together around the crossover point, and does not cancel or peak.

Information on setting the phase will be provided, where appropriate, by the manufacturer of the speaker system. When you've gained some experience using a particular speaker system, it may also be done by ear.

Don't be tempted to use the phase control to deal with the room's acoustic problems. If you misuse it, the result could be embarrassing, with the unpleasant sound of a response peak at the crossover point suffered everywhere except where the FOH engineer is standing.

This control is not found on digital crossovers,

as it's eclipsed by a better facility – namely 'delay' (which we'll come to shortly).

Sub-sonic (infra-sonic) filter

This button and/or knob may be very useful to sound quality and the level at which you can drive the PA – or it may have no effect at all, depending on the particular music at hand.

If it's a button, it switches in a 'below-bottom-end' filter that simply removes frequencies that the sub-bass (or bass) cabs can't reproduce. If there's a knob as well, this tunes the 'roll-off' frequency below which signals are filtered away. If this sounds familiar, it is – it's the same kind of HP (high-pass) filter that's available on the house graphic (see 'high & low pass filters' earlier in this chapter), and also used for individual sources on the mixer channels (see Chapter 2). But this time it's for the benefit of the PA speakers.

Don't be tempted to use the phase control to deal with the room's acoustic problems

If the sound contains a lot of sub-sonics (low bass that's more feelable than hearable, and which also uses up valuable PA power), and if the cause of that can't be traced or disposed of at the mixer channels, then the HPF can be switched in – assuming this sub-bass isn't a desired part of the sound. The HPF may even be switched in as a matter of course.

If adjustable, it should be set at minimum (lowest frequency), then adjusted upwards after switching on, until you notice it just starting to reduce the wanted part of the bass. Then (as we saw in Chapter 2), back-off an octave – that's to half the frequency. This provides a fair compromise between retaining the useful parts of the bass or sub-bass while saving the PA from using most of its power and valuable headroom amplifying what may be a meaningless 'muddy' rumbling.

Note that if the preceding house graphic has a HP filter that's in use, this control may have nothing to contribute. Both controls do the same thing, and whichever one is best for the job may be used – although it can be easier to use both for their different roles: the crossover one for the speakers, and the graphic one for acoustic matters. Both could be used at once, but the one with the higher frequency setting will appear to do most of the work, unless its slope is lower – say only -12dB per octave, rather than -18dB/oct (see earlier section on 'Slope'). Then, for example, a 'lazier' -12dB/octave filter's action set at 100Hz, say, would eventually be overtaken by a -18dB/octave filter's action, even if the latter starts at a lower frequency of say 70Hz.

Sub-sonic filtering is much the same with digital crossovers, with the benefit that the frequency is likely be variable with fine numeric steps (of maybe 1Hz at a time), and also the effect occurs at exactly the frequency the readout shows (ie with greater precision). There may also be a choice of different alignments – one parameter which the FOH engineer can experiment with at leisure, as there is no particularly 'wrong' choice that would cause harm.

Crossover limiters/limiting
Chapter 3 introduced and dealt with limiters generally, and focused on their use (with compressors) on individual instruments, to smooth out level variations and potentially improve (tighten-up) the sound.

In the real world, limiters also have to be used on the main feeds (left & right, or left, centre & right) to protect the PA's drive units from occasions when the PA is overdriven – even if only momentarily.

A limiter is essential for two reasons: first, accidents can happen, even with the best engineers in well-regulated situations. Second, speakers can easily be blown-up by such accidents.

Overall, a PA system that doesn't have limiters is like a small car with a huge engine and a light throttle (accelerator pedal) that can send the revs high enough to blow up the engine. As with a fast car, a lot of expensive damage can be done in a few thousandths of a second – faster than the FOH engineer (the 'driver') can react. Unlike the fast car, at least no one will die – but it might kill the show.

The kind of things that limiters in the crossover protect the PA speakers against include:

● A mike being dropped. The impulse signal (which we hear as a loud bang or thud) when it hits the stage, could well be 100 times (40dB) higher than the vocalist's loudest scream. The compressor on the vocalist's channel might help reduce the excess level a bit, but not always enough to prevent damaged speakers.

It's the same if a mike on its stand is hit by a half-filled soft-drink can. Or even if the vocalist sneezes or coughs into mike (bearing in mind this can involve 'wind' speeds of typically 60 mph).

● A 'whoops' button gets pushed (that's anything that makes the FOH engineer say 'whoops' – or something rather more explicit). For instance going from AFL (after-fade listen) to PFL (pre-fade listen), so the sound level suddenly leaps up by 10 or 15dB (about three to five times).

● Feedback. When this happens it subjects some parts of the speaker system to a continuous tone-type signal which tends to rise up swiftly into clip. The limiter prevents this part, giving the engineer a few moments to drag down the level enough to stop the howl outright.

It's useful to be aware that the limiters will 'let through' one-off (but not fast repeating) peaks of excess level briefly, before they act. This means the speakers do still need to be able to handle the full level of the amplifier being used, at least for a few thousandths of a second. Generally (and perhaps surprisingly) this is not a problem if the amplifier's power rating is as much as two to five times the drive-unit's continuous power rating (RMS). A typical situation might be:

150W rated amplifier – feeds an HF driver rated at 50W. Limiter operates at equivalent of

---

Latency – the digital down-side
When using all-digital (DSP-based) crossovers, equalisers and/or delay lines, you'll need to be aware that some digital units (particularly older or cheaper ones) can have 'latency' (signal delay) of one to several milliseconds. (As a comparison, analog electronics pass signals almost instantaneously – within a few millionths of a second). This latency can upset live musical performances.

At this stage in the PA the keynote is to avoid directly combining signals that are pre and post-DSP equipment – unless you want an echoing effect – or at least be prepared to check for this before use.

50W, but occasionally allows the full 150W to reach the driver, momentarily. Note also that the crossover's limiter, although not intended to prevent the amp clipping (overloading), has this effect.

Crossover limiters vary widely in what they sound like when they're in use. It's also quite common (if also quite wrong) for crossover limiters to be driven 'hard' all the time – for instance at dance events. This is not ideal, but if it has to happen you'll want a crossover with limiters that 'sound good'.

Limiter threshold

This is a vital setting that protects the PA speakers by fixing a signal level above which the music's peak levels will not be allowed to increase. If set too high, there won't be any protection. If set too low, the sound will too readily become unbalanced (for instance bassy, or middy) when the PA is driven to higher levels.

> # It's quite common (but wrong) for crossovers to be driven 'hard' all the time, as at dance events

This knob is vital to PA safety, but only appears on crossovers with built-in limiters. If the crossover doesn't have one of these, either don't use it for PA, or you could add a limiter afterwards – on every band. That could prove costly if the system is three-way or above, and should be compared to the cost of starting afresh with a crossover with built-in limiters.

Because the limiter's threshold control will need re-setting if a PA system gets re-configured with different gear, it'll need to be a panel control, even if it's a 'dangerous' one to tamper with – meaning that tampering could leave the PA's speakers vulnerable. By comparison, on less suitable crossovers that are really intended for fixed installation, the limit level may be set by plug-ins inside the unit, or else a screwdriver-operated control round the back – neither of which are handy for use 'on the fly'.

On digital crossovers the limiter threshold works just the same. Digital benefits include being certain that all the bands' limiters will work in sync when they're all set to the same level (due to digital gear's intrinsic high accuracy and matching). This won't always be the case with analog crossovers – but to help this, see the tip on 'knob aligning' in Chapter 3.

Fixed EQ (horn EQ, CD EQ)

This will be needed whenever 'constant directivity' (CD) horns (the type in most modern PA systems) are in use – usually only ever in the high-mid or high frequency. Note that not all makes are necessarily called 'CD' – but all such types have noticeably straight-sided ('triangular') flares, rather than the rounded flares of older radial horns (see Chapter 6 on horn directionality).

The EQ comprises a steadily increasing boost with frequency, like that of an HF shelving EQ before it shelves (see Chapter 2). In reality there is no more actual energy than before, it's just distributed more widely. The HF energy's density (in space) is 'thinned', so the EQ boost is needed to recover a decent, flat response.

In most analog crossovers this fixed EQ is set by a card (a fibreglass circuit board) that's inserted inside the crossover – it's supplied by the crossover maker and is intended for certain makes and models of CD-type horns and not others. The maker may also supply blank (unfinished) circuit boards along with instructions for making-up any CD-type horn EQ likely to be needed – with a table of resistor and capacitor values that shows what parts you have to obtain and solder into place.

With digital crossovers, the required EQ may be available, much as described here, but more likely it will have to be created using the PEQ facility. The PA speaker maker should give instructions for this.

Variable EQ knob(s)

Some analog crossovers have variable EQ built-in, usually intended either for 'CD' horns (a more 'tweakable' version of the above) or else to help recover some response in the HF's top octave, above 8kHz – where the drive unit always ultimately has a reducing output, and where the air 'soaks up' the sound level quickly with distance.

Typically such EQ allows up to +4dB to be applied at 10kHz, with a moderately high Q of 3. If using it, bear in mind that this will leave less headroom: 4dB boost is equivalent to reducing the amp's power to less than half. Also bear in mind that with high treble the effects of overload and resulting severe distortion may not be very audible until the sound suddenly goes dull... because you've blown a tweeter. This can be avoided by watching the crossover's metering (which we're just coming to). See also parametric EQ, in a moment.

Metering

The signal levels are usually shown by LED bar meters at the band outputs. Some crossovers also have input level metering, which can be useful – since, as with graphics, the levels at the input and band outputs of the crossover can be different. That's largely because the treble band, for instance, may comprise little of the signal content. And also the case at high levels, where the limiters on one or more bands are operating.

On more sophisticated crossovers the output metering is arranged to be relative to the limiter settings. This means you can see the headroom – meaning how many dBs are left before the signal hits the limits. In some cases there are just two separate LEDs for the limiter section, one of which lights to show the onset of limiting, with another to show that limiting is reducing the level by, for example, 4dB.

Extra controls

The following control features are mainly (but not always) exclusive to crossovers employing digital signal processing:

Attack time

The attack time of many crossover limiters is preset and not adjustable. On some units (particularly digital ones), a wide range of times can be 'dialled up'. For example, the rate of attack is limited by the filtering – meaning that fast attack isn't very useful on a bass output.

Delay

Often called 'time delay' (although obviously it's the signal that gets delayed 'in-time' rather than any physics-defying time-shifts taking place).

This setting is universal on digital crossovers. It's less feasible (but not impossible) in analog crossovers because you'd have to pay for A/D and D/A converters, to go to digital and back again, all just for the delay.

Delay is used as an enhancement of, and replacement for, the phase control on analog crossovers that helps smooth the blending of sound between speakers working in adjacent frequency ranges, and the synchronising of the soundwaves arriving from different drivers. All digital crossovers have delay on one or more bands. (Delay is dealt with in depth in a moment.)

Parametric EQ (PEQ)

Some crossovers – mainly digital types – have 'full-blown' PEQ on-board. (The use of PEQs is described in some detail in Chapters 2 & 3.) In digital crossovers, the convenience and accuracy is particularly welcome, in terms of precise frequency, Q (bandwidth) and dB.

Another useful feature of digital EQs is having lots of bands. Four-bands of PEQ are powerful enough, yet the best modern units may offer as many as 30 bands. These are intended as powerful room EQ'ing tools (see Chapter 8 for more on this), and may even be used instead of the house graphic EQ (as described earlier in this chapter).

With digital PEQs, very high Q settings (very narrow, select EQ) can be made with complete stability. Very high Q filters aren't often seen on analog gear, because they're hard to keep stable with temperature changes – they could turn into a 'screaming' oscillator (a problem that's solvable, but a tad too expensive in comparison to the cost of competing digital gear).

With digital PEQ, very high Q settings can be used (for the first time outside of expensive, fussy lab gear) to deal with very specific, bad room resonances, for example, with minimal effect on wanted sounds at only slightly higher or lower frequencies.

Tip from the top - Using delay

If the amount of delay (in milliseconds) is set very slightly behind perfect synchronisation, so it's ever so slightly late, this gives the impression that the sound is coming from the main stage, rather than just from the nearby speakers. The correct amount can be found by slight adjustments back and forth, while listening in a position that's beyond and some way off-centre of the delayed PA stack, so you can not only hear that stack but also the sound from the main PA on one side of the stage.

the drive rack

Setting up a crossover in 10 easy steps

This job interacts with setting the graphic, as the more extreme changes in the sound caused by the room will prevent the crossover being set correctly. But until the crossover is set right, the graphic can't be set properly.

● First: turn off one side (say the right side) of the crossover, using either the mute buttons or the appropriate main fader on the mixer.

● Second: turn off (mute) all the bands on the remaining side, except the HM (high mid) or mid. The idea is to be able to listen to the mid alone to start with.

● Third: use a high quality signal source (like a CD) of some music you know well, that's recorded in a way that's not too wild or weird, and listen to the high-mid (or mid-range), for problems caused by the room. Is there anything ringing in a particular frequency range?

If you think you hear this, find the knob on the graphic that provokes it by boosting one at a time. Once the right knob is found, reduce that frequency band on the graphic by say 1 to 2dB.

● Fourth: if applicable, mute the high-mid band (on the crossover) and repeat the above for the low-mid.

● Fifth: you can now un-mute the other mid-range (if there are two) and combine these – adjust the levels – so the music being played sounds right. Or as right as it can without any bass and top.

● Sixth: you can now un-mute and bring in all the crossover's bands. It's best to do this by introducing the bass next, and lastly the top, rather than the other way round, as this seems to leave a lingering impression of there being more top than bass. Then adjust for the right tonal balance between bands. This will be easier having initially dealt with the worst of any room effects in the mid-range areas.

To get the balance right, it's important to listen loudly enough – as tonal balance changes considerably between medium (say 90dB SPL) and higher (say 105dB SPL) sound levels.

When the crossover controls are set right, the music's different frequency ranges will sound 'connected' or 'seamless'.

● Step Seven: you can now set the crossover knobs on the other side to the same positions. This is correct in theory only. It helps (for a start) if the

knobs are aligned correctly (see Chapter 3). To be really sure, and maybe discover a problem of matching between the two sides of the PA (better now than later), we can perform the following test…

● Eight: to prove or check that left and right sides really are set the same, turn down the right side of the CD's test signal, then pan the remaining channel (on the mixer's input channel) to equal L&R (if it wasn't already) – ie dead centre.

This means that exactly the same signal is sent to both sides of the PA. Beware that on some recordings some of the instruments won't be on the left (or right) side, so either don't expect a familiar recording to be fully present, or else choose music for this test where the two sides are fairly similar.

You can then use the mixer's main faders, or L or R side mute buttons, or the pan knob (the quickest option), to flip between them, to see if the sound is the same over both sides of the PA.

Also listen to both sides at once, making sure the levels are the same – check the metering at the various points from the mixer outputs through to the crossover outputs. If the levels are the same at the crossover outputs, you should hear a central image. If all's well, you can now carry on to step 9.

If the image is lopsided (or there's something funny with the image positioning when you move your head from side to side), it suggests something is wrong in the set-up or wiring of the amps and/or speaker cabs/boxes.

To find out more about the problem – for example, 'Is it in one particular frequency band only?' – mute all but one band in turn. So start with everything other than, say, treble muted. Is that image lopsided or does it have a 'hole in the middle'? This order isn't essential – but working downwards from the top is suggested because the judgment about lopsidedness will get harder at low frequencies and probably won't be discernible at all with sub bass (at these frequencies most people's ears are less good at locating sources).

Once you discover (say) that only the low-mid band is causing the lopsided sound – where the image is off to one side – you know where to carefully check the connections, check the amplifiers are turned on, and the low-mid drive-units are connected, etc.

(You can read more about these in later chapters, particularly 6, 7, 8 & 9.)

If it turns out that all the bands are wrong, this signifies the problem is earlier on, before the crossovers' outputs. If the signal levels before are the same according to the metering, a polarity transposition is the most likely cause.

This can be overcome by flipping over hot and cold connections on one, not both, of the feeds (see Chapters 7 & 9). But it may be a good idea to find out why this has happened first. Note that if the sound image jumps as you move your head slightly from side to side, this is a strong sign of a polarity error – somewhere in the PA's 'back-end' wiring.

If you don't trust your ears enough to set the crossover bands correctly, you could consider the aid of an RTA – a so-called 'Real Time Analyser' (see Chapter 9). This can be used to show if you are 'about right'. Essentially it achieves this by showing a flat response on a screen or display, when the bands are lined up. But it's generally better and far safer to use your ears for fine tuning (see Step Ten, below).

Step Nine: you must 'time-align' (synchronise signals emerging from) the drive-units in the PA cabs, before you proceed to fine-tune the crossover. You synchronise the soundwave at the front of the PA by delaying some of the drive-units' signals by suitable amounts. The amounts will be specified (at least approximately) by the PA system's maker.

If you've read the earlier parts of this chapter you'll know that the cabs can be aligned in two ways, either using a digital delay line or by an older 'phase alignment' method found on analog crossovers. This latter technique wasn't perfect but it was acceptable in the past, when digital delay was expensive and even the best digital units sounded fairly horrible. Nowadays, both digital and analog crossovers employ digital delay, as it has become the superior technique.

Finally, Step Ten: at last, you can now fine-tune the crossover's band controls – doing this by ear with a variety of well-recorded music you know, being sure to do this at an averagely high sound level.

Delays / delay lines

These units can be found both before and after the crossover, or may be built into the crossover. Having read about the crossovers, delays should now be easier to understand and see in context,

We've already encountered delay (in Chapter 3) as an FX device that can be used on individual voices or instruments, either to achieve a repeat echo effect, or as a more subtle thickening agent to 'double-track' a signal. But its presence in the drive rack is either for use on the PA signal as a whole or on different parts of it.

With a properly set-up PA system, delay lines should have no effect on the signal's level

Older delay techniques that used tape and other methods were just too noisy for the main outputs of the PA (compared to use on individual instruments), so for practical reasons signal delay units have only been feasible (at least universally and on any sort of scale) in the past decade or so, since the arrival of acceptable and low-cost digital technology.

Delay logistics

You may have gathered from this section that you might require several channels of delay, even for quite a small PA. For example, if only output delay is used (to align the speakers' signals), then for a stereo, four-way system, you'll need six channels of delay, usually in the form of three stereo units. If input delay is being used, that adds another two channels, or one two-channel delay unit, to the drive rack.

This is an appropriate point to say that it's not necessarily a good idea to try to 'gang' or merge the input and output delays, to save on the number of delay units. For example, you could try adding the input (base) delay to each of the output delays. This is feasible, and is sometimes used.

The trouble comes when environmental conditions change (see 'Delay time'), and the output delays may not need so much change, if any, compared to the input delay. Also, adjustments will be made more complicated and will need more care.

At large venues, and particularly at outdoor gigs, the PA cabs on the left and right of the stage may not be enough to cover the more distant audience with an acceptably high sound level. Increasing the number of FOH PA cabs (and amplifiers etc) helps to boost levels – but only to a certain extent (for a variety reasons to do with combining many speaker sources, explored in chapters 6 & 7). In extreme cases the PA becomes too loud for the front rows while still being barely audible for the crowd further back.

An electrical signal is about half a million times faster than sound

The practical solution is to mount another PA stack (or two or three) further down the field (or stadium). But the problem there is that the signal fed to these (via the speaker cables) will reach the listener before the 'acoustic' sound from the stage-front PA does – resulting in a horrible echo effect. (The speed of sound through air is about 775mph (at sea level) – which seems quick, but an electrical signal is actually about half a million times faster.) That's why delay is needed.

Delays, or 'delay lines', are used to retard one or more signals by exactly the right amount so the outputs of displaced PA cabs can be synchronised or 'time aligned' with each other. According to their placement, the 'delay towers' (also called 'delay stacks' or 'delayed clusters') may be mono (L&R) or individual repeats of the main L or R sides.

A related use is to delay the signal to 'flown' cabs, which will also be further from the audiences' ears. Again, the delay overcomes echoey or other effects caused by hearing the sound sources out of sync.

With both the above uses, the signal being delayed is a branch of the entire full-range signal, taken just before the crossover. Some crossovers, particularly digital types, incorporate delay facilities which may be called 'input delay', or 'base' delay.

Output delay – signal alignment

A second major use is to retard the signals from different drive units (bass, mid, HF, etc). Again, this is to synchronise the sound sources so they

'speak in time'. This makes sense when we realise that some bass horns (or bins – dealt with in Chapter 6) are several feet (one to three metres) long, whereas HF horns can be as short as a few inches (50 to 150mm).

If the mouths (front) of these are lined up for various practical reasons (such as avoiding sound reflections, and mutual mechanical support) then the treble sounds will emerge ahead of the bass (according to the difference between the horn's interior sound path lengths), as most likely will the midrange, by a lesser degree, and so on.

Delay setting examples

For the 'input' type of delay – for a secondary PA stack or 'delay tower' located half-way down a big concert hall or field – a typical setting will be up to hundreds of milliseconds (tenths of a second). The amount is mainly determined by the distance: the rule of thumb is 3ms per pace (roughly one yard, 0.92m). Start by standing under or against the main PA, then pace-out (walk and count) till you're against or underneath the front of the delayed PA cabs – then multiply the number of paces taken by three. That's the time (the amount of delay, in milliseconds) to enter into the delay unit.

Fine-tuning will be required, because pacing gives only an approximate distance, particularly if the ground is sloping or stepped, or one PA stack is way above you and the other is near the ground. And also because the speed of sound is affected by the air conditions (temperature, humidity) and even altitude. At altitudes above sea level, and in warmer or drier air, the speed of sound is slower, so more delay is required than the above formula suggests.

Conditions can change in the course of a gig, too – not only could the the weather turn, but the presence and movement of large numbers of people creates a lot of heat and moisture. Delay line units made for serious PA use are fitted with (or have the facility to fit) sensors to monitor the air's average humidity and temperature, so the delay period can be automatically adjusted to compensate. Without this feature, synchronisation problems may develop as the gig proceeds.

To check settings, use a click track (the sort used by musicians for drum sync'ing etc) or have someone play a snare drum, or anything percussive to create a loud, 'sharp-edged' test sound. Then stand past the delayed stack, and about 20 to 30 degrees off to one side, so you hear both the main and delayed PA speakers.

You can then adjust the delay so the image 'centralises' – meaning it appears to come as a phantom image from between the delayed stack and the main PA.

At higher altitudes and in drier air, the speed of sound is slower

For the 'output' type of delay, there are no typical amounts, except they'll all be a lot less than the periods used for most delay towers. The one thing that's certain is that the bins/horns/drive-units (covering particular frequency bands) have to be delayed relative to the 'reference' one. The 'reference' is the cabinet/bin/horn (working in one particular frequency range) that delivers the signal to its outlet (its front opening) the latest.

It's nearly always the sub-bass cab – this is because most serious PA bass cabs are horn-loaded, and the lower the frequency a horn is made to handle the larger its dimensions, including its length, will be. Hence the soundwave produced by the drive-unit inside starts further back than in the higher frequency horns (see Chapter 6 for more on speakers).

To set the right delays, you need to know the figures recommended by the PA cab makers. For example, maybe a delay of 3mS between the HF and bass cabs. The maker's figures will usually assume the cab or horn fronts for each frequency band are placed in a vertical row. But there will be tolerance variations – not all cabs are identical in size, and the cabs may not be stacked perfectly in line, although they appear to be.

Use the recommended figures, then maybe try tuning up to ten per cent either side (while playing some music from a test CD again) to see if you can find a point where the sound 'clicks' into 'seamlessness' or 'focus'(particularly between the frequency ranges concerned).

You'll have to do a bit more work to find the right settings if the PA cabs aren't vertically aligned or are substantially vertically separated. For example, if the sub-bass cabs are tucked under the stage, or have to be placed, say, three feet (0.9m) further back than the other cabs fronts. Or else the mid and HF parts of the PA are 'flown' (hanging down from above).

In the first case, the bass would be even more delayed than normal – by about another 3ms, so all the settings (HF to sub-bass, HM to sub-bass, etc) would need to have this amount added. In the latter case, the higher frequency cabs would be further back (from the audience) than if they were down below (not flying). So several milliseconds would need to be deducted from each setting – essentially the opposite of the previous process. As negative delay settings aren't usually possible, if this is required it indicates you have to re-determine which (different) band is the latest arriving, and re-reference all the other delays from there.

Delay time setting

First, bear in mind that the delay unit's display doesn't necessarily read the real delay, but just what the delay should be for the apparent settings. It's unlikely (though possible) for the readout to read a time that's different to the actual delay you are getting – if the unit has a minor fault, perhaps. Trust your ears: they are experiencing the reality, while the readout is likely disconnected from this by one or more steps.

Digital settings

As well as using digital electronics internally, delays often have digital-type control surface features – meaning keypad entry, sparse multi-controls (one knob does all jobs, for instance) and nested menus on necessarily small LCD screens. These features can obviously keep down the size and cost of the units, but (although they are getting better) they are not always automatically intuitive and easy-to-use in the real world. Make sure you run through how you will use them, preferably before buying the unit, as there may be unseen snags for your preferred method of working.

For example, LCD screens need to be the sort you can see at many angles, in the dark and in bright sunlight, and also when the air is freezing cold. Another thing to watch for is 'depth' of menus: a unit won't be easy to use if some adjustment that you'll be needing to use a lot is located 15 steps down. (For more hints on buying PA gear, see Appendix A.)

Secondly, for those working on a low budget, and maybe using second-hand gear, you'll want to avoid any 'not made for PA' units where the delay isn't easy to set and alter. For example, on a few older, low-budget delay units, delay was set by depressing a number of DIL switches (looking rather like miniature piano keyboards, and used on some computer equipment for fixed settings) – these give different, 'binary-weighted' combinations of delay times, such as 1, 2, 4, 8ms. (So you would depress the first and second button to get (1+2 =) 3ms. Such units are made only with a view to use in low-cost fixed installations in pubs, churches etc.)

Remote operation saves running back and forth to check results

Slaves & cascading

It's common for each of the delay lines' inputs to be able to deliver signals to two or three separate outputs. The purpose of this is so each of these can have different delays, or some none at all. These independent delays were once called 'taps'– which dates back to the days when short delays were created by passing the signal down long lengths of cable (possibly coiled up) and then 'tapping' or accessing the signal somewhere along the total length to get intermediate delays.

The more sophisticated units allow these separate outputs to be ganged when required, so that when one delay is adjusted the others will track it. With some delay lines this is taken further, with the ability to 'slave' additional units from the first delay line unit, designated as the 'master'. This is rather useful when you need more channels of delay.

Remote (wireless) operation

This sounds like a luxury feature but it's actually more than useful when it comes to trying to set up input type delays (which are of course at a distance). The alternatives – running back and forth to check the results of your adjustments, or trying to talk someone through the control tweaks

via a walkie-talkie or mobile phone – are much slower and also prone to human error of the 'Chinese whispers' sort (where meanings get confused as they're passed down a chain). With remote-control operation, delay settings can be tuned and directly proved out in the field – and at speed. Like focusing a camera lens, getting a PA delay set correctly should cause the sound to suddenly become clear.

PC control

As we've just seen, the use of delay in larger PA set-ups can require a lot of settings to be adjusted and monitored, so the more sophisticated delay line units allow you to download the settings into a computer – preferably a portable laptop type. The computer gives you far more versatile control (using a mouse or trackerball, and with multiple, scrollable and zoomable screens) than any tiny, upright equipment panel can. Note that the audio signals aren't passed through the computer at all.

Gain settings

In an ideal world, with a properly set-up PA system, delay lines should operate at unity gain ('what goes in comes out the same size'). In other words delay units should have no effect on the signal's level.

In most instances the signal to the different input/base delays is taken from the mixer's L & R outputs, and is then fed into all the required crossovers and delays. The signal to these is controlled by the mixer's main faders. But you may sometimes want to control the levels from the different base/input delays from the mixer. As things stand, though, if you were using an analog crossover (which has no overall level control), you'd need to fiddle with and re-balance all the crossover band controls, or maybe use the gain control on the graphic – which is really a misuse, and could upset the gain structure.

The solution is to use an aux stereo or matrix output or, at a pinch, a mono (central) main output (see Chapter 2), which are fed from left or right outputs, and so will track the main faders.

None of these will normally be necessary if you use a digital crossover, since these usually have an overall gain control. This is also the case with some digital delay lines that have a gain control – but only if you don't mind having to turn to the drive rack.

In the real world, anything less than state-of-the-art digital equipment can also require manipulation of its levels (or, strictly, gain structure) for the best sound quality. Much digital equipment still hasn't got quite the dynamic range that has been the norm for analog audio equipment built to pro standards. This equipment can operate over a range of about 140dB – meaning the highest signal levels can be up to 140dB louder than the background noise.

By contrast, basic 16-bit digital is limited to about 90dB, which is 50dB less, or a factor of 300 times noisier, or else with less headroom (you can view it either way). Higher bit systems (20-bit, 32-bit) get progressively closer to analogue's 130dB range – except that there are sometimes doubts about whether all the 'bit' figures claimed are fully 'real'.

Added to this, digital noise can be far less pleasant sounding and 'noisier' than analog noise. This is possibly because the conventional measurement of noise ignores the spikiness of the peaks, or the effect on people listening, of any repeating patterns embedded in the noise.

Expect the unexpected and make no assumptions

Manipulation of levels (aka 'headroom control') means that a digital delay line unit may automatically boost low-level signals at the input then put the level back where it was by the time the signal reaches the output socket. And vice-versa. And can also vary this amount of boost continuously with the signal level, so the signal-to-noise ratio is always kept high (more signal, less noise) but without overdoing it and risking overload.

In cheaper units, input and output gains may be manually adjustable to maintain unity in/out. With these, unless the in/out gains are set about right, there's a risk of either noise (signal low, input level set too low, or output level set too high) or premature sound break-up at higher levels (signal high, input and/or output set too high).

Note that earlier (and cheaper) digital delay equipment may not match the headroom standards of professional grade analog audio, which typically handles maximum levels of at least +18, to +24dBu (about 4 to 9 volts). Units which can't match this are best avoided since their use would mean, among other things, that input levels would have to be dropped typically 10dB in level. And this 10dB of lost level wouldn't be recoverable in the unit. Instead, a succeeding unit would need to do this – either the crossover or the power amps. This sort of complexity is best avoided by sticking to pro-standard gear.

Overall, expect the unexpected, and make no assumptions. Watch for overload (usually very audible with digital) and if in doubt, allow time to check out the optimum operating levels by listening to different gain structure settings.

Other than manipulation of the delay units' operating levels, it's also possible to reduce the sensitivity of the connected amplifiers – see 'Panel Gain controls' in Chapter 5.

In emergencies – to bypass or not bypass?

With a delay unit used for input delay, if the unit itself fails in a major way, or its power connection fails, you'll lose the feed. Bypass isn't provided because the echoey sound that would result would be more confusing than hearing no signal at all from the delayed PA. So be especially careful that power plugs don't fall out of the back panel, and that you carry a spare mains fuse etc.

The same precaution goes for crossovers. If either of these units fail, spares will need to be found and wired and set-up before the show can go any further (see also Chapter 9).

Delay unit metering

Very accurate '1dB before clip' red LEDs are required. Or even more accurate peak metering of the top 10 to 20dB below clip and quite fine increments of, say, 2dB – again with a red LED showing just before clip starts.

Multi-processors & system controllers

While the sonic performance (sound quality) of 'pro-grade' digital processed units (crossovers, delays, graphics, PEQs) have improved greatly since they were first introduced, there's still a sonic price to pay when using them due to their analog-to-digital and digital-to-analog converters (A-D/D-A) –

the drive rack

especially if the signal has to go through several such conversions in a row on its journey.

Each instance of conversion generally introduces unwanted and sonically degrading effects such as ringing (which 'smears' the HF parts of the signal in time – see sidepanel below on 'HF roll-off'). So it's a good idea to minimise these conversions. For instance, if delays are used before or after other items operating in the digital domain – namely the graphic and crossover, as in this chapter – it makes sense to create a direct digital connection between them.

The more integrated approach to this is a 'processor' type unit that combines the delay and crossover, and even the graphic, or else provides PEQ. This is even better as it avoids or at least reduces possible sonic degradation due to jitter, caused by digital signal reflections and/or noise pickup in imperfect digital connecting cables.

In recent years multi-processors and system controllers have been used more and more in PA drive racks. These may appear superficially similar, but there are major logistical differences.

Multi-processors combine and 'compact' the equipment discussed in this chapter – EQ, an active crossover, comp-lim and delays – all into one box. Generally full (or at least wide) control flexibility is allowed – users are expected to know what they're doing, just as if they are using the separate units. In recent years, using 'soft' digital controls has made this possible in a compact box – though ideally this shouldn't be at the expense of a well laid-out control panel.

This 'compacting' has enabled the PA's drive rack to shrink. This isn't a big deal in a small PA –

but, for instance, at Roger Waters' Berlin 'Wall' concert (held in the early 1990s before such compact gear was the norm), the PA ended up with four, gear-packed drive racks for the FOH sound.

A valuable feature of some of these digital 'wonder boxes' is the ability to store the drive-rack settings. This gives the FOH engineer the ability to try 'what-if' settings; maybe have a baseline EQ (etc) for venues, stored from the last gig there; and also deal more readily with any changes in the arraying or even the model or make of the PA cabs being used. The value of this will be appreciated with experience. Your ears are still the ultimate judge, as always, but this kind of help can give you a head-start when time is short at a gig set-up.

System controllers are always the creation of PA speaker cab makers

System controllers – also known by such aliases as 'processor-driven' or 'electronically controlled' systems – are made to suit specific makes and models of PA cabs. Indeed, such units are always the creation of the cabinet maker. The way they operate can vary. There are fundamentally three sorts:

First, in the most purist cases, the unit is likely to be called a 'system controller' or 'speaker management system', and is no more than a multi-processor set-up to make best use of specific cabs – for instance the PA cabinet maker will have

---

High frequency roll-off

Whenever you use analog-digital and/or digital-analog conversions, all frequencies higher than the ones required must be very rapidly filtered away. This filtering isn't user adjustable – it's hidden away inside – and is very carefully set up to give the best sound or performance figures.

It may not seem that this should affect the sound, because the filtering barely starts before 20kHz, which is higher than most adults can hear. But a side-effect of the very steep (rapid

roll-off) filters required for digital is that they do cause a degree of 'peaking', and also ringing – which means they continue to affect each signal over time (an effect we met in Chapter 3). As well as being directly audible to some listeners, and adding to existing acoustic reverb effects, this behaviour will boost (to some extent) any higher harmonics in the sound, and these may be heard as added bite or grittiness.

The ringing will also 'smear' the acuity of HF transients, meaning instead of

discrete 'zings' you hear multiple, echoey, fluttery 'zizizings'.

Another side effect of low-budget digital delays is distortion at high audio frequencies. So listen to the top end sound carefully before purchase. Older, earlier digital delay lines also made weird sounds when you swept a test signal upwards.

A/B testing (alternating between delay connected, then bypassed) with a wide spectrum of musical styles is the best way to check for any effects you don't need or want.

provided suitable system-specific settings for the crossover frequencies, limiting, any EQ, etc. These settings are then locked (in digital units, relevant menus are greyed out or absent; in analog units, relevant knobs are absent), so that users who are inexperienced can't ruin the repute of the maker by coming up with bad sounding settings.

Such system controller units are specified for use with particular amplifiers (or whole amp racks) whose gain is known. This is necessary so the controller can also include suitable pre-set limiter thresholds. If the amplifiers' gain was higher (say users bring in some other, 'unofficial' power amp) then the limiter would operate too late. And, frustratingly, even an experienced user will likely not be able to reset the thresholds to overcome this situation.

On the helpful side, these kinds of 'system controller' units are capable of driving any number of amps and cabs, in the normal PA configurations covered in the rest of this chapter. This isn't the case with some other system controllers, as we'll see in a moment.

A feature that can be common to both the other types – namely 'processor systems' and 'electronically controlled PA cabs/speaker systems'– is to employ a form of speaker protection (avoiding overdrive abuse and possible damage) that's an alternative, or is additional, to the conventional limiting covered earlier in this chapter.

Typically this involves sensing the actual levels that the speakers are receiving. Such a signal may then be used to drive conventional limiters, or some other related means of control over excessive levels, eg compression and/or a gain 'leveller'. This is more foolproof than limiters as ordinarily used.

The reason this sort of approach isn't routinely used with mainstream PA systems is that it comes with the complication of extra wiring, to communicate (send) the drive level at this point back to processor/controller unit. In some units this is achieved by passing each amp's output back through the processor.

That makes the wiring easy, but means the processor practically has to be sited in the stage wings (or under the stage), with the power amp racks. Otherwise the speaker leads would be unfeasibly long – taking these back to the FOH (where the drive usually is) is just not practical. And it's not very practical for the FOH crossover to be placed in the stage wings, except for a very small

and basic level of PA.

The conclusion is that such systems (which are usually for FOH cabs) can only be applied to the stage monitoring part of the system – where the crossover/controller is in the stage wings already. So they are probably best avoided unless you want a small PA that's very easy to set up and won't ever need to be expanded.

Other then assuring a certain ruggedness ('unblowability'), the overall aim with such an electronically-controlled system is often to get the greatest sound level out of a given PA cab. It might be cynical to mention that this is also often a low-cost speaker-maker's agenda, using a direct radiator cab that has limited loudness in the first place (and possibly other corners cut). While such systems allow the speaker maker to push a low-cost cab to its limits, thus proving economic, sound quality is not always a top priority.

These digital 'wonder boxes' can store the drive-rack settings

The third and distinctly less-good 'processor-driven system' uses a method which may be called something like 'spectrum power transfer'. This operates quite crudely as follows: to get the greatest sound level out of a (let's say two-way) PA cab, the low/high crossover frequency is increased as the level gets higher. This saves the HF driver from excursion strain (the strain of jerking back and forth further than the mechanism is designed to do), which is the way most HF drivers are 'blown'.

This seems useful, except it also inevitably changes the sound quality at high levels (described as a 'spectral shift'), and alters the PA's directivity and coverage – something that's been fine-tuned in a decent system.

The acid question is: if the higher crossover point is acceptable at high levels, why isn't it used all the time? This is a distinctly bad idea in an arrayed system (see also Chapter 7), because it will cause lots of narrow, 'comb-like' peaks in the sound at most locations, which will sound horrible.

If such a processor/controller unit effectively protects drivers from getting blown, then provided it's not needed to be used at higher levels, and it isn't creating bad sonic effects at lower levels, it may be reasonably acceptable in a small PA system. It's particularly suited to situations where ultimate sound levels and quality are less important than compactness, speed of set-up, and 'skill-free' ease of use.

A similarly less-than-ideal method of protecting the speaker drivers, used in some 'processor' type of 'controllers', is to 'pull in' the bandwidth. This means to reduce the low bass by moving the bass band's high-pass frequency upwards and, similarly, cutting out the top of the HF band by reducing its low pass filter.

It may be cheaper to buy better speakers in the first place

This removes some of the stress, while only disposing of some of the 'less important' signal content – which has the 'least-bad' effect on most music. This is typically achieved by filtering and EQ'ing that are 'level-conscious' (meaning they vary as the drive level to the speakers varies), together with frequency-conscious compression – meaning compression that's applied more strongly at some frequencies.

Artistically, this processing trio isn't very acceptable – because it alters the sound of the performance too much, and a skilled sound engineer is left fighting their effect. Again it seems weighted more in the interest of selling a low-cost speaker with profitable electronic assistance than in aiding musical performance.

Units of this type are providing something akin to severe limiting, which destroys the dynamic of the music over a wider range of sound levels, more than is the norm in PA systems that don't use this 'push a speaker to it limits' approach. For instance, the low bass might be rolled off and gradually compressed as levels get higher.

In other situations, pumping is audible – and if that happens, unlike an outboard compressor, you can't turn any knobs. These processed systems are made to be automatic – there are no knobs to turn. You can't even bypass these unwanted effects.

Ultimately, ruined music means fewer customers. The flexible, automatic EQ inside such processors can certainly produce an impressive sound at low levels, but its ability to protect at high levels falls away, often leaving the user with a shrieky sound. This needs careful assessment – you really must hear the processor bypassed to realise perhaps what a poor or mediocre quality of cab design is being made to sound better and 'juicy' by EQ'ing.

Another down-side of electronically-controlled PA cab systems' is they lock you in to using specific cabs, and possibly power amps. Also each controller commonly drives only one cab, or a pair of cabs. If you need to expand the system it can soon get expensive. It may be cheaper to buy better speakers in the first place – ones that aren't so dependent on processing. (This is not a problem with monitors, as there we already need lots of channels with their own crossovers, limiters, etc.)

Drive rack for monitors

This drive rack will be much the same as that used for FOH, with a few differences. For instance the graphic EQ is used mainly to deal with feedback and not tweaking the house acoustics. And crossovers for monitors are rarely more than three-way, and often only two-way.

Also input (base) delay lines aren't always required but, if used, will be arranged so the wedge monitor (see Chapter 6) is delayed relative to the sidefill cabs. Similarly, output delays may be used on the sidefills much as on the FOH cabs, as well as on the wedges – if the added complication is deemed worthwhile. Monitors have worked for years without this modern refinement (brought on by the fact that the delay is ready-to-use inside the digital crossover unit) – so don't feel pressurised to do this unless the 'time aligning' effect really helps the vocalist's confidence, singing voice etc.

Note there are a lot more channels of drive rack for monitoring – one per monitor mix. So the compactness of combined digital crossovers/delays/PEQ units is quite a blessing, as long as you can access the controls fast enough – to subdue feedback, say. If you use a digital crossover's PEQ for this, rather than a separate GEQ, it's best to do it via a computer screen.

CHAPTER 5

power amps
power delivery, amp rating, fault protection etc...

Quick view

Usually sited at the side of (or underneath) the stage – as close to the speakers as possible – power amps are fed with the signals from the drive rack and then boost them to a level suitable for the PA cabs. There may be several amps – depending on the number of frequency bands created by the crossovers (bass, mid, high etc), and also how much overall power is required to achieve the desired level in the venue. Compared to most PA gear, power amps are plain-looking units with few controls, but they are the unsung heroes of live sound, and deserve more attention, even respect...

Up to this point, the signals in a PA system can be quite 'lightweight'. They are low in both voltage (electrical 'energy', 'force' or 'potential') and current (the flow of electrical charge through the system). In order to get speakers to deliver the sound levels required at a gig, the signal has to be much more powerful – in other words speakers require a certain rate of energy transfer, or amount of 'horsepower', to drive them. This is where power amplifiers come into the picture.

How is power amplified?

Strictly speaking it's not 'power' that's being 'amplified'. Power amplifiers are signal enlargers or boosters. They generally use/absorb AC (mains) electrical power and deliver audio (music signal) power. In the process they convert the AC power to relatively smooth, noise-free DC (akin to battery power), then they use this to create a more potent version of the low-power incoming signal. Power amplifiers are in effect 'power converters'.

A power amp produces gain (boosting) in both voltage 'swing' (which we'll explain in a minute) and the availability of current.

Electrical law states: voltage multiplied by current equals power, written in symbols as $V \times I = P$. (Chapter 10 has more on basic electrical theory – see p182.) Power is 'the capacity to do work', measured in watts. It's a dynamic quantity – you can't store it. It's either being delivered or absorbed – and with PA power amps it's mainly making either sound or heat. The higher the voltage and/or current able to be delivered by an amplifier, the more watts can be delivered, which generally results in a larger (and louder) signal, up to the capacity of the speakers being driven.

Voltage (V)

The first step to getting high power delivery is a high enough 'voltage swing'. This refers to an ability all audio gear has – one that's exaggerated

power amps

in high power amps – where the signal can 'swing' (change at a whim) from nil to high levels and back again, typically spanning a range of a million times in the process.

An amplifier's voltage swing is elevated by a factor of x10 to x50 (ten to 50 times) compared to typical maximum line levels of 2 volts.

The key equation for finding how this affects the power an amp can deliver is: power equals voltage squared, divided by resistance (or $P = V^2 \div R$).

The fact that voltage is squared – multiplied by itself – shows that a small increase in voltage increases the amount of power we can deliver a lot more than you might imagine.

Assuming the load resistance (of the connected speaker/s) remains the same, the power delivered into them will be doubled when the voltage is increased by just 141 per cent – or 1.4 times bigger.

Similarly if an amp's voltage swing is doubled (x2), the capacity to deliver power is ideally quadrupled (x4).

Note also that we divide the resistance into the voltage part. This means that a lower resistance – an attribute of the speaker we'll be connecting to the amplifier – allows the speaker to receive more power.

In other words, this formula shows that for any given voltage from an amplifier we can (ideally) increase power delivery by using lower and lower resistance speakers. In practice the resistance of a speaker is quite a complex quantity, so we call it impedance (which is abbreviated to the mysterious-looking letter Z).

Also, if the resistance (or impedance) of the speaker is made too low, we may end up wasting valuable power in the wiring, which begins to have a large resistance by comparison.

Another drawback with attaining high power through low impedance speakers is an amplifier's current capability – read on.

Current (symbol I)

An amplifier's ability to deliver current (measured in 'A' for Amperes) governs how low an impedance (or resistance) the amplifier's voltage can drive.

As we saw earlier, power is the product of both voltage and current. Both are needed. Note that an amplifier can have high current capacity or capability, but current only flows when it is forced to, according to the amount of voltage and the 'lowness' of the speaker's resistance. So the amount of current flowing is dependent not only on what the amplifier can manage to produce, but also on the voltage the amplifier can make available, and also the resistance (impedance) of the speaker(s) being driven.

This is summarised in the key formula: current equals voltage divided by resistance ($I = V \div R$).

If an amplifier doesn't have enough current capability for the (low) load impedance (resistance) it is driving, the voltage can't be sustained – it reduces or collapses. The amplifier is under strain, and the sound quality gets bad.

A lower resistance allows the speaker to receive more power

So current capability is a key rating of amplifiers. But it is normally described indirectly, as the lowest (speaker) impedance (or resistance) that the amplifier can drive – so many ohms, typically between 8 and 1Ω (Ω stands for ohm).

How does speaker impedance affect power delivery?
When the impedance of the speaker connected to the amplifier is halved, the current flowing (for a given signal level) increases, and the power that can be delivered should (ideally) be doubled.

For example, if an amplifier delivers 200 watts into 8 ohms, then it would be expected to give 400W into 4 ohms. This is true up to a point. The relationship breaks down gradually as small voltage losses accumulate – the amplifiers' internal parts have resistance, and the losses in these become proportionately greater as a lower speaker impedance forces more current through them. This accelerates as the amplifier concerned runs out of current capability.

Equally, in the power supply (see the 'Amplifier technology' section, p95), the output stage's power transistors and other parts handling the high energy all have their limits, and you get either saturation and overheating (think of a busy road at rush-hour: above a certain flow rate there are limits, and cars and tempers will heat-up), or some other form of limiting occurs that's purposely built-in to protect against damage.

So for example, a 'real world' power amplifier typically delivers, say:

- 200W (watts) into 8Ω (ohms) – the usual reference point.
- 390W into 4Ω – not quite double, as the higher current causes increased losses in wiring, transistors, the power supply, etc.
- 675W into 2Ω – quite a way from double (should be 800W). The amplifier is getting hot, and this and other limits are starting to restrict the current that's available. Also, the sound quality won't be very nice.
- 76W into 1Ω – the amp is protecting itself (see 'Adverse load protection' later in this chapter, p109) so the power is now a fraction of even the 8-ohm rating. Also, by now, the sound quality will be total distortion.

For impedances above 8Ω, problems with losses diminish, so the power decrease gets closer and closer to a perfect halving.

For most PA situations, the total speaker or cab impedance (also called 'nett' impedance) connected to the amplifier channel's output can range between 16 ohms and 3 ohms, and occasionally as low as 1 or 2 ohms. It depends first on individual drive-unit impedance, and then on how many drive units are connected. (In PA cabs, they should always be wired together in parallel.)

In this case, the nett (overall) impedance is a lower figure than the individual impedances. (An explanation of this is given in the 'Input impedance & loading' section, later on.) To work out the exact nett impedance figure, you'll need to know the number of drive units, and then their impedance. If they all have the same impedance (as is the norm), the nett value is easy to compute – you simply divide the impedance value of one speaker by the total number of speakers.

For example, if there are four drive units, each with an impedance of 8 ohms, the nett impedance is:
8 divided by 4 = 2 ohms. Similarly, if there are two speakers each rated at 15Ω, the nett impedance is 15 divided by 2 = 7.5Ω.

If there are two different impedances, you need to work each out separately: for instance, say there are two drivers with 16Ω impedance (16÷2, or 16/2 = 8Ω) and also two drivers of 12 ohms (12/2 = 6Ω). Next, we need to know what the 6Ω and 8Ω in parallel amount to.

To work out the exact nett value of these, you'll need to use 'reciprocals' – a mathematical

---

Good amps & power efficiency

PA amplifiers need to combine the delicacy of a good hi-fi amp with the robustness and reliability of a farm tractor, blended (increasingly) with the low weight and compactness of aeronautical gear.

Good-sounding power amps (ones which add minimal coloration or distortion to the signal, purely making it louder) require great sophistication to enlarge and deliver the signal very precisely over a wide 'canvas' of levels and frequencies, while also delivering high currents and voltages.

And these quantities are not delivered into docile power-absorbing elements, but instead into speakers – which, as we'll see in more detail in Chapter 6, are quite complex and 'reactive' in the way they interact with the amplifier.

No power amplifiers are 100 per cent efficient – even the best manage only about 80 per cent in reality. The best speakers, meanwhile, only approach 25 per cent efficiency. Best overall efficiency is consequently about (0.8 x 0.25) = 20 per cent. (If you're not familiar with multiplying percentages to get these results, see 'Power supplies & true amplifier efficiency' later in this chapter, on p94)

The average overall efficiency figure is more often between five and ten per cent. Taking ten per cent as an approximate figure, this means to get a certain amount of acoustic power – in other words music at a suitable sound level – in the room, we have to provide about ten times that power from the electricity supply. And so this is the amount that an audio power amplifier has to handle and 'process'.

We'll also want to have some power capability in reserve – since inadequate power results in amplifier overload and bad sound. In general, erring on the side of over-rating is better than under-rating.

And remember that the relationship between watts and loudness isn't proportional in the way you might imagine (as we explained in the book's Introduction). As a reminder, a rule of thumb is that you need to increase the power delivery into any particular speakers by at least tenfold (x10) to attain about twice (x2) the audible level. This appears on a sound level meter as a 10dB higher SPL (sound pressure level) – so, for example, if 100W gives 90dB SPL, 1000W will be required to increase the level (where nothing else is altered) up to 100dB SPL.

In short, much much more power is needed than you might expect.

power amps

operation that solves this kind of problem. This means turning each number on its head (so 10 becomes 1/10th, one-tenth), then adding the numbers in this form, then changing the resulting number back to ordinary form.

This is easiest with a scientific calculator. If you don't have a scientific calculator, you could arrive at the figure closely enough for most purposes by observing that (in this case) a 6 and 8-ohm paralleled pair must make up a slightly lower impedance than two 8-ohm units – which suggests a figure slightly below 4 ohms. So you might guess around 3.5 ohms – near enough.

There is no point in being *too* particular about exactness in PA cab impedance calculations, because speaker impedances are nominal averages. The impedance varies quite widely over the range of frequencies, and even increase when the speaker is driven hard. Equally, amplifier ratings are often – or at least should be – conservative, with a fairly generous leeway.

Power ratings for music

Except for some special, extreme and momentary conditions when 'blips' of high signals slip past the (crossover's protective) limiters, a PA amplifier is rarely used to 100 per cent capacity – and also doesn't even ever normally approach this in the same fairly steady, continuous way as (say) a generator's diesel engine.

Although the full power may only be used for a small percentage of the time, that will be very important to some part of some music at some stage – it may be heard as 'dynamic contrast' – so we still need it. (Music is made up of continuously varying levels. Even when music is at its loudest, it is filled with louder and quieter parts.)

Recognising different power amps

There are various different formats, and also various ways to classify power amps, from a practical as well as an academic perspective.

By size or number of channels

Two channel (2 ch) units are the most common. Sometimes called 'stereo', the two 'sides' may be used in a variety of ways, but almost never as left/right – as in home hi-fi, or small studios.

A few multi-channel (eg 3, 4, 5, 6 ch) units have been created. These are useful and technically valid for PA, offering the possibility of a simplified set-up (using multi-way input and output connectors) as well as reduced weight, and possibly also reduced size and cost.

For example, a three-channel 'tri-amplifier' unit suits three-way speaker system – for bass, mid and high. But such packaging advances have been less widely accepted and employed. One reason is

---

Audio power – terminology and facts

'Power rating' is shorthand for 'power delivery capability of an amplifier into a given rated impedance'.

'Power' is defined as 'energy delivered, averaged over time', and in electrical terms is measured in units called watts (symbol W). Amplifier power is delivered, not made. PA power amps do not generate or 'have' power and, unlike engines, the energy/power source isn't usually kept next door to them in the form of a fuel tank – except as relatively puny reservoirs called capacitors. These can keep a high power amp going for only about one second when the electrical supply is turned off. The energy – and power – is ultimately supplied by a power station or generator (see Chapter 10 for electricity basics).

Specifying power ratings

In fully professional amplifiers, 'Watts RMS' or just 'W' in specifications nearly always means genuine average watts. Because an amplifier's power rating has a large bearing on its cost – and also, no less importantly, its weight and how compact you can make it – there's long been a temptation for makers to be 'economical with the truth' about the wattage. For example, a maker might add up the power ratings of the transistors used – which has little bearing on the power rating for PA use. The maker needn't claim anything untrue, but the large number – '3200 watts total dissipation' looks impressive

for a 150W-per-channel amplifier...

Other makers have measured the power for very short musical bursts, which helped to give high figures – in the same way that a sports car engine may manage higher revs for just a few seconds.

Some more dishonest makers have used arbitrary round figures, knowing that most users will never measure the exact power – which might turn out to be only 75 per cent of what's claimed. In other areas of amplifier sales (outside of pro PA), makers have used (and some still use) 'phony' watts, such as 'music power' – giving figures typically up to 150 per cent higher – and 'peak power', which gives double the true average power rating.

reduced flexibility when requirements change or things go wrong. For example, a fault that stops the entire amplifier working could 'take out' one whole side of a small PA. And you're more likely to be able to carry a spare two-channel amp than a spare 'half-a-PA'.

Another reason is related: the individual amplifier channels may be right for a particular make or model of PA speaker, but less so for others. Other than this, there's nothing much to say against using them. Such units have usually been developed either by PA loudspeaker makers or by associated amplifier makers.

Single-channel amps ('mono blocks') aren't used much either nowadays – except maybe some behemoth multi-kilowatt (into below 1 ohm) amplifiers, often built and used for reggae & dub sound systems. There's no problem using second-hand amps of this single-channel type, so long as you can afford the rack space.

By 'U' size (panel height)

The space an amplifier takes up is primarily measured (in PA circles) by the units of vertical rack space it occupies (see Chapter 3 for more on rack-mounting).

A PA may need to use a lot of amplifier channels. At the same time, the number of and total volume of amplifier racks that are used has some bearing on the PA's hire-ability and cost of running. Makers are therefore under competitive pressure to make compact amplifiers. Naturally, this also has to be done without sacrificing any power delivery. The upshot has been a trend towards amplifiers that occupy as few 'U' of rack space as possible with the biggest power. Little is said about depth. The result is that the modern generation of PA amplifiers are mainly slim and sometimes almost unfeasibly deep.

Most '1U' amps – assuming they are two-channel – are limited to about 50 to 100 watts per channel (W/ch). Exceptions include a range of conventional amplifiers that use ordinary technology particularly cleverly, and some emerging Class D amplifiers (sometimes wrongly called 'digital' – see 'amplifier technology' sidepanel). In both types, maximum power delivery lies between 250 and 550 W/ch.

Most power ratings that are required for PA are widely available in 2U form. But to achieve this the casing may be as much as 19-inches deep.

And the higher-powered designs are increasingly likely to be cutting corners somewhere – probably where the user can't quite see or pre-judge.

For example, the parts are likely to be very tightly packed inside, and the airflow path only viable if it's kept clean and the fans work only at high speed, and this may reduce lifespan and increase service costs.

Similarly, the higher power ratings are commonly achieved at the cost of thermal capacity in the amplifiers' heat exchangers. This means that when tested with continuous signals, some of the higher power-rated 1U and 2U amps may overheat after a few minutes operation, and shut down (see 'Excess temperature' in 'LED signals' section later – p106).

But this won't happen with music signals – unless they're overly compressed. For example, the signals from instruments making up the overall signal for the PA's bass speakers may all be quite heavily compressed, as a matter of course. As a purchasing tip, it's a good idea to use a mix of these sorts of signals to help evaluate the suitability of such very compact amplifiers .

3U permits more breathing space for reliable conventional amplifiers, with ratings up to 1-2kW per side. Sizes of 4U and above are usually multi-channel amps or else medium power amps of an older era, when rack space wasn't such a concern.

Why were older amps so large? Because in years gone by road fuel cost proportionately less (particularly in the UK), touring budgets were proportionately greater, delayed stacks and central clusters weren't used, and the 'back-end' used maybe more speakers connected to a smaller number of amplifiers and bands. And the world was a generally more relaxed place. So it's generally the case that older (second-hand) amplifiers will have lower power ratings for any given rack size.

Amplifier technology

There are many technical differences between power amplifiers which may be loudly trumpeted for marketing purposes, but which aren't particularly visible or appraisable. These include:

Amp classes

Amplifiers are classified solely according to the

power amps

different schemes used to handle power in the output stage, which are concerned with getting better (or just good) efficiency, and nothing else.

It's easy but misleadingly superficial to confuse this kind of classification with something like air or rail travel or postal services – where 'second' or 'B' class is inferior. In fact, amp classification actually says more about how hot an amplifier runs for a particular workload, and how compact or lightweight it can potentially be for a given power rating, rather than anything particular about its sound quality or general ranking or worth. There are many factors other than class that determine sonic quality.

It does just so happen that the best sound quality is easiest to obtain from the first but most inefficient scheme, called class 'A'. It's also true that if set up wrongly, its easy to get less-good sound quality from any of the other classes. But in an ideal world, it should be possible for the different schemes to perform about equally. Some of the common amplifier classes used for music are as follows:

Class A – not useful for PA as it's inefficient: too heavy, and runs too hot, burns far more power than it delivers most of the time. This class was first used when the earliest PA amps were made, around 1917.

Classes B and A-B – both types are similar and the most widely used for PA amplifiers and in most other audio amps, except high-end hi-fi. Efficiency at full power delivery is reasonable at around 70 per cent (vastly better than a car engine) and little power need be wasted when the amplifier is ticking over or only reproducing levels well below maximum.

Class B was first invented in the 1930s so people could play music when they went for a picnic outdoors, with a valve (tube) radio that could run for more than a few minutes on large 'accumulator' batteries. Today it's the dominant species.

Class D – so-called 'digital' amplification uses a radically different technique compared to the other classes, which involves 'chopping up' the audio signal. The method used is different to ordinary digital audio – it uses PWM or 'pulse width modulation', as opposed to PCM or 'pulse code modulation' as used for CDs etc. This type of amplifier amounts to using very rapid, accurate switches that turn power on and off to create a high power mimic of the musical waveform. The switching gives high efficiency – up to about 96 per cent in theory – as the switches (in practice, transistors) are either on or off, so little energy can be wasted.

The idea is elegant in theory, but over 50 years after their invention, Class D amplifiers still aren't widely used, and modern products still have problems with radiating large amounts of RF noise, with debatable sound quality and poor flexibility into different speaker loads. The biggest incentive for making them (in today's fiercely competitive industry) is that in theory expensive parts aren't required to achieve high power delivery.

Classes G & H – these are a development of ordinary (analog, as opposed to 'digital' class D) amplifiers, and may be combined with a Class A, B or A-B amplifier 'core'. The idea is to allow high power ratings without wasting energy on this when it isn't needed. This higher efficiency is achieved by quickly switching to a higher supply voltage to allow more power delivery at the instants when this is needed, then reverting to a lower power capability when it isn't. This approach can save weight and cost in theory, but amplifiers using this technique, while certainly efficient, do end up quite complicated.

Class A amplifiers are too inefficient for PA – they often burn more power than they deliver

Class G was invented in the US for military electronics in about 1962, but wasn't used in PA amplifiers until around 1980. Class H (a useful variation on Class G) followed about two years later, as the PA industry's own innovation – invented in the US by Quilter, founder of the QSC company.

Manufacturer's claims of 'high efficiency' from Class G, H and particularly class D power amplifiers must be tempered by the kind of power supply employed.

Power supplies & true amplifier efficiency

Power amplifiers, like most other audio equipment, are powered from the AC power supply. The supply is converted into the steady, smooth DC voltages (like battery power) that the amplifier and other parts require by the amplifier's own PSU (power supply unit). This unit, which isn't readily discernible as a separate part inside most amplifiers, accounts for about half the cost, bulk and weight of ordinary (analog, class A-B or B) amp units.

The real efficiency of an amp is the product of the efficiency of its power stage multiplied by the efficiency of its power supply. Neither can exceed or even get very close to 100 per cent. The significance of efficiency is that it determines how much money will need to be spent on heatsinks and cooling gear (fans) to get rid of the wasted power (as heat), and how much bulk and weight these will add.

Inefficiencies magnify each other... if one part is far less efficient, it penalises the other part

The crucial point is that the product (effective sum) of any two per centage efficiency figures is lower than either of them. For example, if the raw amplifier part has 90 per cent efficiency (say it's Class D) and the power supply part has 60 per cent efficiency (an ordinary figure), then the nett product $(90/100 \times 60/100) = 54/100$ – which means an efficiency of just 54 per cent.

In other words, two inefficiencies magnify each other – and if one part is far less efficient, it penalises the more efficient part. This means an efficient power stage (using say 'Class D' methods) will not offer anything like the expected improvement if it is partnered with an ordinary, relatively inefficient power supply. The upshot can be (and has been) a so-called 'high-tech' amplifier that weighs hardly any less and is hardly any smaller than a well-designed conventional amp.

Power supplies have their own equivalents of Class D amplifier, known as 'switching supplies'.

These switch the AC power at high frequencies – typically 70,000 times a second. The high switching rate enables a far smaller transformer to be used.

Power amp input connections

Balanced XLR inputs are the norm. Use of quarter-inch (phone/jack plug) or other unsuitable connectors isn't recommended at all. Power amplifiers bearing such sub-standard connectors are best either not used or should be modified.

For connecting each feed (crossover band output) to more than one amplifier channel, see also the 'Multiple amplifier connections...' section in a moment.

Input sensitivity

This is the amount of signal input (in RMS volts or dBu) that you need to 'put up the input' to be able to drive the amp to its full capability – in other words to the point where it clips. High sensitivity means only a small voltage (low signal) level is required to do this. The amplifier's gain control is assumed to be set at max.

You might imagine that high sensitivity is best. Actually, equipment offering -10dBu sensitivity (which includes some supposedly 'pro-looking' amps really made for bedroom DJs) is really only suitable for lightweight, disposable 'consumer hi-fi', and should be avoided for PA.

Higher sensitivities – anything below 0dBu – are not generally a good idea in PA, as residual noise is increased and, if used without due care, the amplifier's input or the speakers can be seriously overdriven, even though the PA may have correctly set limiters.

For example, the amp's input stage (before its gain control) can be overdriven if the amp's gain control is backed off and the input driven harder to compensate.

A few amplifier makers cite the input sensitivity on the amp itself. For most PA amplifiers it will lie somewhere between 0.75 volt (@ 0dBu) and 2 volts (@ +10dBu). These are RMS (average) values.

Sensitivity cannot ever be an exact figure, because the amp's full capability – where it clips – varies with the speaker load and also the mains supply voltage. The speaker's load can vary in turn with the frequencies of the dominant components at each instant, and also, if the drive-units are being driven hard, the music levels' recent 'history'.

power amps

It follows that a PA amplifier's 'rated input sensitivity' is purely 'nominal', typically within a range of 0.5 to 1dB either above or below. The sensitivities of amplifiers handling the same signal and connected into the same sort and number of drive units should, however, be much closer to being identical.

Of course the amplifier's sensitivity changes – it reduces – as soon as you back off the gain control knob from maximum. See the 'Panel gain controls' section later on this page.

If the control is calibrated with a dB scaling you can easily work out how the nominal sensitivity has changed. For example, if the amplifier's nominal sens is 0dBu, and you back the gain off to the -10dB position, then the sens is now +10dBu. That's to say you now need 10dB more level to get back where you were before you dropped the level going in by 10dB.

Knowing the nominal sensitivity of the amplifiers is useful as a point of comparison, and to understand (for example) why one part of the system needs to be driven a lot harder to work at the same level as other parts. Knowledge of amp sensitivities can also be required when large PA systems are planned using any sort of CAD (computer-aided design) program.

Sensitivity is often only cited in small print in the maker's printed specifications, which are rarely read. So it can be worth labelling amplifiers (beside their input sockets) with their nominal sensitivity rating in both mV or volts, and also in dBu.

Sensitivity and power delivery

An amp's input sensitivity says nothing about how much power an amp can deliver, or indeed anything in particular about its output capability at all.

In fact typically (and perhaps counter-intuitively) less sensitive amps are often the ones capable of delivering the highest maximum powers. This occurs, for example, when the sensitivities of a 'logically-designed' range of power amplifiers (from a given maker) have input sensitivities scaled in line with the increased power delivery capability.

Another common, logical arrangement is that a maker may make all the amplifiers in a given range have the same sensitivity – for example 0dBu.

Some amplifier ranges that aren't fully professional PA units can have less than logical,

almost random variations in sensitivity compared to the full power rating. This is probably a sign of gear that is best avoided. At other times, sensitivities can sometimes be changed by a few dB either up or down – consult the amplifier maker's technical department.

It can be useful to keep these different approaches in mind when dealing with (and troubleshooting) a PA that's using mixtures of different amplifier models and makes. If you don't know or are not sure of the sensitivity of a given amplifier, see the sidepanel on 'Using clip LEDs', later in this chapter (p104).

Sensitivity and gain

The gain of an amplifier is the amount by which it magnifies the signal voltage, from input to output, with the 'gain' (setting) control assumed to be set at its maximum position. This helps us understand the important relationships between sensitivity and power delivery (as well as helping set up limiters – see Chapter 4).

An amplifier's gain is altered only slightly by the speaker load, maybe by 0.75dB at most. Unlike sensitivity, gain isn't affected by the mains voltage at all.

It is gain that determines an amplifier's sensitivity, along with its rated output power (which tells us the maximum voltage its output can deliver). The maximum voltage is divided by the amplifier's gain to get the amplifier's sensitivity – ie the voltage needed to get full output. For instance, if the maximum voltage is 30V, and the gain is x10 (also expressed as 20dB), the sensitivity must be 3V.

Panel gain control

Also known as volume, level, attenuator, or sensitivity control, knob or pot. Normally found on the power amp's front panel. These controls rarely really change the amp's gain (amount of amplification available). All they do is reduce the levels of signals reaching the amp's later stages. But they do alter gain from the wider perspective of the whole PA system.

As mentioned already, the 'gain setting' knobs (two on a stereo amp) should generally be set at maximum, unless there's a good reason for backing them off (effectively to modify by reducing, or match sensitivity). Having them set at maximum means you can be sure identical channels' gains (on one or more identical amplifiers) will be closely matched – without any need for measurement or level calibration.

But once the controls are backed off for any reason, you really can't rely on visual matching, as the knob calibration and various part tolerances are variable and unknown. Particularly when there are lots of amps in a PA, this situation adds an extra tier of complexity and uncertainties that's not very helpful.

If you must back off controls (typically to reduce background noise levels from digital delays and the whole PA's front end), you can overcome part of this problem by making sure the knobs are mechanically aligned (see technique in Chapter 3). Once this is done, it's then worthwhile using a piece of card (say) that's cut out to line-up against the knob's pointer and any other reference point. This can be used to quickly check and align lots of controls to a better degree than relying on panel markings alone.

On some amps, gain setting is by 'stepped' attenuators. These provide exactly the same sort of attenuation, but in discrete round numbers of dBs. With these, in theory, the benefit is that levels are known to be matched regardless of settings, and levels can be set or reset in the dark, by feel alone – simply 'count the clicks'.

In practice, in most amplifiers with stepped controls, a low-cost method has been substituted – the so-called 'detent' control. This type of low-grade control is produced mainly for 'consumer' products, and comprises a cheapest-of-the-cheap carbon-type pot, widely known to degrade sound quality as well as being subject to rapid wear. Its centre 'detent' mechanism might seem like a useful guide, but because carbon pots are highly imprecise components, the dB levels marked on the panel can differ quite widely from what's really

happening on different amps. This is just a cheap imitation of a real stepped attenuator which uses switching and fixed resistors, and offers typically 100 times more accurate setting.

Other gain controls, ranging right down to maybe -60dB (pretty quiet) or less, may be found on some not-fully professional PA amps, but shouldn't be needed for PA. A backing-off of only 20dB amounts to a notional 100-fold reduction in the power level for a given drive level. It seems unlikely you would use a heavy and expensive 1kW amp then turn it down so it could deliver less than 10 watts.

Knobs should have clear, high-contrast markings

Of course controls can be backed off because of too much level, but it's unlikely to be 20dB higher than what's required. If so, alarm bells should ring, as something may be amiss – perhaps faulty wiring, blown units, something covering the HF speakers, crossover wrongly set up, or someone's stolen the bass drivers...

The control knob action of a well-designed PA amp shouldn't be so light that the critical settings are easily altered if the knobs are lightly brushed against. Some amplifiers, made for fixed installations, go too far the other way, with lockable, screwdriver-adjustable settings. PA gear doesn't usually need that level of security – on the fly adjustability is often more important. Knobs ideally should have clearly visible pointers with high contrast white on black, or even fluorescent markings – helpful in the dark or dim lighting behind and under stages, or

Impedances in context

A high impedance number means a light load. A possible analogy is with a road slope: a '1 in 300' hill is a light climb even for a small-engined car, whereas '1 in 3' is steep and would require far more engine power. Whether an impedance is a light or heavy loading depends on context, as seen in this table (it also shows how widely the 'medium' range varies).

	Low impedance (heavy load)	Medium	High impedance (light load)
Line	below 1kΩ	10kΩ	above 100kΩ
Mike	below 200Ω	1200Ω	above 5kΩ
Speaker	below 8 (or 4)Ω	8Ω	above 8Ω

(There's more about speaker impedances in Chapter 6)

power amps

outside at night.

Amps that are properly designed for PA should also have gain controls that focus most of their span on the top 20dB (or so), while the setting also goes as low as -∞ (meaning minus infinity attenuation = off).

This 'off' or 'mute' setting can be invaluable for troubleshooting and system set-up. The best-designed amps may also have a separate mute button. This is handy since a given control setting doesn't then need to be disturbed at all. In that case, a gain control that reduces the level by no more than -25dB would be fine.

When amplifier's gain knobs are simply turned down, they can become a bit dangerous, since if they're later accidentally turned up (after the limiters at the crossover are set – see Chapter 4), there will be no effective protection from the limiters to prevent the amplifiers being strongly overdriven and 'clipped' – which may well destroy the drive units. The simple solution is to tape down the controls with gaffer tape.

Connecting multiple amp inputs & daisy-chaining

Daisy-chaining is also known as 'connecting in tandem', or 'linking through' – this is the primary method of connecting lots of amp channels to the same source signal (aka 'hanging' the amp channels 'off' the crossover outputs).

It saves having special 'fan-tailed' cables – many cables joined to a single plug connected from the source. Or 'Y' cables, where (in this case) two XLR male (output) plugs are connected to one female plug. This would be repeated many times to connect a row of amp channels.

Both these special cables would be less flexible when system configurations are changed, and more vulnerable to failure – as XLR plugs' cable inlets cannot handle more than one cable of 'stage-proof' thickness – that's about 8mm. Instead, daisy-chaining allows as many amp channels as you like to be connected using standard XLR-XLR line cables.

Daisy-chaining is made possible because professional power amps have a male 'link-out' socket mounted next to (or near) the female XLR input connector for each amp channel. This is simply wired 'across' the input socket's connections (along with the input into the amp itself) and is no more than a convenient sturdy Y splitting staging post, making use of the power

amp's casing for mounting and shielding. Other female/male connectors are possible.

As well as being a pest, amps with jack sockets are not a good idea because, as there is no input/output distinction, two signal sources (crossover bands) could easily be plugged into the same line connection. This could cause some confusion and/or even damage, and something you can do without in a high-pressure situation like backstage at a gig.

Some peculiar power amps have quarter-inch (phone/jack) sockets just for their link-out connections. This means for daisy-chaining you'd use the jack sockets as the feed to the next amp's XLR, which requires you to build your system with odd-leads and sub-standard connectors.

If you encounter this, or any other similarly messy situation, or the absence of any link-out sockets, then it's probably worthwhile having the amp(s) in question modified to use XLRs throughout, as the de facto standard.

If there isn't enough room on the rear panel to add the extra XLR connectors, XLR line plugs can be fitted on short (say 5in/125mm) leads (aka fly leads), passing through the panel inside a gland (a circular cable gripper).

Input impedance & loading

Input impedance (at this juncture in the PA) is a measure of the loading presented by each power amp to the preceding signal source, in this case one of the crossover's outputs. With all normal PA amps, input (load) impedance is moderately high, typically 10kΩ (10,000 ohms).

A few amps may have input impedances lower or higher than this. Ones with lower impedances will tend to exacerbate any loading problems (which we're coming to in a moment); while ones with higher impedances help reduce loading but will exacerbate the chances of environmental noise pickup in a worst case (but not unknown) scenario.

For instance if the returns multicore cable (snake) gets unplugged at the far end – at the drive rack (see Chapter 4) – while the PA is up and running. Then, if the cable shield's grounding connection isn't correctly organised (because it depended on a connection only at the drive rack end – which is actually recommended by some crossover makers), there could be a very loud

power amps

buzzing and/or shrieking from the speakers. It's the same ugly sound effect as removing the plug from an electric guitar and letting it fall to the floor.

Power amps could readily have 1MΩ (1 million ohm) input impedances. This would greatly decrease the loading conditions we're about to discuss. But it would also make amps far more like a typical 'screaming' guitar amp in their susceptibility to the kind of noise pickup that's possible when 'live' leads get unplugged.

10kΩ sounds like a light-enough loading, considering that professional grade PA gear should always be rated to drive its full level into a load as 'heavy' as 600Ω, which is 16 times lower (see sidepanel, p99)

But connecting more than one 10kΩ (or any other medium value of) load across the driving equipment – in this case, any one output of the crossover – causes the overall impedance (as seen by the crossover's output) to reduce in proportion to the number of separate loads – amp channels – that you're adding.

For example, feeding one crossover output into eight stereo (two-channel) power amps (that's 16 channels in all) is going to load it with 625 ohms. This figure comes from assuming the amps' input impedances are 10kΩ, so 10,000Ω/16 = 625Ω.

With normal power amps, input (load) impedance is moderately high

---

This is barely above the professional standard, which is that all line level outputs should be capable of driving a load of 600Ω. Also, once we include the extra loading that's caused by the capacitance of the long-length cabling between the crossover and amp inputs, the total (nett) loading turns out to be actually below this 600Ω limit at the higher frequencies.

(Note that the cable's capacitance is almost invisible as loading at low frequencies – because in general the impedance of a capacitance is high at low frequencies – but appears as a lower and heavier loading with increasing frequency.)

The result is that, with 16 amp channels, the sound may not be so good on the top end. (See

'Driving lots of amps from one source' section, in a moment.) The answer is generally to 'derate', and so use up less of the 600Ω capability – at least on the top and HM (high midrange) connections. Too much will waste too much capacity, and too little won't reduce the distortion much. A derating to half the loading, hence 1200 ohms (a maximum of 8 x 10kΩ power amp channels) will be about right.

Driving lots of amps from one source

When you have to drive more than ten or so amp channels from one crossover output (assumed to have the professional standard minimum load rating of 600Ω) you can avoid overloading the crossover in several ways.

First, use a 'line driver' unit (an electronic device, not to be confused with a speaker's 'drive-unit'). This contains one or more 'line amps' – specialised amplifiers that boost or 'buffer' the strength of the signals from the crossover output, so lower impedances can be driven, without increasing the signal's size (voltage level). Beyond this specific PA application, line-drivers are general-purpose units, that are particularly widely used in some broadcast work.

A typical line-driver package contains two or more 'line amps' that are simply inserted (in the line – hence the name) after the crossover outputs needing to be boosted. Such a driver may either drive some way below 600 ohms (typical ratings are 300, 150 and 75Ω) or else split the job into several parts, for instance four outputs, each of 300Ω capability. The former is superficially easier, while the latter is more likely to give a good performance. It also offers diversity – if one of the four outputs goes wrong, the others – maybe 75 per cent of the show – will keep going.

To use this arrangement, either you will need enough extra 'pairs' (channels) on the return or main multi (snake) – since each crossover output is being split into four 'feeds'. Alternatively, the line amp(s) can be placed at the other end of the returns multi/snake, where the amp racks are sited. This might mean using more units.

When working out how many power amp channels you can connect to a line amp, bear in mind that (just as with the crossover itself) the sound quality may be surprisingly better, particularly the top end, if you don't load it all the way. If you can manage, loading to only 50 per cent of the maximum is ideal – so a line-

Power amps

driver rated to drive 300 ohms would be best not loaded with below (meaning any load heavier than) than 600 ohms.

The reason is that the long cables being driven down, together with amplifier inputs, present quite a high capacitance. In practice, with typical runs of a hundred feet/30m or so, the capacitance presents an added load to the line driver's output that's tiny and unimportant at bass frequencies but gets increasingly significant at high-mid (HM) and treble frequencies (HF). This loading effect is rarely mentioned by manufacturers.

There are two other, not-as-good ways to avoid excessive loading of the crossover's outputs.

The first is possible with a few PA amps that have a small line-amp fitted at their inputs. This is connected so that it boosts (or 'buffers') the signal that's passed-on ('daisy-chained') to the next amp. If every 'link-out' was to be buffered before sending the signal on to the next amp or channel, then the loading seen by the original source (and by each succeeding line-amp) should be no more than the impedance of the first input in the chain (say 10kΩ), plus a large part of the total cable capacitance.

The snag here is there'd be a gradual but inexorable and needless extra build-up of noise (hiss, hum, buzz) and distortion. A better way to use this scheme, if feasible, is to 'fan out' the crossover's output into to a reasonable number, say five stereo amps (5x2ch = 10 ch, 10k/10 = a 1kΩ load), then from the output of one of these take another 'fan tailed' or 'Z-Z' lead (1 XLR female plug wired to five destinations) on to the next ten channels. This can of course be repeated several times by taking additional set of five 'Z-Z feeds' from the other four amp channels in this first tier, if required.

Although it needs special 'Z-Z' or 'fan-tailed' leads, this has two advantages. First, the signal to the 'furthest' amps down the chain has only passed through one line amp (or if not, far fewer than in the preceding example). Second, the signal is rapidly diversified, so a cable fault may only affect a small fraction of all the amps. The one thing to watch is that the amps in such a scheme must keep their line-driving (aka small signal handling) circuitry working – even if/after the rest of the amp has shutdown for some reason.

As another, more desperate option, you might consider using additional crossovers, if they're

available. But bear in mind that each additional crossover output will usually provide just another 1.2kΩ (1200Ω) worth of drive capacity, based on the 600Ω standard, less the derating to half, as discussed above.

Also, this is not a good idea with most analog crossovers because the matching between all the settings – levels, frequencies, slopes, alignments, horn-EQ etc (see Chapter 4) need to be accurately matched, and this matching must be readily maintained when any control setting is adjusted. If the various crossover characteristics are slightly different, it will make it all the harder, if not impossible, to get good sonic quality and have the sound 'click into focus'.

The answer is to use digital crossovers. With today's top professional models, the settings can be uploaded into a PC (a laptop running from the main AC supply is suggested for convenience and reliability). The settings can then be adjusted from the PC screen, and any settings that have to differ can still be ganged, so their relative differences remain.

Amp protection

Power amplifiers are more vulnerable to damage than their (often substantial) weight might suggest. The primarily vulnerable part is the electronics that drives the output.

> # Power amps are more vulnerable to damage than their considerable weight might suggest

To drive PA speakers with musical signals and not degrade or distort these, amplifiers have to be able to deliver a 'dance' of high voltages and high currents in varying permutations, within millionths of a second and without hesitation, yet also quickly sense (notice) any excessively high current flow that will threaten the amp's life.

Distinguishing between the two is hard. If the amp's protection is set-off by real music, the sound is likely to suffer unnatural distortions, such as breaking up. The other option – shutting down – is not much better. The upshot is that better-sounding amps are often not quite so ruggedly protected, and that many very rugged amps are commonly noted as being less good sounding.

Without protection, if a power amp's output stage goes wrong, a large amount of energy can be unleashed very quickly. This can result in violent miniature firework displays and quite loud 'cracks' of miniature explosions – and sudden connection of a high-current-capable DC voltage to the speaker cabs, which can fry the speakers' voice coil in a few seconds.

Serious fires are mercifully rare, but to survive real-world PA use and 'reasonable' accidents, power amps are fitted with a variety of protection features – for their own sake, as well as protecting speakers from expensive collateral damage. Some kinds of protection stop an affected amp altogether, while others operate more stealthily, without even any indication.

Indicated protection

Some protection functions (and sometimes other problem situations) are indicated by LEDs. To be useful in a PA system, the LEDs should adhere to the almost universal colour-coding system that every PA user on the planet is already familiar with – from travelling anywhere by road or rail:

- Red – Stop. Shutdown. Serious error.
- Orange – Warning. Take care. Check something.
- Green – OK. Normal. All present and correct.

This leaves the two other common LED colours – yellow and blue – for other kinds of messages.

These conventions are not always precisely adhered to. One large PA maker has used red LEDs to indicate 'safe, OK' which causes endless confusion (and occasional panic), but was done just to save a few pennies.

The other important feature of protection LEDs is their visibility. On the better amps they can be seen in bright sunlight, and also when viewed from the side.

In many amps, some of the indicated kinds of protection are latching – meaning they 'stick on' until you sort something and reset. It means the user has to intervene by either pressing a reset button or, more usually, turning the amp's power

switch off then on again. It's a bit like 'rebooting' a computer. The amp should then operate normally – but only if the protection system is able to detect that the fault or problem has been cleared. The need for a user to 'intervene' is no bad thing if the fault is a genuine and serious one.

For some sorts of conditions, such as overheating, protection systems can also reset automatically at the instant the problem is over. This isn't possible with most other kinds of faults, as these cannot always fix themselves, so the protection could cycle on and off, eventually causing damage, as well as strange sounds. To overcome this, auto-reset may be controlled – for instance by only attempting to reset a limited number of times, at preset intervals.

Amp LED signals

Signal Present

Unlike other 'warning' LEDs, this shows that a signal of 'reasonable' level is reaching the amp. Some LEDs flash with the signal, others light steadily. Absence of illumination may mean no signal, defective wiring, or a defective amp. Or possibly none of these – since what constitutes 'reasonable' signal level varies widely. Some sig-present LEDs won't light up if the amp is driven more than 20dB below full output.

So before deciding a signal isn't reaching an amp or amp rack, and ripping the intervening wiring apart, first try raising the drive level so it's above 20dB below clip (full output). Use the meters on the mixer (see Chapter 2) to set this level.

Clip (peak, 0dB)

An important LED that shows when the amp has reached or is exceeding its 'operational ceiling'. Occasional flashing is fairly harmless, but anything more (certainly steady illumination) can rapidly destroy loudspeaker drive-units – particularly HF units – and also hurt most people's ears (once again, this includes your paying customers).

Well-designed amps use 'true clip' indicators. These show actual clipping by directly sensing that the amp is overloading. In comparison, less well-designed (often lower-budget) amps indicate at the level at which clipping should occur – but then only under particular and ideal conditions.

One detail that's rarely mentioned in this

context is that a signal only has to exceed clipping by 0.2dB or so to cause damage. At the same time, the point at which an amp clips can vary by over 1dB above and below the norm, depending on the speaker load, the signal frequencies involved and even the mains voltage.

So primitive clip indicators can read typically up to 1dB early or 1dB late. Lighting up 1dB early is no bad thing, but 1dB too late (1dB above actual clip) can result in unexpected damage – the LED might just be flashing occasionally while drive-units are being destroyed. (This is as frustrating as the oil light on most cars, which lights-up to say there's a problem only some time after the engine has been damaged by lack of oil.)

Using clip LEDs

This is a good place to mention that, most of the time during a gig, the amps' LEDs aren't being observed, at least for the FOH PA, because the amps are out of sight of the FOH engineer.

But when a system is being set up (certainly at rehearsals/soundchecks) you should check that the threshold level at which the crossover limiters are set (see Chapter 4 for more on this) is some way below, rather than enough to light any of the amp's clip LEDs.

To determine the level at which the first/any clip LEDs just start to light (or flash), the limiter thresholds should be temporarily increased gradually until the LEDs just start to light up. This should be safe provided the test is only done momentarily, long enough to note the threshold knob position at this point.

It will obviously be more approximate with a clip indicator that doesn't show true clip. Make sure that the full complement of PA cabs is plugged in because that governs the result if the indicator does show true clip.

The amount (in dBs) to back off the threshold control is then determined primarily by variations in the AC power supply (which can change with the venue, time of the day, and day of the week). This is because the output delivery of nearly all PA amps directly depends on this. If the AC supply is ten per cent low, the amp's clip point will also be about ten per cent low.

To allow for the AC voltage being up to 15 per cent lower in real use, the crossover's limiter thresholds need setting at least 1.5dB below the level where the first of the amplifiers' clip LEDs

light. If a PA system is being used between countries with supplies centered on 240V (in UK, Eire) or 220V (mainland Europe), so bigger AC variations of up to 20 per cent are possible, this leeway should be increased to 2dB (see Chapter 10).

'True clip' or not

To determine whether an '0dB' LED (on any amp) is showing true clip or not, use a CD player to repeat a musical passage with large transients (something with perhaps loud trumpet stabs, or kick or snare drum hits) through the amp under test, at first with no cabs connected, and arrange the levels to get the clip LED(s) to just start to come on.

A well-designed clip indicator will stretch illumination

Next, back off the gain control by a fraction so they just don't light. Then connect one or more cabs, up to the maximum allowed. If the LED lights up more solidly when you do this (plug and unplug the cabs to be sure, while the music levels vary), then the amp almost certainly has true clip indicators. If the indicator isn't affected by the extra loading at all, the LED is almost certainly working like a meter – it indicates the absolute level perfectly but is useless for showing the clipping you need to know about.

What about very brief clipping? In general, momentary clipping (under a few milliseconds) is not very audible – you could experiment with a limiter's attack time, between 3 and 30mS, to demonstrate this. But somewhat longer clipping (still under a third-of-a-second) is audible.

At the same time, the LED illumination it causes appears dim and fleeting to the eyes. A well-designed clip indicator will stretch illumination so any audible clipping is appropriately brightly represented by the LEDs. You can also use the clip LEDs if you don't know the sensitivity of an amplifier. To do this, you'll need to also use another amplifier you do know the sensitivity of, or else one you want to compare it with.

Connect the same speaker load – number and

type of cab(s) – to each amp's output. Then drive each with the same signal (not L & R signals) with the gain controls on each amplifier set at max. Now raise the source's level – usually using the mixer's main fader until one or other clip LED starts to flash. If both LEDs flash at the same setting, the sensitivities are the same. If not, disconnect the amplifier that has begun to show clipping – this will prevent damage to the speaker.

Also note carefully the position of the mixer's fader on the adjacent dB scale. Say it's at -6dB. Now push the fader up until the second amp starts to clip, with its clip LED flashing to the same degree. Note this position too. You can now turn down the level. If the second amp does this with the fader set at -3.5dB, you'll know the sensitivities differ by (6 – 3.5 =) 2.5dB. It's then a simple matter to determine whether the second amp is the known amp, in which case the unknown amp's sensitivity is 2.5dB higher.

LED level metering
On many PA amps there is no LED level metering at all. Paradoxically, quite expensive pro-standard amps omit metering to save pennies, while lower budget 'semi-pro' amps are more likely to have ordinary LED metering (see Chapter 2) – probably because it brightens up the power amp's otherwise dull front panel.

On the better pro-standard amps that do have LED bar metering, true clip indicators are integrated as the '0dB' LED at the top of the display. This means they light independently of the rest of the display, although you wouldn't readily know this – at least without a 'true clip or not' test, like the one suggested above.

In a few 'blue chip' amps, the entire metering is 'clip referred'. This means the '-3dB' LED lights at 3dB below true clip, etc. Unlike an ordinary meter, these levels show the actual headroom that's left for the amp under the actual operating conditions at each instant. This is really far more useful than ordinary metering at this stage.

Anti-clip (clip eliminator) & soft clip
Instead of clip indicators, some amps have LEDs that show the operation of built-in limiters. These can prevent hard clipping (hence the name 'anti-clip'), replacing it with less harsh distortion and making any amp less readily able to destroy drive units.

In effect, anti-clip acts as a 'catch' that can save the gig (and some very expensive drive unit failures) just in case the PA's main limiters (in the crossover) miss the leading edge of a 'spike' of loud sound, or get defeated or mis-set.

The presence of anti-clip inevitably increases an amplifier's distortion just (say 1dB) before clip. This can look bad in a written specification, but operationally it is not a real problem, since the limiters associated with the crossover (see Chapter 4) should be set to kick-in at or below this level (called -1dBr).

An alternative technique is 'soft-clip'. This means the amp enters overload gently. But soft-clip begins to affect the signal's purity a long way before clip occurs. For example, if the soft clip starts to alter the signal at 10dB below clip, as is the norm, then an amp delivering 1kW will only deliver 100W (as -10dB is one-tenth of the power) before the sound quality starts being altered – if not degraded – by the anti-clip. Soft clip is somewhat akin to compression, except there are no time constants or related settings.

Excess temperature – warning & shutdown
Also called 'thermal shutdown', 'overtemp', 'hi-temp'. In the real world, amps get hot because they generate significant heat (because the conversion of electric power into music is only about 20-50 per cent efficient at best), then that heat generation is concentrated into confined spaces (PA gear has to be compactly cased for efficient transportation).

Many amps give no warning of overheating before they shut down

In ideal conditions, amps will run cool enough. But overheating can occur rapidly for a variety of reasons: a cooling fan inside may fail; the cooling ducts inside may get blocked with dust; or the 'cool' air needed at the air intake may be unusually hot (and not so good at cooling), due to the combined effects of stage lighting, a sweaty audience, plus the hot air discharged from surrounding amp racks. Or someone may throw

power amps

have thrown a coat over the rack, smothering the airflow.

Driving the PA hard with compressed signals, and/or loading the amps with all the speakers they can bear also increase(s) temperatures in and around the amp racks. Overheating in a PA amp can cause expensive damage to the amp's innards, and possibly collateral damage to the speakers. It might even lead to a fire (fortunately likely to be highly localised and not emitting too many toxic fumes, as pro-power amps are mounted in metal, not plastic cases).

For these reasons, when the maximum temperature of parts inside a power amp are exceeded, the main 'programmed' response is to turn off the power to the parts of the innards that produce heat. A secondary response may be to reduce or mute the incoming signal, which may be at least partly the cause.

Many amps give no warning of overheating before they shut down. An LED labelled 'thermal' or 'overheat' or 'high temp' then lights up to say, "I have turned off the power and/or signal and/or have disconnected the output so I can cool down a bit". A few better-designed amps have a warning LED which lights-up when the temperature approaches the shut-down point.

Automatic reset is the norm with thermal protection. This is feasible because, firstly, this problem can be 'self healing'. If an amp is observed to be 'thermalling-out' (cutting out due to overheating) the problem can probably be

rectified (say by removing the flightcase lid that someone has placed over the rack's front air intake), or mitigated enough (by turning down the level or compression, or both, by a few dB) to see a recovery.

But however good the amp's cooling (usually fan-assisted), it will take time (at least a few minutes) for the temperature to fall. Also, when an amp's thermal protection resets, it will take a minute or two, at least, for the temperature to rise back again – assuming the problem is still present. The significance of this is that an amp's temperature can't 'move fast enough', meaning that there's no worry that the thermal protection might cut in and out at any high rate.

If you look after your amplifiers and racks, thermally-speaking, you may never see thermal warning LEDs light up.

DC fault / DC protect

(Aliases: DCF, DCP, DC, DC error, or 'fault'.) This is an LED that mainly says, "very serious fault", and either, "I have turned off" or, "I am still on but have disconnected the outputs". Mostly you'll never see it lit.

If lit up, there is a chance that this has been falsely triggered by a whopping on/off 'thump' received from preceding gear, or by some very low, loud bass. So the first thing you should do, if there are no distinct damage signs (like a burning smell), is to attempt to reset the protection by turning the amp off, waiting for a few seconds,

RF pickup and instability

Something that equipment makers almost never ever mention is that some conditions and connections can occasionally upset or somehow provoke power amplifiers, causing them to generate high frequency signals. A related situation is when they're fed such signals, which have arisen for similar reasons somewhere earlier in the PA's audio path.

Usually these are at frequencies well above audio, known as RF (radio frequencies). These can also be picked up from large local radio transmitters – often by speaker leads as much as the many preceding cables and wires.

These sort of RF problems aren't at all common, but when they strike they can cause havoc and/or a lot of head scratching and wasted time if you don't know about their existence and behaviour.

If an RF 'attack' is affecting your amplifier, you won't hear anything directly, since even if they're converted into sound by any of the HF drive-units, signals at these ultrasonic frequencies would be too high to be heard. But you may well notice indirect effects – such as inexplicably distorted sound quality, strange noises (sometimes like those heard from radio receivers), and the power amp

may well be running warmer than you'd expect for the work it is doing. Another symptom is blowing a lot of HF drive units, where there's no simpler cause, such as an incorrectly set limiter.

Most amplifiers have no active protection (meaning no shutdown or muting), and no indication either, of any kind of RF problem. With some amplifiers, a generic 'error' will be indicated, and shutdown will occur, leaving the user protected but usually none the wiser.

With a minority of more sophisticated models, a 'DC/RF' or even a dedicated 'RF fault' LED will light.

then turning it on again. But before doing this, best unplug the speakers so they aren't unnecessarily put at risk (just in case).

A well-designed clip indicator will stretch illumination

If the 'DC' LED doesn't show again straightaway after re-powering, then it's almost certainly safe to reconnect the speakers. This makes sense, because the kind of serious fault that the DC protect light should indicate is normally one that won't ever repair itself by any amount of re-powering of the amplifier.

Beware, though, that with some amplifiers the protection can be re-triggered falsely if you are in too much of a hurry. So if the DC fault LED does show again after the first resetting, it's sensible to make one more attempt, leaving the amp switched off (to thoroughly discharge) for about one minute first.

If DC protection fails, or if you use an amp without any DC protection, there is a risk to most of the speaker drive-units. In circumstances where the drive units are directly connected to the amps, HF units are most vulnerable, followed by mid, then bass drive-units. It's down to the mass of wire in the driver's voice coils (see Chapter 6), and how long this amount of wire takes to overheat when subjected to the DC voltage coming from the amp with a DC fault.

The good news is that in the better PA systems, DC protection or 'blocking' capacitors ('caps') are fitted in line with the HF drivers, and sometimes in line with the mid drivers too, and occasionally even the bass drivers. But many LF drivers are not protected by caps – because of the large size and high cost of the caps required for bass protection, and also because of sound quality concerns.

Otherwise, these series-connected caps largely protect the drive units from amplifier failures even if there's no fault protection on the amp, or if the amp's DC fault protection itself fails – which is not unknown. They protect by stopping the DC reaching the drive units, although they do still allow an interim 'thump' when the DC first 'hits',

and also if there is any further step change in it – which is why they are not 100 per cent foolproof.

Bear in mind that, with some amplifiers, DC protection is provided but there's no separate indication. Instead, indication is by way of a general 'fault' or 'error' LED, that will likely also light up when clipping or overheating. This is obviously less useful, possibly confusing – and also rather miserly, considering the low cost of providing another LED

RF protection

Like DC ones, RF faults (see the 'RF instability' sidepanel opposite) can have dire consequences. For example, HF drivers can be damaged as the RF signal 'rides' on the audio signal, which readily leads to hard clipping. But this isn't the case all the time, and it's possible to have a lingering RF fault without knowing anything about it.

A few pro PA amplifiers have a dedicated RF fault indicator. In others, the two indicators are combined into a single 'DC/RF fault' LED. The RF protection LED says, "I have turned off the amp because the amp is unstable at RF, which suggests something is wrong with either the amp or the wiring. Either way, I'm playing safe rather than let this destroy the HF drivers or even the amp".

If an RF fault is indicated, unplug the input and speaker wiring and then 'reboot' the amp (turn off, then wait a moment, then turn back on again) to see if the fault stays. If the amp has the problem on its own, then it is faulty. If not, replug each cable in turn to see which causes it. (See also the 'RF instability' panel.)

Other fault protection

The following protective features may share indicator LEDs with others. Some may not be indicated at all, working 'behind the scenes' to safeguard the power amps (and the rest of the PA). Some may just be indicated by a general 'fault' LED that says, "I have turned off – but you will have to find/figure out why...".

HF Protect

This is a valuable feature for PA amps, though it only appears on a few model ranges. The idea is to provide added protection for HF drive-units – which are far more delicate than the others – for situations that even correctly set-up limiters (on the crossover – see Chapter 4) cannot discriminate against.

This feature enables high-power amps – ones with high voltage swing that can deliver their power into 8 ohm (and pairs of 15 or 16-ohm) drivers – to be safely connected to HF drive-units having continuous power ratings that are four to ten times lower. This arrangement gets the most out of a single drive-unit, and allows treble transients to be cleanly handled.

It's also not a problem for HF drive units, because the average power in the treble frequencies is (unless the signal has been highly compressed) very much smaller than the peak level. So we can have 80V peaks, equivalent to 400W into 8 ohms, but the average power level will be closer to say, 20 to 50 watts.

The trouble comes most often when there's acoustic feedback at treble frequencies, giving a fairly continuous tone. Or if any instrument or sample contains continuously high levels of steady signals at HF – anything above 2kHz. Or if anyone feeds a continuous tone into the PA by accident – such tones shouldn't be used for testing speakers (particularly HF drivers).

If any of these things happen, the HF driver's innards will be 'fried' and destroyed quite quickly, probably within a couple of seconds – depending on the amplifier's power delivery. Damage will occur even if the crossover limiters are set correctly, since their threshold doesn't need to be exceeded for the damage to be done.

The HF protection feature prevents this, by distinguishing between high-level continuous tones at HF and musical signals which aren't continuous – or if they are, at least they come and go in bursts so the average level is lower.

As soon as a high frequency signal has been detected and determined as being continuous, either the amplifier is prevented from passing any signal at all, or the HF response is reduced so there can be no harm. Reset to normal working may be automatic.

Power on/off muting

Without this feature, when amps and other pieces of gear are turned on or off, they can make random noises, some of them violent and/or alarming. These 'glitches' or 'power up/down transients' might also damage (or at least hasten

the wearing-out of) HF or bass drivers – quite apart from scaring the daylights out of artists, crew and public.

Some amplifiers have a 'start-up' (or similar) LED which lights to confirm that the amp's outputs are muted – usually for a few seconds during and straight after switch-on. When an amp is powered down, muting has to act swiftly to avoid the same sort of sharp cracks, bangs and zhitts. But this is not normally indicated.

Amplifier inputs should be muted when an amp isn't powered

Strictly speaking, amplifier inputs should also be protected ('muted') when an amp isn't powered. One definite way to achieve this is to disconnect the returns snake/multicore before powering down the rest of the PA.

For this to do more good than harm, you'll need to first make sure the cable shield is tied (connected), through XLR pin 1, to the equipment's casing/earth at the other end, otherwise there's a risk of hum. The two halves of the PA – amplifiers versus the rest – can then be powered down in any desired order.

Adverse load protection

Amplifiers are vulnerable to being abused. There's no 'amplifier police' to stop you from connecting an amp's output to too many speakers (cabs, boxes), or making other kinds of accidental connections which demand power deliveries that are outside of what a PA amplifier is capable of.

These conditions may be categorised as 'adverse loads' – as opposed to the 'legal', rated load impedances that are specified for a given power amplifier (the ones it's designed to drive into).

Either most PA amps will simply find adverse loads difficult to drive and not give their proper performance, or they will be instantly damaged. Alternatively, protection circuitry may operate, and/or the amplifier will shut-down and may need resetting (after the problem has been sorted).

The following list starts by categorising 'legal' rated loads (for most PA amps), then increasingly adverse loads, finishing up with the most severe and dangerous:

No load, no problem (usually)

- Driving an amplifier with signal, with nothing connected at all to the output is not stressful to PA amps – this only upsets amps using valves (tubes), such as instrument (eg guitar) amps.
- Driving as above, but with speaker cabling connected (but nothing beyond that) also shouldn't be at all a problem. But a few less well-designed amps may be upset by some cables, depending on the type and length – upset means they may shut down or even, at worst, self-destruct. This is fortunately rare.

Rated ('legal') loads

- Driving an amplifier into speakers having any impedance above the rated minimum is fine. There is no maximum. The rated minimum is commonly 4 ohms or less, down to 1 ohm in a few cases. With some amplifiers the figure given – in the manual and often repeated on the rear panel – is the actual limit.

  With others, lower impedances (down to half what's rated) may be driven without the amp overheating or shutting down. But this arrangement won't help sound quality. Indeed, this sort of leeway ('conservative rating practice') has been provided expressly to ensure that the amp is always used with some 'current headroom', and so stays sounding good under all conditions.

Potentially stressful & damaging to the amplifier

- A 'low load' – meaning anything well below the minimum impedance the amplifier is rated to drive. In most cases this means impedances below 4 ohms. Such low impedances commonly arise when more than two drive units or cabs have to be connected to one amp channel. There can be pressure to do this when an amplifier channel 'goes down (develops a fault) or when, for other reasons, not enough amplifier channels are available.

  Although misusing an amplifier in this way can't be condoned, if it has to be attempted, it is less stressful for any amplifier if the drive (output) levels are kept down as far as possible, and also if the signals being used aren't subjected to much in

the way of compression.

When an amp is loaded with more cabs (and so a lower impedance) than it is rated to drive, the symptoms will generally include overheating – or at least, running hotter than normal – and also reduced sound quality, with more distortion.

Sometimes it's discovered that a particular amp rated at, say, 4 ohms will apparently happily drive particular speakers having a somewhat lower impedance without any sign of stress.

Usually there is a good explanation for this – the amplifier maker may allow some leeway in the minimum impedance rating to allow for speakers which exhibit impedances below their ratings over some parts of the overall frequency range. But some speakers don't exhibit these dips, so they can exploit this leeway capability without overloading the amplifier.

For example, most 4 ohm speakers will dip to no more than 2.8 ohms. So many amplifiers rated for 4 ohms will actually operate down to 2.8 ohms – provided the load exhibits no impedance dips below this. In this case a pair of speakers rated at 6 ohms, giving a total of 3 ohms, could be connected with no problems.

The opposite of this situation is that some speaker cabs exhibit load that's more stressful than their impedance figures suggest – perhaps because the 'impedance' rating of a speaker is only a nominal figure, really an average. The actual impedance can be higher at many frequencies – which is fine – but also lower at other frequencies. Usually the frequency range where it is lower is quite narrow – usually around the lower end of the drive unit's range. If the music doesn't have continuous signals at these frequencies, the amp won't 'see' and suffer this low impedance so much.

When an amp is loaded with more cabs than it is rated for, it will run hot

The adverse loading problem is aggravated by some speaker designs that aren't made for professional PA but which are widely sold for low-cost installations, and which the unsuspecting smaller PA user may end up with. The

power amps

troublesome type involves 'ported' (aka 'reflex', 'vented' or 'Thiele') bass cabs – and particularly developments of these, commonly called 'double tuned' or 'bandpass'.

The trouble with these types is that there can be more than one area of impedance dip, so there's all the more chance of the musical signals being at frequencies where the cab's impedance is low. This means the amplifier sees, on average, a lower load. Speakers of this type can be identified by their hidden (internal) drive units and lack of horn-loading (see Chapter 6).

Stressful or damaging to most amplifiers

The following conditions can destroy the innards of amplifiers that aren't properly protected. Otherwise, protection (including shutdown) may be triggered.

● Bad or loose speaker connections are rarely mentioned as stressful loads – meaning connections that are 'on the edge', alternating between a good contact and not. The fact that one bass driver out of several is pulsing on and off may well go unnoticed in use, but such intermittent connections can cause tiny sparks – so-called 'micro' arcing, which can be stressful, or even instantly damaging, to some amplifiers.

It can be avoided by regularly checking for solid connections – and, in particular, that all screw-held cable conductors are secure. Better still, avoid screwed connections, or at least ones with only single wire-clamping screws.

It can also be beneficial to lightly tin (lightly soak in solder) all stranded wires before they are screwed into place, as this will prevent 'whiskers' – fine, individual strands of wire – splaying out

to cause the intermittent connections in the first place. (For more on troubleshooting and maintenance, see Chapter 9.)

● Short circuit (or 'a short') directly across the output terminals – meaning between the hot (+) and low (-) side. This might occur anywhere between the amp's output terminals, along the cabling, or inside the cab, and even at the drive unit's connecting terminals.

Typical causes are untidy wiring, often where strands ('whiskers') of wire make accidental connections to the 'wrong' side of the output. Short circuits can happen less often when cables are crushed or split, and also when drive-units or other parts inside cabs are repaired or serviced and not re-connected correctly.

● Capacitive loads. This means that the load presented to the amplifier involves a fairly large amount of directly connected capacitance – something over $0.5\mu F$ (half a microFarad). 'Directly connected' means the capacitance is seen by the amplifier without any resistance in line with it to moderate the effect in any way. This capacitance may or may not be in parallel combination with other loads, such as ordinary speaker drive-unit loads.

Only slightly larger capacitances – much above $1\mu F$ – cause extreme stress to most power amplifiers: in some ways even worse than a short circuit.

A common cause of capacitate loading is a wrongly wired or connected passive crossover (see Chapters 4 & 6) or DC protection capacitor inside HF and other cabinets. Most PA amps have protection against things like low-load, short circuits and capacitive loads. This differs

---

Lowest ideal loading

To avoid low load problems – discussed in the 'Adverse load protection' section – establish what the actual minimum impedance rating of the amp is, and also the actual minimum impedance of the cabs it will be used with. Work out whether you can, in theory, connect one, two, three or (rarely) four drive units or cabs to each channel's output. Then try this number of cabs with a wide variety of music, and at high levels. Go to one

cab too many – one cab more than the rating suggests.

If the amp doesn't get too hot and 'thermal out', measure how hot the heatsinks or outgoing hot air gets (a domestic electronic thermometer can be used), and also check the sound quality. Then take out one cab or drive-unit, and check temperature and sound quality again. Then step back once more, so you're using one drive-unit or cab less than the legal rating allows.

If with either of these lesser loads the

amp runs quite a lot cooler, and/or sounds considerably less distorted, and 'clearer', then this shows the loading with which this particular amp is most comfortable with the particular speakers used. If you use other speakers, the experiment will need to be run again.

It sounds complicated, but after a while users build up a mental library of ideal matching amp and cab configurations, which may be different from the manufacturer's recommendations.

power amps

depending on the type of output ('power') transistors used.

In amps that use bipolar output transistors (and also a minority of amplifiers that employ MOSFET output transistors), the protection against these is provided by a method called 'V-I load line limiting'. This causes the output transistors to be abruptly starved of signal drive (to save their lives) if the combination of the voltage (V) applied to them and current (I) passed through them oversteps a line on a graph.

This protection is 'dynamic' – able to cut in as fast as there is danger, and out as soon as it is safe again. This saves the transistors lives, but there may be some audible effects, such as snapping, popping, rasping or other 'sonic break-up' sounds, and/or a considerably reduced output level when the protection is triggered.

Oddly, few if any PA amps have any indicators to show that this sort of protection is occurring. As a result, real users cannot so easily learn about which cabs present difficult loads, or what might be causing grungy sound. Instead, experiments must be conducted.

Most power amplifiers that use L-MOSFETs (a type of MOSFET transistor – see Glossary – specially made for audio amplifiers) do not require V-I protection. Instead, the MOSFETs protect themselves, and also do it in a more gradual, musically acceptable way.

A few power amplifiers use V-MOSFETs in their output stages. This type of MOSFET is not made specifically for use in amplifiers and is not self-protecting. Instead, they rely on thermal sensing (since, if there's an adverse load problem, it will usually quickly show as overheating) followed by rapid switching off of the supply voltage if the problem persists long enough to threaten the amplifiers' innards.

Earth/ground lift

This facility is provided on a few amplifiers, mainly semi-professional types, to help eradicate hums caused by earth connection loops. If the amplifier has a balanced input (which should be the case for any serious PA use), and this is correctly connected, there should be no need for

it, and as discussed in Chapter 3, this facility shouldn't be used unless it's absolutely necessary, as it can detract from electrical safety.

Pin 1 lift

Related to the above, but safer, some PA power amplifiers have a 'header' or 'jumper' plug inside that alters the connection to pin 1 of each input's XLR socket. The connection should normally be direct to chassis. But sometimes RF pickup problems (see earlier, and also 'RF troubleshooting' in Chapter 9) can be reduced or averted by opening one of these connections – usually on one input, or all but one input for amps with more than two channels.

This is simply done by moving the header plug's position. But it needs some forethought and logistics planning, otherwise the screen may not be connected to the adjusted input(s), if the cable at the other end is unplugged. This could (at worst) cause damaging, high level buzzes, RF noise or audible shrieks.

Switch-on surge (inrush current)

Most PA amplifiers can draw high 'inrush' (or 'surge') currents when first powered up. The extent to which they do this depends greatly on the power supply they're connected to. The surge current will be highest when the amplifiers (or amp racks) are connected directly with short and stout cables to a 'beefy' high-current capable (low impedance) AC supply (for example, one rated at 100A or more in the UK, or about double this current in other territories).

Most PA amps draw a high inrush current when first powered up

As this is also the ideal condition for getting good sonic performance from the power amplifiers and the PA, it is not a helpful solution (even if it were practical and safe) to power the PA from longer, thinner cables, and lower current outlets. So the surge needs to be handled separately.

How big the surge is also depends on 'pot luck' – whether you push the 'on' switch/button at exactly the wrong moment, when the mains voltage is (cyclically) at its highest peak value. This occurs for a few thousandths of a second, a hundred or so times a second.

In worst case situations, where there's no control over the surge current, and the supply impedance is quite low, and several power amplifiers are switched on at once at the instant of the mains peak voltage, the momentary current draw can exceed 1000 Amperes.

Whether the 'inrush' current is this high or not, it can be enough to drop the AC/mains voltage for a few cycles, which will dim ordinary incandescent lights momentarily, and possibly crash adjacently connected mains-powered computers and instruments with volatile digital 'brains' – as well as certainly tripping any circuit breakers in the AC supply line that don't have enough 'anti-surge' (surge-handling) delay built-in. (There's more on electricity and safety in Chapter 10.)

This high momentary 'switch-on' current is drawn by nearly all high-power-capable audio amplifiers to charge their large reservoir capacitors. It also 'charges-up' the magnetic circuit of the power transformer, which isolates the equipment from the electricity source for safety, and changes the AC voltage to a variety of voltages required by the amplifier's innards.

Soft-starts & staged power-up

Some amplifiers are fitted with inrush current surge suppression (aka a 'soft-start' method) that employs thermistors – resistors that reduce their resistance with the background heating caused by passage of the current drawn by the amplifier.

This method is simple and fairly reliable, but may not reduce the surge sufficiently, because this type of part can usually only vary in resistance by a certain amount. Surge suppression with thermistors also slightly reduces the amp's performance – notably its power delivery. And once the amp is warmed up, thermistors offer reduced (or no) surge protection if the amplifier is turned off then straight back on again.

In the more effective kind of soft-start found in fully professional PA power amplifiers the power is initially connected to the amp through a high-power handling resistor that lowers the surge by a factor of a hundred times or more, making it insignificant (provided you aren't switching on dozens of amps at once).

After the surge is over, the resistor is bypassed by a relay or triac (electronic switches) to enable the power amplifier to work normally.

Even this can occasionally go wrong. In some cases, the soft-start's current limiting resistor fails – the amplifier tries to start without it, and the resulting, uncontrolled surge soon destroys the bypass switching. If this uses a relay, the contacts may be welded in the 'on' position. The result is an amplifier that apparently starts normally, but draws a high surge current.

Again, this condition won't usually be clearly indicated – it's something the user has to spot. One symptom may be if an amplifier with soft-start is occasionally blowing mains fuses at switch-on.

You can reduce the stress placed on the amplifiers' soft-starting facilities, so they're less likely to go wrong, by never applying a high-level signal to the amps when they're first powered up.

For example, if the amps' power was detached when the PA is running, turn down ('dim') the PA's drive, to at least 20-26dB below full level. At this sort of setting you'll still be able to hear the moment the amplifiers are running again, but the power current surge stress at turn-on won't be made any worse by having the amplifiers passing high current into speakers at the same time.

You can reduce amp switch-on current by turning on one at a time

A higher-level 'system solution' to reduce amplifiers' switch-on current surge is to turn amplifiers on one-at-a-time, or one rack's worth at-a-time. This can either be done manually, or you could use 'sequential switchers' that will do the job automatically – once they're connected in line with the power supply to the individual amplifiers, or as many amplifiers as can be switched on at once without problems, and subject to the switcher's current capacity.

The downside is generally having more gear to go wrong, added wiring complexity, and using up more valuable amplifier racking space.

power amps

What neither of these techniques deal with on their own is that the larger (higher-power-rated) amplifiers can individually draw very high currents at switch-on, enough to cause problems when connected to less sturdy electrical supplies. So you still need amplifiers with soft-start.

It's worth noting that if you have otherwise good power amplifiers that have no soft-starting facility, it can usually be added ('retro-fitted') by a qualified technician.

Other (less useful) amp features

Remote control – great for permanently-installed PA in clubs, pubs etc, not so practical for touring PA, though evolving all the time.

Integral (active) crossovers – are intended for a simple, low-budget install. They could also work in a very small PA (where only one channel is used per frequency band) and possibly for stage monitoring, but are otherwise not very practical, certainly not for FOH use. The reason is that you would have to set up multiple crossover frequencies identically, when using more than one amplifier channel.

You'd also have to use the amplifier gain controls to set the band levels, and match each every time you wanted to change one. Also, these controls would be sited behind the amplifiers, not at the FOH position.

Sub-sonic protection – this is another name for a high-pass (HP) filter, which removes signals at frequencies that are too low to be useful. As with other HP filters in the preceding stages of the PA, the slope may be -12, -18dB or more rarely -24 dB per octave (see Chapters 2 & 3).

It's not a very useful feature on PA power amplifiers for logistical reasons – lots of separate adjustments are going to need to be managed and these will be physically remote, at least from the FOH engineer. In any event, the house graphic and/or crossover in the drive rack (see Chapter 4) already offer this facility in a handy form.

Amplifier fan-cooling

There are two main types of fan-cooling system – both have advantages and disadvantages.

One type draws cool air from the rear and exhausts hot air at the front (with this kind of amp, anyone getting close enough to set controls or read indicators also gets 'physical feedback' on how hard the amp is working, thanks to a face-full of hot air).

The second type draws cool air from the front. This can keep amps cooler – especially if the space at the rear is more enclosed, which is often the case in a flightcased rack of amps. The downside is that 'waste air' pumped out at the rear can be hot enough – under worst case conditions, hard driving etc – to soften PVC cables or even melt them.

'Dressing' (positioning) the cables to one side is a must – or, better still, use all-rubber-sheathed cabling.

Front and back exhaust types should be kept separate from each other

There is another type of amp – less often seen nowadays – which exhausts hot air at the sides, and this needs either a rack with side exhaust grille or else a chimney up the sides.

It's important that the front and back-exhaust types be kept separate from each other in amp racks – otherwise hot air will get recirculated, possibly with show-stopping or permanently damaging results.

If the different types have to be stacked with each other, place the one emitting hot air at the rear *above* any drawing in cool air from the rear (as hot air rises).

There are two other notable problems with fan-cooling: one is that it tends to draw in lots of dirt and dust from the air, which can be pretty disgusting in some venues (see Chapter 9 on relevant cleaning and maintenance); the other is acoustic noise.

Fan-motor droning may be less of a problem at relatively noisy live events (as opposed to, for instance, a recording session), but it's still something to bear in mind when positioning fan-cooled gear in relation to microphones.

power amps

Bridging

Bridging (also called 'bridge-mode operation') means connecting between the hot (positive, red-labelled) sides of two amplifier channels' outputs. It's the way to get a usefully higher maximum (approximately doubled) voltage swing from the output. This in turn can deliver more power.

In practice, bridging is most feasible and useful for delivering higher power into higher impedance cabs and drivers – namely 8, 15 or 16 ohms – rather than 6, 4 or 3 ohms; or else any combinations of drivers (in parallel) giving 8_ and above. This is useful, if not the 'only way to go', because few power amplifiers have individual channels that can deliver as much 'per channel' power into these impedances.

To avoid disappointment with bridging, we ought to remind you again that there's 'no free lunch'. Power delivery *can* be doubled, or more – which is what you'd expect when combining two amplifier channels to one output. But the bridging action could force the amp to deliver more power than it is strictly rated to do, if rated solely to cope with the per-channel power delivery.

Nearly always, this will be the case into lower impedances (6Ω and particularly 4Ω and below), and you shouldn't expect to drive these. You might get away with this in borderline cases (say into 6 ohms), but there's always a chance of creating bad sound (especially noticeable in mid and HF), and such 'bridge-abuse' will certainly be stressful to the amplifier (heating it up) if the signals are strongly compressed.

Bridging works best into low impedance cabs

The limits of bridging show up in spec sheets where the minimum impedance you can drive is doubled. So an amp that's good to drive 2 ohms per channel is 're-rated' to a minimum of 4 ohms when the amp is bridged.

The significance of bridging is that it's relatively simple to build 'beefy' power amps that deliver higher and higher power into lower and lower impedances. This is rather like developing a car that can go faster and faster down steeper slopes...

With bridging, you can end up being able to deliver more power than is safe or desirable into low impedance drive units (or lots of parallel drive units giving a low impedance). But you can also focus more power into individual 8Ω or higher impedance drive units – as are usually used in PA speakers.

A few power amp designs can effectively 'change gear', by re-deploying their power source's voltage and current capacities, as an alternative to bridging. Notably, this method doesn't reduce two channels into one. A switch marked 'low' or '2-4Ω' and 'high' or $8/16\Omega$ is provided for each channel. When set 'high', the amp channel operates with a higher voltage, lower current supply. When set low, the power supply is 'gearshifted' to provide double the current but half the voltage.

Setting-up & using bridge mode

With most PA amps, bridging (or 'bridge mode' operation) is achieved by moving a switch to the position labelled 'bridge' (or 'mono bridge', 'power bridging' or just 'br'). Or sometimes you might see some strange symbol that suggests a (river) bridge.

But before you flick or set this switch, read on... When you set bridge-mode, most amps are reconfigured so the signal you connect to Channel 1 now drives both outputs. Both outputs will continue to work individually, but the signals will be opposite in polarity, ie one output will be 'blowing' while the other 'sucks'.

So if you leave existing speaker connections to the individual channels intact, the signals' opposing polarities will largely cancel each other out, which is pointless, and will also usually lead to poor quality sound.

Bridging is most usefully set up when an amp rack is readied for a PA job. Other than making the bridged connection, you have to make other arrangements for the signal that would have used Ch2. Any signal that's left connected to Ch2 will go precisely nowhere.

Another reason for not setting-up (or changing over to) bridging 'on the fly' applies in small PA systems, where Channels 1 & 2 might often be handling different frequency ranges. Say, for example, Channel 1 was handling LF and Channel 2 was handling HF. This is a common enough situation. Switching to 'bridge' might damage the HF drivers, as they'd be fed with bass.

If you're lucky, the DC protection capacitors, as mentioned earlier, might provide some protection, but to save this happening by accident or careless curiosity, it's a good idea to put white or silver gaffer-tape over any amps with a bridge switch, boldly marked 'Do Not Use'.

Back to bridging: having connected the required feed to Ch 1 and switched the amp to work in bridged mode, you finally have to reconnect the speaker(s) between the two channels' positive outputs (+ to +).

Bridging can also push the limits of the speakers

Bridged output connections

On a few better-designed amplifiers there's a dedicated bridged outlet – which may be a Speakon socket on more recent models, an XLR socket on older units, or else dedicated 4mm sockets/binding posts. If not, bridge connection can require some fresh logistical thought.

The trouble is that otherwise, the Channel 1 & 2 '+' outputs require unusual, individual (single-wire) connections. Even if they are individually accessible as 4mm sockets or terminals (binding posts), the Channel 1 & 2 '+' outputs will be typically widely-spaced apart. They may even be on opposite sides of the panel.

In many PA racks, connections to these terminals are made with dual (twin) plugs, which aren't directly suitable to make a bridge connection. So, a bespoke/customised wiring adaptor will need to be made to ensure a neat and safe connection. This might comprise a 4mm socket which plugs into the half of the dual plug you can't insert. This socket would then connect to a plug on a suitable length of single-core flexible wire, which is then connected to the opposite channel's '+' output, wherever that is on the panel.

Likewise, if the amplifier has only 'per channel' Speakon (or XLR) outlets, the adaptor lead you'll need to construct comprises a specially-wired line plug. With this, one side (likely pin 2 if an XLR, or pin 1+ if a Speakon, or whatever other pin goes to channel 2's '+' output) is not internally connected, but is extended with a suitable length of flexible wire and arranged to emerge out of the rear of the plug. The other end is then connected to a second plug (to pin 2 if XLR, or 1+ if a Speakon plug) to

make contact with channel 2's '+' terminal.

When making bridged connections, care is needed (as ever) to preserve the correct polarity ('phasing'). Channel 1 '+' output is normally '+', and so Channel 2's '+' actually becomes '-', once the bridge switch is operated. (You can read more about polarity in Chapter 7.)

Effects of bridging

As the output voltage has been doubled, the sound level (dB SPL) from the connected cabs should (or could) increase by 6dB – but that's based only on assumptions and simplistic theory.

First, as noted already, the bridged amplifier's power delivery capability is increased up to a natural maximum of quadruple (x4) into the same speaker impedance. But more often, the amplifier is limited in the lowest impedance you can drive, or in the power delivery increase it can handle.

So, in practice, power delivery increase may be lower than x4, and is often only x2 in conservatively specified (read as 'cautiously rated and tightly designed') amplifiers. This then amounts to an increased SPL of nearer to 3-4dB.

Of course, this extra capability could also be left latent, and kept in reserve as headroom. This would be done by turning down the gain on the particular amps (or crossover band etc) by 3 or 4dB, so there was no nett change or difference in the acoustic output's level.

Also, even if the amp's voltage swing is exactly doubled, the bridged sound level increase won't be quite 6dB, due to a quirk of speakers known as 'power compression'.

When the power the amplifier is delivering to the speaker exceeds the heat-removal capacity of the driver's voice coil, the temperature rise causes a resistance (and hence impedance) increase. This effectively absorbs the extra power being delivered, and defeats some of the anticipated sound level rise.

In other words, if you try to push the speakers too hard with extra power, there will come a point where they just won't go much louder. That's also a sign you're getting near to burning them out.

This problem, incidentally, is not solely caused by bridging – it's just that bridging puts higher power delivery at your disposal, so you'll be pushing the limits of the speakers.

speakers
drivers, horns, impedances...

Quick view

Our PA signals – split into left, right and maybe centre channels, as well as into several frequency bands – have now been amplified and finally arrive at the speakers. This is where they're turned back into sound, after their long, eventful journey (which has of course taken less than a nanosecond – it's all instantaneous as far as the audience and performers are concerned). Speaker cabinets contain 'drive-units' which must be carefully matched to the amps, as well as suited to the venue and audience, so the music is delivered with clarity and projection to everyone who wants to hear it.

the word speaker (short for loudspeaker) is often used casually to refer to the whole unit (or box) that ultimately emits the sound, the end product, in a PA system. In fact it can be broken down into more accurately-named parts. The exterior casing is called an enclosure, or cabinet (cab) – this will be designed and shaped to project the sound in particular ways. But the working heart of the speaker, the bit that essentially produces the sound, is generally called the driver, or drive-unit.

A driver is a transducer – like a microphone, except it works in the reverse way, converting electrical signals into acoustic soundwaves. Most drive-units used in PA systems work on a moving-coil principle, like a dynamic microphone – though there are also other kinds of speaker which parallel other mike types: for instance, an electrostatic speaker operates in a similar way to a capacitor mike, while a ribbon mike uses the basic concept of the mike of the same name (see Chapter 1).

But it's the moving-coil/dynamic type that's most common in live sound – largely because it's

most economical and relatively efficient. Simply put, it consists of a large, powerful magnet in which there's a gap where the magnetic field causes the 'voice-coil' to vibrate.

This coil is attached to a cone, or 'diaphragm' – usually made from paper pulp, but can be other materials – which can vary in diameter from about three to 24 inches. Larger speakers generally handle deeper, more bassy sound, as we'll see shortly.

The fragile cone is held in place (but allowed to vibrate) by a metal chassis, which gives its name to the main, multipurpose work-horse type of driver used in PA work. 'Chassis' drive-units are responsible for most of the bass and mid-frequency reinforcement. (Bass drivers are sometimes known as 'woofers', and come fitted in 'bins').

High-frequency speakers (sometimes called 'tweeters') need a different type of driver. For high-mid and particularly treble ranges a very lightweight but rigid 'cone' is required, and the paper type in a chassis driver is not so suitable.

Instead, the answer is a tiny cone turned inside out and rounded, becoming a mini dome-shaped diaphragm, made from either a light, rigid pressed metal or plastic-impregnated cloth.

This forms the core of specialised 'compression drivers', designed to be used in conjunction with a 'horn' (a flared tube attached to the front of the diaphragm dome) to efficiently direct top-end-frequencies at high levels. This is the sole practical approach for delivering high frequencies in a controlled way, and is universally accepted.

Behind the speaker grille

As with microphones, a good speaker design tends to stay around and develop gradually (compression drivers date back to pioneering work in the mid-20th century by companies like Altec and JBL).

They are also the most high-profile and publicly visible parts of a PA, and are often individualistic in looks and the sound they produce. Because of this, particular brands of speaker are often seen to characterise a given PA system – deservedly or otherwise.

In other words, the PA speakers become a shorthand for 'describing' the PA – for many artists and audience members, the speaker cabs are 'the PA'. Few people mention the names of the mike makers, or crossover maker, or the type of multicore/snake – though all are, strictly speaking, equally vital.

This shorthand is naive but also rather inevitable. As we've said, it's partly because PA cabs are physically and visually dominant, and also because differences are more obvious to the 'untrained eye' than with those boxes of electronics behind the scenes.

Though a theoretically simple mechanism, speakers are in fact about the hardest part of a PA system for the maker to 'get right'. Compared to most other elements they remain some way off any sort of perfection.

PA speaker makers have to satisfy many conflicting requirements – if they make one part work well, another area may be compromised. The basics of speakers, their complexities, and why particular approaches are used or have been found to be optimum are too complex to put into one paragraph. Instead, they are worked through over the following few pages.

For many artists and audience members the speaker cabs are 'the PA'

If you buy and use modern 'packaged' PA speakers, you probably won't need to think about what's inside them until they go wrong (see Chapter 9 for more on this).

For users with more restricted budgets, there's a tradition of DIY (do-it-yourself) cabinet building

Speaker impedance & load matching

As we've said before, impedance *impedes* the flow of current in an electrical (AC) circuit, such as an audio path. It's basically the amount of resistance every device offers to the signal it's either sending or receiving (everything offers some resistance).

Every speaker has a nominal impedance – the exact impedance varies at different frequencies, and depending on how the particular speaker reacts to particular sounds. PA speakers are usually rated at either 4 ohms or 8 ohms.

You may remember there are two sorts of impedance: source (or output) and load (or input) impedance. The figure commonly quoted for a speaker is its 'input' impedance, which indicates how much of a 'load' it presents to the amplifier that's feeding it.

As we mentioned in Chapter 5, the destination 'load' impedance should be at least ten times higher than the 'source' impedance – in this case the amp. Power amps usually have very low output impedances (below 1 ohm, and typically around 0.02 ohms), which means the average speaker impedance of 4 or 8 ohms safely satisfies the 10:1 ratio guideline.

The higher the load impedance figure, the less the actual load, so the easier the speaker is for an amp to drive – in theory at least. In fact power amps have a maximum output current and usually a safe minimum load that they should be connected to.

If the speaker impedance is too low, and the amp is driven at a high level, the amp will either protect itself by shutting off, or underperforming, or else overheat, start clipping and damaging the speakers, and possibly expire.

(See also 'speaker impedance & power delivery' in Chapter 5, and also the 'power handling' panel over the page)

and the recycling of older cabs – putting new or 're-coned' drivers into them etc.

Full DIY advice is beyond the scope of this book – some drive-unit makers offer plans or booklets of basic designs, or you can check in relevant magazines or websites. But it is still useful to know about different types of drive-units and horns, to have a basic feel for what's behind or inside the units you are buying or using. The following serves at least as an introduction.

Drive-units/drivers

The drive-unit is the main 'motor' part of the loudspeaker. It's required to make sound from electrical signals. The most practical, universal type is a paper-coned chassis drive unit.

At very low sound (SPL) levels, a single paper-cone drive-unit (or 'chassis driver') can just about cover most of the audible range – for example in earpieces, headphone drivers, small transistor radios etc. But as more sound level is demanded from a paper-cone driver, it gets strained at the low frequency end. That's because any cone has to move disproportionately further to reproduce lower frequencies at a given level. First distortion rises then, if pushed too far, the cone hits the stops, or rips.

Most drive-units, when worked hard, can cover a maximum of about three octaves. Exactly where these octaves lie (in other words whether it's a treble, mid or bass driver) depends on the design. If you bear in mind that the average human hearing range is between 10 and 12 octaves, you can see why most speakers struggle to reproduce

more than a fraction of the full range.

Most high-quality, high-power-delivery sound systems use different drivers in at least three or four separate frequency bands (the frequencies are divided by means of a crossover, as described in Chapter 4).

In the light of this, the use of just two drive units in many domestic hi-fi speakers can be seen as economic and idealistic. It explains why they break-up (or simply break) when any high level – particularly bass – is attempted.

Once it's accepted that drive units work at their best over a three or four-octave range, the search is then for the best technology for each range. There is some consensus on this, based on many years of experimentation and technological advances.

Most speakers struggle to reproduce more than a fraction of the audio range

Mid-range (mid, mid-frequency, MF) is the most straightforward to reproduce, because there are no extremes frequencies to deal with. Here, a small-to-medium-sized (5–12in) paper-coned drive unit is unbeatable. Other approaches have been tried, such as compression drivers – but sonically these are best kept to the upper-mid or, even better, purely high frequency.

Other types – electrostatic and ribbon/panel

Power delivery & handling

As we mentioned back in the book's Introduction, it's a misconception to believe that speakers 'have' power. Even in so-called 'powered' or 'active' speakers, it's a built-in amp that delivers the power, rather than the speaker itself. When we talk about 'powerful' speakers, we mean to say they have a high power-handling capability – they can cope with a powerful amplifier.

Logic might seem to suggest that a big powerful amp would inevitably cause

damage if it was connected to under-rated speakers. But in fact, as long as you're not excessive with the level, and as long as the impedances are matched sensibly (see Impedance sidepanel), the speakers will be happy just to use what power they need, and the amp will operate with a comfortable 'headroom'.

What tends to be more often a problem is driving a large, power-hungry speaker system with a weedy underpowered amp. A voracious load like that demands to be fed with high amounts of power,

which a low-powered amp cannot deliver. The result is very much an inadequate sound and a damagingly overheated amp.

In fact, a generally acknowledged rule-of-thumb among PA professionals is that amplifier power-delivery ratings should be between five and six times higher than the speakers' average power-handling ratings. So if your speakers are rated as able to handle a total of 400 watts, your amp(s) should be capable of delivering at least 2kW.

drivers, as used in top-end hi-fi – are direly inefficient, can't handle much level at any very low frequencies, and don't lend themselves to horn-loading (which we'll explain in a moment).

For low frequencies (LF, bass, bottom end), larger cone drivers are generally the optimum. These range from 15-inch (nominal diameter) up to a goliath 24in. Large cone diameters are required to curtail the otherwise extreme distances the cone has to travel to create useful sound levels at lower frequencies.

For instance, to produce a given sound level at a particular bass frequency, a 12in cone speaker would need to travel about three times further than a 24in unit.

But the distance of travel is limited by practical mechanics, including the stress on the surrounds (the 'necks' at the inner and outer edges of the speaker cone). 'Long-travel' of a speaker, as seen on cheap hi-fi drive-units, also creates unpleasant distortion, disrupts the power-handling capacity, and actually makes the driver more vulnerable to damage from any excess motion (as caused by the noise of dropped microphones, breath pops, etc).

The crucial features used to describe a speaker are size and power-handling

---

Instead of trying to get low bass from a 10in driver unit – which might have to move 0.5-1in (12-25mm) to give useful low bass levels – the PA approach is to use the largest drive-unit that's fast enough for the frequency range concerned. A 24in driver, say, would need to move only about an eighth of an inch (3mm) for the same result in the sub regions (say below 70Hz).

For the larger sizes of drivers (18in and above), a cone material other than paper – such as plastic – may be used to improve rigidity while not increasing weight/mass, which would not help acceleration. On the other hand, higher up in the mid-range, plastic cones don't work.

An alternative approach to delivering low-frequency sound is to use multiple smaller drivers (as with a 4x 12in guitar or bass guitar cab). This creates a 'faster' overall bass drive-unit, since the individual cones are lighter. On the down-side, it costs more for a given result, creates more chance of mis-wiring or other errors, and can occupy more cabinet frontage (take up more space at the sides of the stage and in the trucks).

Chassis speakers (cone drive-units)

Chassis drivers are used in all sorts of cabs – horn, vented, and direct radiator. Although chassis drive-units all look similar at first, there are countless variations in magnet size and strength and the magnetic circuit shape, plus cone profiles, cone surfaces, surround elasticity and damping, voice-coil diameter and length and materials, and the width of the gap the voice-coil moves in.

The crucial features, and the ones generally used to describe a speaker, are size and power handling. Commonest sizes are 8, 10, 12, 15 and 18in. The voice-coil sizes may also be described – diameters range from 2in to 6in and more. The larger sizes generally handle more power.

When looking at 'continuous power handling' figures (expressed in watts) you need to be wary – this is an area where there's scope for makers 'being economical with the truth'. It should usually be rated according to either the AES or

---

'Power' compression

When drive-units are worked towards their maximum power ratings, their voice-coil temperature is raised high enough to increase the wire's electrical resistance. This increased resistance (which lasts as long as the wire stays hot) 'taxes' the energy supplied to drive the cone. The upshot is that, at high levels, the sound level doesn't increase by as much as you'd expect when the signal (drive voltage) is increased.

In effect, when pushed towards its limits, every drive-unit acts like a 'soft-knee' compressor. The overall result is not necessarily un-musical, it just frustrates those who need to squeeze every last dB of sound level from the PA.

Although not harmful for short periods, you are approaching the driver's limits. The best you can do is reduce compression levels, so the signal is less dense, which will permit higher levels (in bursts at least).

speakers

EIA standard (the test signal should be pink noise, treated to simulate fairly dense music).

'Program' power rating has to be derived after the continuous rating is known. The result varies with the nature of the music, so it's not very useful on its own, at least without qualification as to the assumed 'crest factor' or peak-to-mean ratio (see Chapter 2 & Glossary). Generally it is taken to lie somewhere between three and ten times the continuous rating, based on crest factors (PMRs) of between 10 and 20dB.

As a measure of efficiency, sensitivity is around about 100dB SPL @ 1 watt @ 1m for a direct radiating 15in speaker. This rises to about 115 to 120dB SPL when horn-loaded.

Compression drivers

Compression drivers are usually compared by their throat diameters: from small – 1.75in to 2in (51mm) diameter; to medium – 3in (75mm); and large – 4in (100mm). Generally the wider-throated drivers have large voice-coils, handle more power, and give lower distortion for a given sound level. But smaller-throated drivers may give a wider or smoother top-end response.

Even so, many driver-horn combinations may exhibit a gradual roll-off of typically 3dB per octave above 4kHz. This is readily corrected with EQ, provided you have the headroom, hence amp power capability.

The best compression drivers are about 30 per cent efficient; a typical 'high-end' hi-fi cab, by comparison, is about 0.3 per cent efficient – 100 times less. 1W will typically give between 115 to 125dB SPL @ 1m distance. Of course in all cases the driver must be coupled to a matching horn.

As well as being affected by the size of the driver, the sound is also characterised by the diaphragm material. Metal ones can sound 'metallic', as you might imagine. Phenolic (plastic-coated cloth) types can be more musical sounding. Unfortunately, for most units only one of these types can be fitted.

Horns & horn-loading

So far we've looked at the reasons why certain drive-units are used in PA systems. But making a

usable PA speaker system requires some other rather major steps to be taken.

We could simply take those optimum bass and mid drivers, choosing the ones with the most power-handling, fit them into overgrown hi-fi-like enclosures, and plug your power amp output straight into them. But what would happen?

As well as needing an unfeasibly large number of outsize cabs and racks of amps to drive them, the dispersion of the sound would vary wastefully over the frequency range, from being omni-directional (wide-beamed) at the low end of each driver's range, narrowing to nearly a torch beam at each driver's upper end.

Of course, with domestic hi-fi these problems do not exist, or just don't need to be faced. Home listeners can choose the best seat in the room, can use any number of huge amplifiers if they wish, and in any case wouldn't expect their speakers to reproduce concert volume levels (without breaking).

Horn-loading improves speaker efficiency by at least ten times

This is where horns are needed. A horn is a flared tube or passage that creates a perfect coupling between the drive unit's radiating surface and the venue's mass of air. (The idea is adapted from the basic horn speakers used in cinemas back in the early days of 'talking pictures'.)

Horn-loading any driver means it can be controlled better, and improves 'efficiency' (the way a speaker converts electrical power to sound power) by a factor of at least ten times.

The cost of doing this is far, far less than trying to increase the system power handling by ten times (by using amps of higher power delivery capability, and either drivers rated to handle the higher power or, even costlier, more drivers).

Horn-loading has another very useful property: dispersion or 'beam-width' control. This means as we move through the three octaves that a typical drive unit is asked to cover, this coverage, while not perfectly stable, varies far less than with a direct-radiating (non-horn-loaded) cone. This is particularly true with mid and HF horns, which can be designed with beam 'angles' from 30-90 degrees .

Another way of looking at horn dispersion or beamwidths is 'short versus long-throw' – wide dispersions (90 degrees and above) are naturally short throw; while long-throw dispersions are generally 30 degrees or below – though these could also be 'spot' used for more local, narrow infilling.

Another feature of horn-loading is that the driver works under compression. As a result, whether the driver is a cone type (chassis driver) or a compression type of driver, the radiating surface will only move very slightly compared to the movements it would make in a direct-radiating situation.

Compression drivers made solely for horn-loading aren't designed to make much excursion at all – since they don't need to. For this reason compression drivers can't (and shouldn't ever) be driven at any level without being first connected to a horn.

The small movements help reduce several inter-related kinds of speaker distortion (Doppler, intermodulation, cone break-up sub-harmonics, and ordinary harmonics). This can result in sound so clean that it probably doesn't seem as loud as it really is.

Because bigger horns are needed for lower frequencies, by the time we get to the bottom end of a PA stack, normal horn-loading would result in an excessively large and unwieldy bass unit. So 'bass horns' sometimes use an internally 'folded' horn system to reduce the depth of cabinet/bin required.

Spread versus focus

The dispersion of sound from a driver affects not just what's heard at different parts of the venue but also the overall power. If sound is radiated in all directions, not only is much of it wasted (often annoyingly, such as when unwanted spill affects musicians on stage, or angry local residents outside), but also the maximum sound level in any direction is lower, as there's less sound to go around. In short, it's an inefficient use of a speaker's energy.

As we've just seen, horn-loading offers a useful dispersion pattern (with a choice of angles) which is relatively constant over each driver's range. This also means the speaker system's efficiency is controlled and fairly consistent, as we're able to point sound mainly where we want it (at nearly all frequencies). Effectively horns offer far better projection than direct-radiating cabs, delivering focused sound power.

You can find out about the directivity of individual horns and PA cabs from 'polar plots' (similar to the ones used for microphone response patterns) on makers' data sheets – either ask for them or see if you can download them from a website. These show changes in the relative sound level in all directions from the horn's mouth.

Of these, the horizontal coverage (the dispersal width and distance the sound is 'thrown') is generally the most important. Indeed, in many venues, only a few degrees of vertical dispersion is required, as most people will be standing or sitting at roughly the same level.

More vertical coverage is obviously required if the floor is 'raked' (getting higher towards the back, as in a theatre or cinema) or if some of the audience members are on one or more balconies.

SPL (sound pressure level) at distance

When looking at data sheets of cabs and/or drive units, you'll see that the sensitivity of individual drive units (in suitable enclosures or horns) is commonly specified by citing the sound pressure level (SPL) at a distance of one metre in front of and on axis to (directly in line with) the radiating surface or horn mouth, when using one watt of electrical drive (this is written as XdB@1W@1m).

But for systems with more than one sound source (more than one PA stack, for instance – see sidepanel on p122), the level has to be tested at a greater distance. This needs to be the nearest point beyond which it is judged that the wavefronts have come together. A fair guess is about one-and-a-half times further from the cabs than the distance between the speaker stacks.

This all makes it easier to understand maker's system specifications – for instance when they say, 'Max SPL @ 30m (or 100 ft) = 110dB continuous, 120dB peak, coverage angle 60 degrees (h) x 40 degrees (v)' (where h is horizontal and v is vertical).

121

speakers

You might then check (by walking around at a listening demo) to see how well that coverage angle and the SPL hold out at the frequency extremes, particularly the bottom end – since dispersion control is harder to achieve at low frequencies. This is because both bass horns and direct-radiator cabs radiate increasingly widely at low frequencies.

Eventually the radiation becomes omni-directional (spherical). Even so, bass bins always have more focus and direction than direct radiators at any given frequency.

Dispersion control is harder at low frequencies

When two or more bass bins are set together, the effect is quite synergistic – in other words they mutually reinforce each other, and also become more focused. PA companies can take advantage of this effect by using bass bins with smaller mouths than they should ideally have for efficiently reproducing soundwaves below 100Hz.

In other words, when four cabs are stacked together the overall increase in mouth area means the system's low-frequency capability would be extended – effectively doubled in this case, say from 100Hz down to 50Hz. That's quite a major improvement.

It also shows why you might use four single bass bins (or two doubles) in a small P, to achieve high-level low bass – even if you don't need to use all the power-handling capability of the cabs.

PA size

Even though, as we've seen, small speakers stacked together can be more efficient than one huge speaker, PA users still face the unavoidable fact that in order to get adequate sound levels for a large gig or venue you will need lots of good-sized speaker cabinets, of all types. In other words, multiple sound sources.

Here's the problem: when cabs are stacked together to get more power, the resulting sound can often be horrible, and possibly even a lot quieter than you'd expect. It's a classic dilemma of PA, which has not yet been fully solved – although it's an area where there are vast differences between systems.

The terrible sound and power 'fallback' are mostly caused by mutually destructive interference between each cab's soundfield (also called 'phase cancellation' and 'comb-filtering' as the peaks and dips can look 'comb-like' on a response printout or oscilloscope). Unlike solid matter, soundwaves can be cancelled out, and mutual interference ultimately leads to this.

This in turn leads to a wild and ragged frequency response, which varies as you move across and past the field where two or more sound sources overlap. The quality only recovers in the areas where only one sound source is contributing.

The outcome? One third of the audience hears reasonable sound while the other two thirds have their ears hurt and/or go home hating the act and 'that sort of music'.

It's also incredibly wasteful of power, and money... 'Power fallback' means it is possible to pump kilowatts of sound energy into the air and

---

Single & multiple-source

In the context of speakers, a 'sound source' can refer either to one speaker cab or a group of them. To complicate matters further, 'single-source' does not have to mean strictly just one speaker – it could refer to a stack, if it's below a certain size:

- At HF (top-end) – if there's only one drive-unit and its horn per side of the PA.

- At mid (MF) – only one horn + driver, maybe two very close-seated drive units per side (at a pinch). Preferably driving into one horn, maybe two if very close seated, and then only at a pinch.
- At bass (LF) – as a maximum, a compact array of, say, two or four cabs per side, with no air gaps between cabs. Each cab could have two or more close-seated drive units.
- At sub-bass (below 100Hz) – a

maximum compact array of say eight or nine cabs per side (2x4 or 3x3).

More units than this means it's a multiple-source PA stack. As you'll notice, the limits on what constitutes a single sound source are relaxed at lower frequencies, where a number of cabs can count as one, provided the horn (bin) mouths are very close together relative to the lengths of the soundwaves involved.

for most of it to be cancelled out – it means you can increase a PA rig from 2kW to 10KW and find it's still not a lot louder.

This was a common drawback of the brute-force 'wall of sound' PAs of the past. Even backline technicians have learned this – so in fact today's walls of instrument amp cabs may well be mostly empty 'dummy' cabs, for appearance only.

The problems are worse when cabs are widely spaced, and they get worse the more sound sources there are trying to 'talk' into one sector. In fact it increases exponentially with larger and larger PAs which need more and more cabs. (The possible exception is when some of a PA's 'power enlargement' is achieved with adequately-spaced delay stacks – see Chapter 4 on delay lines.)

When cabs are stacked together, the sound can lose quality and level

There are particular ways of avoiding these problems, and making multiple sound sources actually work together to increase levels in each sector – for instance: closeness of cabs to each other, relative to the frequency band, with no gaps between cabinets or complex surfaces between sources; curving the cabs in one plane (a curved array); or curving the cabs in two planes (omnispheric).

The last three approaches were pioneered and tested over a number of years by Turbosound, starting at the UK's Glastonbury Festival from the 1970s onwards, closely followed by the first curved point-source array in the US, by Meyer in 1981. The idea of a curved array is to arrange multiple speaker cabs so they resemble a segment of a sphere, and by doing so they simulate a single-point source of sound originating some way behind the PA.

With hindsight it seems an obvious solution to the multiple-source nightmare, and nearly every major maker across the world has since taken up with their own interpretation of this idea.

But 'arrayability' means different things to different manufacturers – it's not enough just to create attractive shapes using any number of partnering cabs. Yet as there are no world standards in this area, and little independent testing, once again the best judge of effectiveness has to be your own ears

Compact PAs

To be useful, PA speakers – however sonically good – also have to have practical attributes. Size and manageability are among the most important, though these shouldn't be pursued at the expense of all else. For instance, it's pointless making a PA cab more compact (maybe just involving a slight squeezing of the cab design) until you have made it as efficient as possible (which may involve a major design change).

'Compact PA', particularly for larger systems, took off when some makers (originally in the US) began to integrate horns or drivers for different frequency bands into one cabinet – creating what are known individually as 'full-range' cabs, and collectively as a 'one-box system'. Before then, all PA systems had comprised stacks of horns covering distinct ranges. Some were very untidy, some just complex.

Rather more ramshackle (or 'visually interesting', to be charitable) examples of old-style horn PA technology can still be seen at beach parties and countryside raves in various countries.

Getting the best from a small PA
Here are a few practical things you can do to make the most of a small FOH PA, mounted either side of the stage:

● Place it as high above the audience's heads as possible. The higher the frequency, the more this matters.

● Point the mid and especially the HF horns slightly downwards into the audience. The angle need only be subtle – say 15 to 20 degrees – and can be achieved with shallow wooden wedges, taking care that these don't make the cabs physically unstable.

● Make sure both sides of the PA are pointing fairly squarely at the audience.

● Make sure the cabs don't rattle or shake due to a wobbly floor surface. If they do, solve this temporarily using beer mats, or more permanently with a hammer and some shallow wooden wedges, or shims.

speakers

The one-box system – putting the bass + mid + HF horns into one enclosure – may not actually have saved much space or volume, but it did make a whole PA system quicker to ship, erect, tear down, pack, etc. This in turn saves labour costs – especially in US cities, where gear 'humping' labour has always been less 'casual', more organised and Health & Safety aware, and therefore costlier.

Tour budgets can be significantly raised by the length of time a show takes to load in and out again (especially if this involves working 'anti-social' hours). One-box PA systems, along with multicore speaker cables and other time-saving tweaks, have reduced rigging times from a whole day, or even several days for some large concerts, sometimes down to a few hours.

Compact PA configurations

Here are just a few examples of how all-in-one-box speaker systems are put together:

●Two-way = a 10, 12 or 15in + HF horn
●Three-way = 10, 12 or 15in + 10, 8 or 6in or HM horn, + HF horn.
●Four-way = 15 or 18in + 12 or 10in + HM horn + HF horn.

(Note – in some cases speakers may be doubled up in one or more frequency range; for instance 2x10, 2x12, or even 2x15in.)

Small standard FOH system

As we've seen, a PA with the minimum number of sound sources (one horn-loaded source per frequency band) should be the most efficient. But of course it won't be the loudest – sound level delivery is restricted to the power-handling limits of the few drive-units. With current speaker technology this means in practice about 800W of 'program' (music) on HF, 1.5kW on mid, and 2.5kW on the low end (where more than one driver can be used, as described already). In other words, a total system rating of about 3kW (an average of 1.5kW per side).

With horn-loading on all drive units, this is enough to cater for between 100 and 2000 people – depending as usual on the venue acoustics and size, not to mention the maximum sound level required (ie whether it's a genteel folk band or hardcore techno).

Of course, sub-woofer cabs (few enough to constitute a single source) may be added.

Medium-to-large-sized PA

This is the point where we have to move into using more than single sound sources per frequency range, and begin 'arraying' identical cabs. In some senses it's simply a development of what's been said about smaller systems, gradually expanding until you reach stadium proportions, which brings in the concepts of delay towers (see Chapter 4) etc.

At this level, a rule of thumb for flown, 3D point-source clusters is that in every seat/listening position you should be able to look up the mouth of an on-axis horn, at least for the mid and high-frequency bands.

With outdoor FOH systems (for festivals and the like), there's more scope for hoisting heavier arrays (you can hire a bigger crane, whereas you can't replace a weak roof in a venue). You can also hoist higher.

Unless you're in a guaranteed rain-free zone, wet weather always has to be protected against. You can either spray the speaker cones with a waterproofing agent (based on modern 'hydrophobic' chemicals), or use a large enough sheeted canopy above the PA (and mixer) to keep off the worst of any rain.

A neat benefit of horn speakers is that the driver's cones are well protected from occasional rain, hail or snow, as they're set far back or even (with bass bins) round a corner.

If the point needs proving, it's worth mentioning what happened at a major rock festival some years ago, where a well-known British direct-radiating PA (where the speaker cones sit right at the front of the cabs) was caught in a pre-show rainstorm. The rain turned the paper bass cones to mushy pulp. Then the sun came out and started to dry them out. The poor sound engineer, enthused by the sunshine, then drove the system at maximum for the opening bars of a particular heavy metal act – and all 20kW of bass drivers were blown to pieces.

Speaker enclosures

Boxes are put around drive units for several reasons: to help with directing and shaping the emerging sound's wavefront; to avoid the soundwaves being cancelled by external noise; to help keep control of the drive unit's physical movement, and protect the parts inside.

Horns are the most efficient method of controlled projection, but as we've seen, aren't always feasible for low-mid and bass frequencies because of their inevitable size.

Horns aren't always feasible for low-mid and bass frequencies because of their size

Most 'budget' PA makers offer (or base their whole ranges around) vented cabs, also known as 'reflex', or 'ported' cabs. These use resonant techniques to create some added 'pretend' efficiency in the mid-low bass regions. But in the regions below this they have far less bass power capability, and also have abysmal projection. So they will need to be driven harder, which produces less pleasant sound, and they will be more prone to damage as a result. Fairly understandably, makers never mention this.

If you must use this sort of enclosure, consider building a DIY horn to fit on the front. This can comprise nothing more expensive, large or weighty than three sheets of plywood, hinged to provide a 'barn-door flare'. This will increase efficiency slightly, particularly at the low end, and also give a measure of directivity at the higher bass frequencies.

Bass isn't so fussy about smooth corners, so a horn on a bass cab can have quite an angular 'mouth' with a rectangular or square cross-section. But a good bass horn should be completely air-tight – the woodwork must be built to metal-type fitting tolerances, and all screws must be kept well tightened down. Using a bass bin can quickly start to take apart any shoddy woodwork. The higher quality sorts of cabinet (made of birch ply) should stay airtight after years of use and abuse.

Lack of air tightness will also initially wreck sonic quality and, if bad enough, allow the drive-

unit to lose control at some frequencies, leaving the cone vulnerable to being ripped apart.

Well-equipped PA bass cabs come with EP6 or other multi-pin connectors (as well as possibly Speakon or XLR sockets – see Chapter 7 for more on connectors), plus linking fly-cables (to daisy-chain to the next cab), and a compartment to fit the cable in.

At mid frequencies the 'headline rating' (the way the horn is defined) is firstly the dispersion angle and then the size of drive unit that fits the horn's mouth. The better-sounding mid-range horns are like bass bins in that they are coupled to cone (chassis) drivers. Some types are also like bass bins in that the horn flare is made of wood. A few ancient types (copies of a 1930s Altec design) try to be vented (reflex) cabs as well.

Most modern horns are moulded out of a suitably acoustically-dead plastic. The flowing plastic shapes used in some up-market PA cabs are not just for visual effect but are based on years of experiments into alternative electro-acoustics. They sit directly in front of mid-horn drivers, and are superficially similar to the 'phase plug' (see Glossary) regularly placed into the throat of compression drivers. The action is subtly different, though, and more multi-faceted in its benefits – which include smoothing, 'accelerating' and re-vitalising the sonics, and lowering distortion.

Cabs using these devices achieve a higher degree of sonic/musical quality than the norm, as well as good efficiency – it's rare to achieve both together.

With high-frequency horns, again the headline rating is first the dispersion angle (horizontal & vertical) and then the size of drive unit that fits the mouth. HF horns were originally made with soldered sheet metal, with sawdust in-between as a 'deadener'.

Some horns are still made with die-cast metal, which is economic but not always ideal for PA as some types shatter quite easily, and most are quite resonant. In the past 20 years, acoustically-deadening fibreglass 'flares' – pioneered in the USA – have become the most used type.

HF horns are also classed by their shape. The most common in PA – the radial horn – has straight sides with concave flares above and below. The typical dispersion (beamwidth) is 60 degrees x 30 degrees. Other types include the biradial, which looks like a radial on steroids, and is also affectionately known in the UK as a 'baby bum' ('baby butt', in US).

Flying

A major change in PA speaker systems in the past 20 years is widespread 'flying' of cabs. This usually means suspending a single array (apart from sub-bass cabs) centrally above the stage. This has the following potential advantages:

- Wider sightlines (more people can see the stage if there are no stacks at either side) – and consequently more sellable seats in some venues.
- More attractive stage visuals – many PA stacks and arrays are perceived as rather ugly by some set designers, and more 'up-market' clients (eg opera).
- Improved ability to point the sound downwards into the audience, rather than directly at rear walls – where it can create unpleasant-sounding 'slapback' echo.
- Reduced interference between the cabs – notably the low bass vibrations from the sub-bass cabs can't so readily reach the other cabs (where it can vibrate their walls and cones and create sonically unpleasant distortions and sub-harmonics).

Flown central arrays are acceptable because our hearing isn't overly sensitive to the height (vertical position) of a sound source. Alternatively, a conventional (L & R-sided) PA can be flown above the stage wings, with most of the advantages. This technique has even been used at outdoor gigs by suspending PA stacks from cranes.

The flying of cabinets – and arraying them precisely into 2D or 3D point sources – has certainly added to the hardware carried to gigs, as well as to rigging time. But it has also focused minds on safety. To be fair, any kinds of PA stacks are highly dangerous if they fall over – all flown or high-stacked PA cabs should be fitted with properly-rated and certified restraining devices.

Generally, flying involves either chains or steel ropes or cables ('steels') used to hoist and restrain cabs. Speaker arrays and the flying tackle are assembled into position on the ground and the cables connected. The system is then hoisted a few feet up, where it can be drawn into shape (if it's a 3D array) – and a check made that all the horns are functioning – before being fully hoisted up to its position. This is usually done with motor-hoists, and the entire job can be amazingly rapid when a skilled team are working on a well-designed flying system (also see Chapter 7, p144).

Monitor speakers

The purpose of stage monitors is not primarily to reproduce music as accurately as possible (unlike the FOH system), but to help artists on-stage hear what they're singing or playing, so they can perform accurately and confidently.

Another side-benefit, particularly with 'heavier' forms of music, is that performers get adrenalised by hearing themselves at a decent volume – they can 'get off' on the sheer on-stage loudness.

Each musician or singer will generally have their own wedge/floor monitor (more of which in a moment), and ideally a personal mix – see Chapters 2 & 8. Monitors are particularly necessary with larger bands playing loud and complex music, though occasionally DJs or other performers will also want a monitor.

To be useful on a noisy stage – amid competing backline cabs and other instruments, as well as the (confusingly delayed) output from the main PA – monitors need to have a special 'attention-grabbing' quality. This is known to wedge monitor designers as the 'oi' factor – a British exclamation which roughly equates to the American 'hey!', and just happens to be an ideal vocal sound for testing the effectiveness and audibility of monitors (actually much more useful than saying "1, 2, 3, test" – and more entertaining).

Different types of monitor

Wedges (otherwise known as 'floor monitors') are the most familiar type – low boxes with a sloping face so they can be placed on the stage-front with the drivers pointing fairly squarely at the performer. One or two of these are usually allocated to each performer (though occasionally solo singers may have almost a semicircular array on the floor in front of them), with an individual

speakers

mix sent to each (or to each pair/set, for example to backing vocalists). Smaller types (1x12in, 2x10in) are mainly for vocals. Larger types (2x15in) will cope with bass and loud percussion.

Sidefills (sidefill monitors) look similar to FOH PA cabs – some are exactly that, though some are older, semi-retired FOH cabs, and others are given modifications (like a bigger mid-range driver) to give them more 'oi' factor. Others are completely custom-made. (Strangely, sidefills are not usually seen in speaker catalogues.)

Facing across the stage from each wing, sidefills are directed at the band generally – a 'side-wash' effect. So they tend to carry a different mix.

Directivity of the mid and top horns in sidefills should be narrow (under 60 degrees), or else the fills will also project into the audience. This is less important on deep and tall stages. There's rarely anything that can be done about the bass end projecting sideways – unless you just filter out all the low frequencies which are not strictly necessary (using sweepable HP filter &/or bass shelving-EQ cut).

Drum-fills and keyboard-fills are usually specialised, bespoke (custom-made) full-range cabs with anything from two to six drivers, tuned to cut through and be intelligible in among high sound levels. Some drum-fills are notoriously loud – though they do need to be fairly high-level for drummers to hear cues from other instruments over their own playing. (One experimental adaptation involved mounting speakers into a specially-designed drum stool, so the drummer could experience the music from a new angle...)

Monitor drivers

Whatever the shape or type of monitor, the same variety of driver configurations can be found as in FOH cabs. The most primitive wedges have just a single chassis driver (cone speaker). This can be fine for instrumental 'trad jazz' and gentler folk bands, where most of the sound is in the low-to-mid-midrange, and high levels aren't generally required.

Otherwise, driven hard, this type of wedge will cause persistent feedback problems and/or lack of 'oi' factor. The 'oi' factor is particularly needed for vocals, and requires (at the very least) HF horns and partnering compression drivers.

The best drivers and horns are costly, but a low-cost HF 'bullet' horn (or other horn with

integral driver) is an advance, and often quite acceptable in less demanding monitoring situations. The more expensive drivers (both chassis and compression) will have smoother responses – helping you get higher sound levels (if needed) with less time spent EQ'ing. Compression drivers with 2in rather than 1in throats are preferred for their lower distortion at high levels.

Particularly in budget (two-way) systems, but also in multi-way set-ups, the HF part may be 'passive' (using a passive crossover – more of which later in this chapter). This saves an amp channel for every wedge. With some cabs, active or passive operation of the HF driver can be selected – you should generally use 'active' and carry the extra amp channels for louder jobs, and use the 'passive' setting for acoustic folk artists.

With vocals, guitars and brass, 1 x 12in or 2 x 10in drive units are very suitable for monitoring. They have frequency responses that usefully don't

Directivity of the mid/top horns in sidefills should be narrow, so they don't project into the audience

go any lower than the fundamentals of at least female vocals, and generally no higher than the highest vocal harmonics. Effectively these driver sizes filter out stuff you don't want, without using the mixer's EQ or filters.

For reasons of cost and portability, most wedges are nowadays reduced to being two-way active ('bi-amped') – though a few bespoke monitors continue to use up to five or six drivers and frequency bands (capable of delivering 'brain-damaging' levels of up to 125dB SPL between 40Hz and 15kHz).

In wedges, LF drivers are likely to be specialist parts for the job. That's because there isn't room in the box to contemplate using any sort of horn loading. And for a direct radiator, the cabinet volume isn't very big. Yet to allow bass and drum parts to be heard, a monitor must be able to cope with frequencies down to at least 100Hz. To

speakers

handle these jobs as well as vocals, the 'better' (bigger) wedges reach down to 45Hz.

Chassis drivers may also be specialised in other ways. Firstly, the shape of the magnet, as the wedge-shaped space may cramp too oversized a magnet. Second, any elaborately deep chassis fronts may be 'squared off' (cut away) to enable the cab's width to be minimised, or else meet a particular size target (for truck packing, perhaps).

The better wedge monitors also reach up in frequency to 15kHz. This range may not be strictly essential to musical communication and confidence, but is required more to balance the bass extension, and avoid ear fatigue (if there's not enough high treble, the monitor can end up sounding like a loud telephone).

There will usually only be one HF drive unit, and its dispersion needs to match the bass/mid chassis driver as far as possible. Typically, the chassis driver will have an increasingly 'beam-like' dispersion above 1kHz. Usually, this means a compromise, using a horn with a medium dispersion, say 90 degrees.

Commonly encountered types are 'bullets' (developed from the JBL original), the JBL 'pepper pot' (so-called because of its looks) and 'conical' horns. Turbosound have a triangular horn that gives the same effect. Wider dispersion (if absolutely required, though you risk having too much spill) has been achieved in the past using a 'crinkle plate' acoustic lens (slanted slats in front of a speaker). But this only widens the HF dispersion, which is least important as the HF has little 'oi factor'.

For sidefills, refer back to section on drivers for FOH cabs.

Self-mix monitor systems

Various self-mix systems have been created over the past 35 years. One was by the Grateful Dead, who to their credit spent serious money experimenting with PA. They placed their PA behind them, so they could also use it as their monitor system. Different parts of it covered the different players. As they played, an 'arms race' developed – one of the band would turn up, unbalancing the sound for everyone. As the concert progressed, the sound steadily increased in level while being randomly mixed by the individual players, who hadn't a clue what the effect on the overall mix was.

Another, more technically intricate system was created by PA designer Tim Isaacs (co-founder of Turbosound). A 'bus' system (where all aux outs were routed to all stations) allowed each musician on-stage to set up their own mix on their wedge monitor. Alas, the musicians were the weakest link – most couldn't cope with

Musicians can play better when they don't have to worry about things going wrong around them

twiddling knobs to control their mix at the same time as focusing on their own job, either while soundchecking or playing.

Musicians can play far better when they don't have to worry about things going wrong around them. Yet the risk with using any sort of powered monitor with a volume control, even if it is driven from a monitor mixer, is a stream of problems (howls, signals competing for level etc) which any competent regular monitor mix engineers would (hopefully) smooth out without any trouble.

A possible solution in a self-mix system is to set an active wedge's gain knob almost to maximum (9 or 9.5) then reset the system levels to make the monitor's level about as loud as that cab should go. The artist will then be mainly limited to turning their monitor level down.

Placing monitors

With wedges, the drive units need to be pointing fairly well 'on-axis' – squarely at the head of the musician(s) they're intended for. The slope of the speaker baffle (panel) determines how far away the wedge needs to be – for instance, how far in front of a singer's mike stand it is. This might vary depending on the height of the performer.

Steeper-sloped (more vertical-faced) cabs suit larger stages, as the cab has to be set further back – this also gives the artists more space to 'prance about' without going out of range of their own monitor.

Shallower slopes (so the driver points more upwards) allow the cab to be almost under the mike stand/performer. But this is not necessarily the best direction for the ears to receive sound from – nor the best direction for the mike to ignore it...

This is where a microphone's 'polar response' diagram can be particularly relevant: most PA mikes reject sound best from their 'dead rear' – where the lead emerges. Applying this knowledge in a monitoring situation helps keep the feedback threshold as high as possible. (And see Chapter 8.)

Assuming the mike is left in the mike-stand or generally held at the same angle, the cab ought to be some way behind the mike – possibly as close to the stage edge as you can get (at least on a smaller stage).

When placing sidefills, one or more cabs will generally be raised off the floor using a spare flightcase or the like (maybe plastic beer-crates, in a less professional set-up).

In-ear (wireless) monitoring

Although conventional monitors can sometimes work well, there are occasions when everyone concerned wishes they weren't needed. Common problems with traditional monitors include:

● Monitor mixes spill into many of the mikes, reducing the quality and clarity of FOH sound.

● Monitors are heard by the audience near the stage. The different mixes, strong EQ, and the 15mS or more delay in what's heard compared to what's coming from the FOH PA, isn't very complementary.

● If the artists are (by some miracle) able to play with nice quiet monitors, you still can't win – they then get distracted by the different (and delayed) FOH sound, which can produce a hesitant, unsure performance.

Some artists have been able to give up (or reduce their dependency on) conventional monitor by using wireless systems. These effectively operate like radio mike systems, only in reverse. High sound quality with high levels and reasonably deep bass (and stereo) are achieved with pairs of high-quality ear-pieces. At pro level, these are specially designed for each individual artist by first taking a mould of the ears and creating customised ear-pieces that fit neatly into the ear.

Other artists prefer to keep using ordinary monitors (for instance on bass) so they can still get the visceral 'feel' of the low frequencies. (And traditionally the vocal monitor gives a posing frontperson something to rest a foot on.)

Additionally, or alternatively, artists' mixes may include a signal from an 'ambience' mike, which helps overcome the feeling of isolation which using earpieces (or headphones) can give.

Of course among the first musicians to wear headphones on-stage were drummers using click-tracks – often necessary when playing along with sequenced backing tracks or parts.

In-ear monitoring receiver & transmitter

Performers wear a belt-pack receiver, with the aerial/antenna discreetly built into the earpieces' (or headphones') cabling. The transmitter is sited in the stage wings. As this can have a much better aerial/antenna than a radio/wireless mike, the coverage distance is typically wider.

In-ear monitoring brings down overall sound levels and monitor spill

The transmitter's drive-level, typically shown by a three-LED bargraph, can be seen by the monitor engineer from some distance. That's important because the transmitter may not be optimally located, if not sited conveniently at the monitor mixer. Multiple units have to be spaced out at least four feet (1.3m) apart.

The incoming feeds are stereo. That's nice, but might require a monitor desk with more sends, or better, stereo sends.

Graphics or other EQ shouldn't be needed, as there should be no feedback problems. If EQ is needed, it would have to be a stereo unit. The monitor mixer's own send-EQs should cope with artist preferences.

General benefits of in-ear monitoring

● It brings down sound levels on-stage – so you don't get the 'everyone trying to be louder than everyone else' effect.

● Spill from monitors is reduced or avoided.

speakers

- As a consequence, the FOH sound 'tightens up'.
- Avoids the frustrations of dealing with ordinary monitors – the menace of feedback, for one.
- Fewer wedges/floor-monitors on-stage (or none at all) can help visuals for some shows.
- Fewer monitor cabs and amps means less overall weight of gear to transport, making considerable savings on fuel and trucking costs.

There are some (comparatively) minor frustrations to watch out for, though:

- Earpieces are small and easy to lose (and expensive to continuously replace); belt-packs can get lost or stolen when interacting with an audience.
- Plugs breaking or falling out.
- Flat batteries.
- There's also a worry among some people that constantly wearing earpieces may be damaging to hearing, either through long-term high levels or accidental 'zapping' with an excessive blast.

In fact, general SPLs should be no greater than if using monitors (in fact they can be much less, as the signal is more controllable and directable), and as for avoiding sudden 'blasts', radio/wireless transmitters should have effective compression and peak limiting to be legal for use.

(For more on radio/wireless systems, see Chapter 1.)

Passive crossovers

Active crossovers were covered in Chapter 4 – they live in the drive rack, and their purpose is to direct signals at particular frequencies to the drive-unit dedicated to handling that range.

Passive crossovers are a separate species, usually only used with HF drivers, mainly in bi-amped systems, where they create a 'three-way split'. Or they may be used with 'supertweeters', which may make an occasional appearance in five or six-way systems to extend the existing HF driver's uppermost range.

Having read about how great active crossovers are in Chapter 4, you may wonder why passive crossovers are used at all. It's basically because, for HF and super-HF, they're simple (both in the number of parts required and the logistics of using

cabs with them), low in cost, and fairly effective.

They can also just about be suitable for use with high-power PA systems, for the following reasons:

First, the type of filtering required for deriving the treble from the mid+treble signal is high-pass (HPF). This can be far more efficient because the signal passes through a series capacitor, which causes lower signal losses (through resistance) than an inductor in a low-pass filter. Instead, the coil is only connected across the resulting signal, on the tweeter's side.

It's also possible to get away with not filtering the treble from the mid-range signal. The mid-range driver won't produce treble very well, but sounds OK and isn't harmful to the driver.

So, in simple systems, the mid is said to be 'run out' – meaning it's left to its own devices with the high frequencies. This avoids adding losses into the mid speakers' connections.

So there needn't be a (passive) low-pass filter in use, wasting any power, and power losses in the high-pass filter can be low.

Also, any losses are balanced by the fact that, in most types of music, treble signals don't generally have sustained high levels, so what small power losses there are will be concentrated in these peak periods.

Losses are also reduced by the fact that at least one passive crossover will be built into each cabinet, and may be allocated to one or two drivers at most. This means the crossover is used in a circuit loaded with 8 or 16 ohms (assuming 16-ohm HF drivers, which are the most common sort).

This in turn indicates a lower current will generally flow than is the case with the lower-frequency drive-units. So the losses due to the passage of high currents in the passive crossover's series capacitor are less than they could be.

Next, the HF driver's damping isn't so direly affected as it would be with lower frequency units, again because of the typically higher nett drive-unit impedances that are involved (8 to 16Ω).

Damping isn't very much of an issue because experience shows it's not a very audible effect with HF, at least in auditoriums – where perception of the directly-radiated treble sounds is affected, altered and even drowned by the acoustics, far more than in a small room.

speakers

Passive power sharing

The low average power in most music's high frequency also means it is possible for the HF to share the mid-range amp without putting any great added demands on it – such as heating it up, or loading it with too few ohms impedance.

Although connecting two 16-ohm tweeters in parallel to the amp (giving 8 ohms) alongside the mid-range (also 8Ω, say) apparently gives 4Ω, the nett impedance is actually nearer to 8Ω at most frequencies. That's because the capacitor stops the mid-range signal seeing the HF driver and the load it presents.

Meanwhile, the 'run out' mid-driver has a much higher impedance that its nominal 8 ohms at treble frequencies. So the overall load on the power amp stays around 8 ohms at most frequencies.

But you'll need to be aware of one little snag with these – and indeed all passive crossovers – which is that, around the crossover frequency area, the overall load impedance that the drive units and the crossover present may dip to a lower than expected value.

This means you might just hear a less good sound from this area, as the amplifier may be strained. You may well learn to use EQ with particular cabs at this frequency to reduce levels here, so the distortion effects aren't so obvious.

When you get to the stage of needing a -24dB-per-octave slope, it's really best to bite the bullet and 'go active' – which means extra cabling and amplifiers, but also means being able to drive the HF driver closer to its lowest allowable frequency without exciting it significantly below this point.

DC protection capacitors (caps)

These are placed in line (in series) with drive-units. They are superficially similar to the passive crossovers just discussed, but are only required for the drive-units driven from the main (active) crossover. They are not required for drive-units connected via passive crossovers, because the crossover series capacitor can neatly do this job at the same time.

These caps are fitted for two reasons. First, to help protect the drive-units in case someone muddles up the feeds of the different frequency bands (from the amplifiers' outputs, or even between the amplifier inputs and the crossover unit), and connects the bass to the HF drivers etc.

This would normally be a very unpleasant experience, as the HF drivers would be damaged beyond repair after the first couple of loud bass beats (as might some of the mid-range drivers).

This arrangement can't give total protection. You should still bring up levels on the crossover slowly at first, listening at modest levels for any strange sounds affecting whole frequency ranges. Don't 'slam in' to high levels until you're dead sure it's all connected and right and sounding OK.

If you've read Chapter 5, you'll recall 'DC fault indication'. Not all amplifiers have DC protection, and even on those that do it doesn't always work. The occasional amp has a fault – like a relay that gets stuck on – where the protection fails to operate (even though the LED claims it has). So 'DC protection caps' provide 'belt & braces' protection against the rare (but occasional) event of an amp having a DC fault. If this ever happens, you'll just hear a buzzing from the speaker, and no signal. But at least no smoke (from the DC burning the driver's voice coils) or other damage.

DC protection caps are most often fitted to HF drivers and cabs

What they mustn't do is upset the main and highly accurate active crossover filtering. This is avoided by making the capacitors' value larger than it would be for crossover duties. This means the capacitor acts as a crossover at frequencies some way below the actual crossover frequency, so it protects the drive unit from the worst of any much lower frequency signals.

DC protection caps are most often fitted to HF drivers and cabs, less to mid-range, and even less to bass or sub cabs. This is mainly because of physical size and expense, but also because the lower frequency drivers are less quickly damaged by a DC fault. So even amplifiers with DC fault protection that takes a few seconds to work (which is not uncommon) will protect these types OK.

putting it together
connectors, polarity etc...

Quick view

All the (relatively expensive) PA gear we've been looking at is useless unless you have the right connectors (cables and plugs) to join it all together. These often-overlooked items are almost certainly the cause of more problems than any other parts of the PA system. This chapter looks at the type of connections commonly used – plus other vital but easy-to-ignore matters such as how to make sure everything's wired in a compatible way (polarity, or 'phase-checking'), turning this mass of equipment into a manageable, smooth-functioning unit, ready for the soundcheck and gig.

Connectors

Cables and plugs are very often the weakest links in sound systems. However strong and well-made when new, the connections inside will still weaken over time, or succumb to the inevitable abuse and accidents that occur in real-world use.

Just one broken connection inside any cable in the system can be enough to cause a major fault. And also, by its nature, the broken connection will nearly always be hidden (inside a cable or plug), so special skills and or/tools are needed to locate it quickly.

And because cable (or 'lead') failures can happen at any time – often unexpectedly and at the worst possible moment – it's very useful to have spares of each type of lead you are using.

Main types of cable used in PA

Mike and line leads – those used for connecting microphones and guitars on stage have to be very durable and also highly flexible. While they can be quite thick, they can fortunately remain lightweight, as only a small amount of copper wire is needed to pass the signal. Thinner, less rugged line cables can be used between equipment, both inside and even outside of housing racks.

Speaker cables (leads) – used solely to connect between amplifiers and speakers (bins, horns), these are generally quite thick and heavy, as they need to contain quite a lot of copper to handle high power without wasteful losses.

Power/mains cables (cords) – a lot of lightweight PA gear relies on pluggable 'IEC' cables – similar to the ones sometimes used for electric kettles. Some power amplifiers have similar-looking but larger IEC connectors which can handle higher current. Even larger, pluggable cables used to distribute power are commonly coupled together using 'C-Form' connectors.

Multicore cables ('multis' or 'snakes') – are used to connect between the stage and the mixing location. Multicore cables/snakes comprise large groups of line-level leads bound up in a flexible sheath-covered tube. The overall cable is thick, heavy, hard to bend – and also quite easily damaged.

Balanced and shielded cables

The weak, low-level signals sent down line and (especially) mike leads need to be protected from contamination by external electrical noises. The most common of these are hums or buzzes (effectively tones and harmonics of the AC mains frequency), and strong radio (RF) signals. Protection takes several basic forms.

First, shielding of the signal wire(s) with some sort of metal surround. This is routine with all cabling used for mike and line connections. The shield provides some protection against RF, but little against many hums and buzzes.

High-quality cables are shielded with either a solid shield of aluminium foil (for rack use only) or else a tightly-woven wire 'braid' or a conductive plastic tubing (these two types will withstand stage use). These will reject RF noise better than low-cost cables that use a 'screen' made up of a single layer of side-by-side wires twisted around the inner core. With these types, the shielding quickly opens up with use, so it develops holes which will let through RF noise, certainly at high RF (such as TV transmission frequencies).

Another way of helping keep down noise is to twist two or more signal wires. This provides some protection against hums and buzzes caused by pickup from (for instance) nearby mains cables or transformers. But it cannot reduce hum separately caused by earth loops or gear defects.

A more effective option is to use balanced connections – something most professional PAs have done since at least the 1980s. In a balanced connection the signal is sent 'push-pull' down a pair of wires.

Using balanced connections is the only safe and reliable way to get rid of the ugly hums and buzzes caused by (earth/ground) wiring loops. You should *never* disconnect a piece of gear's safety earth/ground connections in the power plug.

In PA systems, balanced connections are generally used in addition to shielding and twisting, and together they help keep external noise to a minimum. A properly-wired modern PA system will employ all three methods – except in speaker cables, which are only rarely shielded.

Using balanced connections is the only safe and reliable way to get rid of hums & buzzes

If some gear in a PA has only unbalanced in and out connections, try to avoid using that piece of gear if possible. If it has to be used, bear in mind that:

- The unbalanced input cannot be driven by a balanced source. To 'unbalance' the source, link pin 1 to pin 3 at the unbalanced input's socket.

- An unbalanced output can drive the next gear's balanced input. Just make sure that the unbalanced output's 0v (low side) is connected to pin 3, and that only the casework of the unit is connected to pin 1.

As every maker follows different connection

Compatibility

When you come to put your system together, it'll soon become clear that the PA industry is not a 'joined-up' business. Makers of the different parts of the system (who may rarely make, or even try, the other parts) may not have been working with other manufacturers' products in mind – or might perhaps make assumptions about how the other gear you need will work. In other words, there's always a risk of incompatibility in any 'mix-and-match' PA system (which of course describes most set-ups). Often the first people to test the uncertain results of connecting units X + Y will be the end users, the poor sound engineer.

strategies, these procedures may not work, and further fiddling may be required to get a noise-free signal.

Because feeding an unbalanced output 'up' a balanced one is much less hassle, this form of half-way house – gear that has balanced inputs if not outputs – is to be preferred, if there's any choice.

Sometimes with balanced connections hums, buzzes or RF occur which can be stopped by breaking the shield connection at one or other end. This is because loops have formed into which mains or radio frequency noise is induced.

In theory, it's better to break the connection at one end for one reason and at the other for another reason. On the returns multi/snake feeding the amps, there is only one end to break it – at the source. Otherwise, if the multi is unplugged at the sending end, it could be deafening, loosing 100 yards of shielding on all the amp feeds.

In other cases shields are best lifted at the reception (load, destination) end. So between two items of gear, a 'shield-lift' XLR-XLR cable is fitted, where the shield connection is cut, insulated and folded inside the male XLR plug.

One feature of balanced wiring – and also the 'floating' wiring of speaker connections – is that if the wires carrying the signal are 'flipped over' (cross-wired), then unlike the situation with unbalanced wiring (as used with most MI and hi-fi gear), the signal isn't disabled, but carries on – although this 'reverse polarity' condition can cause a number of sound quality problems in a PA, as we'll explain later on in this chapter.

Speaker cable

Mostly these anti-noise details don't apply to speaker cables: firstly because the signal levels are generally a lot higher at this stage, and so less vulnerable to interference; secondly because the signals are already at their highest level at this point, so any added noise won't be further amplified.

Occasionally speaker leads are the cause of radio interference, but only because they feed the RF noise back into the power amp they are connected to. The simplest solution may be to change the amplifier or have it rectified or modified so it isn't so susceptible to the problem.

Alternatively, or additionally, you can use shielded speaker cable. Usually this will have an overall shield, like a balanced line cable. This shield has to be directly connected to the amplifier's metal casing – which may cause some headaches when some sorts of plugs are used. It may also help to use speaker cables with very closely-spaced cores, including four-core cables wired for two cores.

Types of plugs & sockets

XLR (or 'Cannon')

The XLR connector (with familiar lay-out of three small round pins/holes forming a triangular pattern inside an 18mm diameter circle) is universally used for mike and line connections throughout PA systems. The family comprises latching (locking) male & female connectors, which either fit on the end of cables or can be chassis-mounted as plugs or sockets. Male is always used for outputs, and female for the receiving end, at every stage in the sound system. Standard cables always have one male and one female end, which means you can readily and neatly join together similar types of cables.

The ruggedness and overall reliability of the most widely-used brands of XLR has been long proven – for instance Cannon (the original), Switchcraft and Neutrik. Low-cost XLR 'copies' are best avoided – there's a lot more to making an XLR connector than copying what it looks like.

XLR pins & wiring

When looking at any male three-pin XLR with the middle pin underneath (so the three pins form a downwards-facing triangle), pin 1 is always on the left. Conversely, on a three-pin female plug, it's the one on the right. Pin 3 is always the 'odd one' – at the tip of the triangle.

Unlike other connector makers, most makers of XLRs are thoughtful enough to put numbers next to some or all of the pins, so no guessing is required.

The world standard XLR wiring convention is as follows:

Pin 1 = Shield (drain or ground wire)
Pin 2 = Signal hot or '+'
Pin 3 = Signal cold or '-'

Some older equipment is wired with pin 3 hot

and pin 2 cold. This polarity flip has no individual effect, but will cause endless trouble in bigger systems where amps with one sort of wiring are mixed with the opposite sort. It is really best to have all inputs rewired to the 'pin 2 hot' standard. Note that the above wiring method may be used even if the destination is not balanced.

XLRs may also be wired for unbalanced use (with line, or sometimes even mike signals) using just pins 1 & 2 for shield and 'hot'/inner. Sometimes pin 3 is linked to pin 1.

Unbalanced connections are not a good idea in modern PA. It is best limited to where some special (archaic) equipment wired this way is being connected, and where the hums and buzzes that unbalanced connections normally cause are not a problem.

It's also a bad idea because amps with XLR outputs and matching XLR speaker cables (see later) are wired in the same way. This creates a greater possibility of damage if any line and speaker level cables get mixed-up in the dark.

In the absence of a two-pin XLR, three-pin XLRS have long been used for speaker connections. Speaker level XLRs are wired like an XLR for an unbalanced line connection, except there is usually no shield wire in a speaker cable, hence:

1: negative/– (black or blue)
2: positive/+ (red or brown)
3: not used at all

XLRs with four, five or seven pins may be used (and should be kept) for special jobs – typical uses include intercoms and control cables. There are no universal wiring standards.

DIN connectors

These austere connectors (a 1950s German invention – the name translates as German Industrial Norm) are no friend of pro-audio. Their use in PA use is thankfully limited to musicians' MIDI cables (and to some intercoms and oddball control cables). MIDI cable wiring is as follows:

Pin 1:	n/c	Pin 2:	shield
Pin 3:	n/c	Pin 4:	+5V
Pin 5:	MIDI data		

Sockets are the same, except that pin 2 ground is omitted on a MIDI input connection.

Even the high-quality makes of DIN plug, which have crude latches, are extremely fiddly to solder, have primitive cable strain relief, limited cable access, and are held together by screws that readily drop out, or have threads that strip easily. Even the pin numbering is disordered.

In short, avoid DIN connectors wherever possible – or at least buy ready-made leads instead of trying to make/fix them yourself.

'Quarter-inch' plugs

Note: in the UK these are commonly called 'jackplugs' – although in the US, where the word 'jack' is used to refer to a (female) socket, they're simply known as either quarter-inch or 'phone' plugs (not to be confused with 'phono' plugs, which we'll come to in a moment).

There are several variations on the quarter-inch-size plug:

Two-pole: also called 'two-circuit phone' or 'mono jack', this makes only two contacts, with the 'tip' and the 'sleeve' of the plug (which is why it's also called a TS plug). The tip is obviously the knobbly bit on the top of the plug, and the sleeve is the rest of the plug, the lower part of the metal shaft.

This type of plug is widely used for 'MI' (musicians) equipment, and wherever low-budget, unbalanced audio connections are required. Because it can appear at mike, line and speaker connections, there is a high risk of causing damage by plugging things into the wrong places – such as mikes into amp outputs.

Quarter-inch plugs are widely used for low-budget unbalanced connections

Three-pole: in the UK called a 'stereo jackplug', otherwise known as a 'balanced quarter-inch' or 'three-circuit phone connector' (for reasons which will shortly become obvious). It can be used either as a balanced connector or for stereo – but not both at once. It makes three contacts – tip, ring and sleeve, so it's sometimes called a TRS or RTS.

The 'ring' refers to the extra section on the plug shaft between two thin insulating (usually black) stripes.

When used as a balanced plug, wiring is as follows:

Sleeve (as XLR pin 1) = shield, drain wire, ground
Ring (as XLR pin 3) = signal cold/negative (-)
Tip (as XLR pin 2) = signal hot/positive (+)

There are two different sorts of three-pole quarter-inch plug which are physically not compatible – and can cause havoc if connected by mistake:

A-gauge describes the familiar plug on many stereo headphones (in the US it's sometimes called a 'stereo cord plug'). It's simply a three-pole version of the ordinary quarter-inch ('mono jack' or 'phone') plug. When used with balanced feeds this type can be inserted interchangeably into 'mono jack' sockets, usually without physical harm. The source (or destination) simply becomes 'unbalanced', because the ring connects to the sleeve.

On some more basic mixers, only the A-gauge type is used for aux send/returns (not balanced). Whether the tip is send or return depends on the mixer. Two single-core leads are used, usually split into two jack plugs at the other end, to access an FX unit.

B-gauge plugs have a smaller tip. This is a more rugged and much higher specification of plug, called 'quarter-inch military telephone plug' in the US and sometimes 'GPO' (General Post Office) plug in Britain.

For PA, this type is widely used in professional patch panels or 'jackfields'. It's also used on some studio mixer headphone outlets, though not usually on PA mixers. You might find it as a balanced aux send or return on the rear of some mixers.

It's easy enough to tell the A and B-gauge plugs apart by comparing the tip, but identifying the sockets is more of a problem – sometimes a serious one.

If a B-type plug is inserted into an A-type socket (as found on low-budget PA gear which may not have XLR sockets), the tip won't make contact, usually stopping signal flow, or probably just creating a loud hum. If, on the other hand, a B-socket has an A-plug inserted into it, the plug won't go in properly and, if forced, it will wreck the socket.

In some cases of intelligent design, B-gauge jack sockets (on mixer panels, for instance) have been marked with distinct colours of plastic 'nut' or surround – usually red, though the colour-coding can vary from one maker to another.

(Note: 'Bantam' jack plugs – in the US called '0.175' phone connectors – are miniature cousins of the B-gauge plug. Also solid and rugged, they are used in higher density jackfields.)

Quarter-inch plugs were originally designed for 'patching' together calls in telephone exchanges (hence the enduring name 'phone plug' in the US), and are even now seen on mixers with patch panels (for highly flexible and accessible aux send/return and matrix connections), and also in jackfields (or patchbays), as well as on some classic synths. This is still what they're best suited for, rather than PA work.

Knowing how they are meant to be used, it is easy to see why quarter-inch connectors are not usually latching (although there are some latching types available).

Quarter-inch 'jack' connectors can cause big trouble if used for hooking up speakers. As well as not really being rated to pass the current delivered by higher-powered amps, they can easily short amps and blow them up, or cause their own contacts to get damaged by 'arcs'. They are also vastly more likely to cause bad or intermittent connections than other, far more acceptable speaker connectors. Also they are not airtight – which can alter the sound of a bass cab by a surprising degree.

The best advice is to avoid using quarter-inch connectors altogether

The best advice is to avoid using quarter-inch connectors altogether. If you meet one on a speaker cab, change it to a decent connector.

Where required, PA-grade quarter-inch plugs (and sockets) are made by companies such as Neutrik, Canare, Supra and Switchcraft. You can tell them by their heavy gauge, all-metal construction, solid terminals with total absence of rivetting, and decent cable clamping that accepts a wide range of cable diameters. Poor quality 'copies' are to be avoided (even standard guitar cable plugs won't really do) unless you want more than your share of loose plugs, bad contacts and plenty of frustration.

Binding posts

Many PA power amps are fitted with red and black 'binding posts', where bare-wire-end speaker cables can be attached to a positive & negative terminal. (Standard spacing for channel output terminals is 19mm.)

The 'binding post' part is rarely used in the intended way, with the wire wrapped around the threads before the knurled end is tightened down. Instead, if bare wires ('tails') are being connected, they are usually passed through the hole then simply clamped. If the wire is clean and located carefully this can be secure enough; but if the wire is carelessly located, so the bared section protrudes, this obviously tempts problems with shorting and arcing, shock risk, and harm to the amp involved.

Also, if the insulated section of the wire is placed under the terminal, so the sheath is clamped rather than the conductor, an unreliable contact may result.

For reliable long-term use – including in semi-permanently arranged PA amp-racks – the knurled knob should really be tightened down a bit more than fingertight. You can use mini-mole grips or pliers. (No terminal or amp maker seems to have had the intelligence to make and supply a proper 'nut-spinner' un-do/do-up tool.)

When clamping stranded wire ends, consider crimping 'bootlace ferrules' over them – this tidies up the ends, making them solid, so they will then last longer and cause less grief. A special tool is required – but make sure all the strands are captured before crimping.

Binding posts usually also accept 4mm 'banana' plugs, more of which in a moment…

Good and bad binding posts

Some PA amps use substandard types of binding post. These may have only a very thin hole for passing conductors through – perhaps suiting wires of 1mm2 size, whereas conductors in many PA speaker cables will require rather larger holes.

Also, the 4mm socket in the rear may only suit US-style 'banana' plugs, which are shorter than the type used in the UK and Europe. The latter consequently don't make very good contact in US-style 'banana' sockets, because the receptacle is too short to engage the centre of their 'lantern' springs, where grip is greatest.

Lesser terminals also may have a greasy, un-knurled and un-fluted plastic surface that's very hard to grip tightly with the fingers. The best designs of binding-post/4mm socket are readily tightened with fingers alone. They can also be tightened with a large screwdriver or coin – more likely to be found in a hurry than some fussy spanner size.

Traditional binding posts mostly don't meet modern safety regulations

Some binding post sockets rely on screws to clamp down 4mm plugs – which is fine in itself, but unless they are tightened only very slightly it ruins the lantern spring(s) on one side, possibly destroying the plug.

Traditional binding posts mostly don't meet modern safety regulations. Safety is less of an issue, though, if the rear of the racks is relatively enclosed, and bears a safety notice like, 'Danger of bare high-voltage connections'.

A weak point of most binding post-cum-socket terminals is that they use brass and will snap off readily if smacked hard from the side. This can happen easily when heavyweight amps aren't racked, or when they are being manhandled in or out of their racks.

It's a good idea to be prepared for this before it happens. Look inside the amp and see what tools will be needed to remove the broken connector and fit a new one. Some amps will have to be taken apart just to get access. Make sure you have all the necessary tools on the road.

4mm 'banana' plugs & sockets

4mm sockets aren't very often used as outputs on

the back of amps, probably because they'd frustrate users who expected to connect bare wires. But they will work fine in systems where 4mm/banana plugs alone are used. (Note that US-version banana plugs look even less like bananas than British ones.)

Though they may appear so at first, standard 4mm connectors aren't much safer than binding posts, as bare metal is exposed when the mating plug is only slightly withdrawn.

The solution is dedicated 'high voltage' sockets, which allow plugs with fixed shrouds to be used. 4mm plugs with spring-loaded shrouds are also available – these are 'self-shrouding' as the plug is withdrawn. Unfortunately both types of plugs are generally only available as individuals, while the US-pattern twin type, which PA users naturally prefer for speedier connections, is usually unshrouded.

Such US-pattern twin or two-pole 4mm 'banana' plugs have other problems. Strain relief is next to non-existent. And no thought has been given to polarity – so it's easy to put a speaker plug in the wrong way round. It's really not what you need in a big system.

Multi-pole connectors
Single and twin connections are acceptable as semi-permanent wiring inside amp racks. But for rigging speed, safety and ruggedness, more foolproof connectors are needed, generally with more pairs, so that several frequency ranges or drive-units can be connected in one 'hit'.

Speakon speaker connectors
These fairly new but widely adopted plastic connectors were designed expressly for use with PA speakers, and are tough enough for all but the heaviest work.

Earlier Speakons (made pre-1996) had a twist-lock rather than a latch. This worked well, though it was not liked by some PA users. Also in earlier versions the 'shoulder' at the rear wasn't flared (tapered) enough, and this could cause snagging when long leads were coiled up and dragged around objects.

Speakon connectors are well insulated against accidental bodily contact, so they allow high power amps (delivering well over 50 volts and possibly over 100V AC) to be connected with little risk of shock compared with the usual

alternatives. The chassis-mounted Speakon sockets are also airtight, which is important for speaker cabs where there may be unused 'link-outs'.

There are three main types of Speakon connector, which allow for most requirements in PA – two-pin, four-pin, and eight-pin (which is physically larger). The wiring is even standardised, with the pin numbers relating to the frequencies being served in the speaker (low-numbered pins connect to LF drivers). All Speakons usually have silver-plated pins, though gold-plated pins are an option.

Speakon connectors were designed expressly for use with PA speakers

Large-ish speaker cable conductors of 4mm² (which assumes stranded flex wire) can be fitted. Connections can be screw-clamped as well as soldered – though if clamped they should be regularly inspected as screws can (and will) work loose with speaker vibrations and metal creep. Nail varnish or some other screw-locking substance may help.

There are several variations in the Speakon family that allow, for example, larger sizes of cable (up to 20mm diameter), or which fit in a particular size of chassis or case hole.

Unlike XLRs, both ends of a Speakon cable have connectors of the same gender (actually male, although the pins are shrouded). But 'couplers' are available, so that standard Speakon cables can be joined end-to-end.

EP connectors
The EP series was Cannon's 1950s' predecessor to the XLR. It is considerably larger, and so the multiple pins can be stouter. But the die-cast metal casing 'shell' has not proved as strong as it appears, at least in the original version. To overcome this a seriously rugged stainless steel 'copy' version may be available in some territories.

A four-pin EP4 connector used with a single driver is straightforwardly wired: pin 1 positive, pin 2 negative, pins 3 & 4 not connected. But for dual drivers or more, wiring may vary from one speaker manufacturer to another, so it's best to check before proceeding.

Six-pin EP6 connectors are widely used for multi-driver connections by a number of leading PA cab and system makers. A de facto wiring configuration, developed by Turbosound and used by some other cab makers, is as follows:

Pin 1	Low Frequency -
Pin 2	LF +
Pin 3	Mid -
Pin 4	Mid +
Pin 5	High Frequency -
Pin 6	HF +

EP8 is also used, but the wiring is even less standardised, so you'll have to contact the cab maker – or else take a cab apart to work it out.

PDN

PDN was Cannon's attempt (before Neutrik's Speakon arrived) at a specialised speaker connector that also addressed safety legislation. While seen in catalogues occasionally, it's not known to be fitted to very much if any manufactured PA gear, and is rarely seen in PA use.

Phono (or RCA, Cinch) connectors

These are substandard equipment, not suitable for professional audio use – but they're fairly cheap and easy to fit, so are seen on much semi-pro and lesser gear. The sooner you can change the connections over to XLRs the better – even if the source (from phono/RCA outlets) is unbalanced.

This can be sorted using 'conversion tail' cables with a phono/RCA plug wired with a short flex cable into an XLR plug's hot and cold pins.

Multicores/snakes

As explained in Chapter 1, all the music signals on-stage – the microphones, the line sources (keyboards etc), and the DI boxes – are usually plugged into a stage-box. From there the signals are fed to the mixer via a multicore cable, or 'snake', as it's known to its friends.

Multicore cables contain many individual balanced 'pairs' – hot and cold wires – with shielding around each. Some multicores are more robust than others, but in all cases it's best to use a cable with several spare 'pairs' available, because the stresses and strains of gigging mean that occasionally one wire will fail – and if one in a pair breaks, the pair is effectively unusable.

The returns (the cables linking the mix position to the stage) may be part of the main multicore or on a separate drum. Either way, make an allowance for at least three or four spare pairs – not just for emergencies, but in case the system requires more frequency bands, perhaps due to a change of speaker system.

Multicores might also be used for FX, amp and speaker distribution. This not only makes setting-up quicker but helps avoid wrong connections.

FX multicores are usually so short (from the rear of the mixer to the rack or racks alongside) that simply to use a cut-down version of a thick main FOH multicore would be unwieldy – it may be easier to have a selection of several 'snakes', some with fewer pairs inside.

Ideally the connection is a neat pair of multi-pin plugs. Quite often, usually at the mixer end, the pairs may well have to be split out as tails, often fitted with one kind of quarter-inch plug. The tails at each end have to be tailor-made to match the mixer and FX rack's connector layouts. The extent to which there's any (re)-patching flexibility will depend on how the cable lengths are organised.

For connections from the returns multi/snake, daisy-chaining through to amp racks, miniature 'fast-connect' (push-pull) multi-pin connectors have proved successful. These can greatly speed set-up and (assuming they're wired correctly) overcome the risk of significant speaker damage that can be caused if low and high frequency drive signals get misplaced. Multicore speaker cables continue this arrangement.

Polarity

This concept is central to the correct operating of any PA system, and is also relevant to all the other chapters in the book. It's about making sure that some parts of the sound system aren't working against the majority, which is an easy way to destroy or throw away an otherwise good FOH speaker system's headroom, sonic quality, and smoothness of coverage.

'Polarity' is the name for the choice of two 'directions' in PA connections and signal wiring. Physically it means that all the drivers' cones should move in the same direction – out or in – as one (at least for a particular frequency range).

Getting polarity wrong takes just one accidental flip of a wire, and can happen at different points throughout a PA system, causing different degrees and types of 'polarity error'.

Wrong polarity in even one speaker cab in a PA can cause:

- big variations in response over frequency range, with different audience positions.
- 'hotspots' (loud areas) at some positions.
- house EQ difficulties.
- apparent 'artist or instrument EQ' difficulties.

(To some extent a well 'arrayed' PA can overcome these problems – see Chapter 6.)

Polarity errors occur most naturally where there are balanced (line, mike) or floating (as in speaker) connections. Hence all PA wiring (except inside equipment) is vulnerable.

Methods of rectifying the problem can vary. To avoid wasting time 'barking up the wrong tree', methodical procedures are required. You need to work through the PA, checking each part for polarity inconsistencies.

Polarity identification (aka 'phasing')

Both compression and chassis (cone) drive-units must be correctly 'phased' (as it's commonly known, though strictly speaking it should be 'polarised' – see panel below). This is simple in itself – about as simple as fitting a battery the right way round – but gets complex with PA systems as it operates throughout on 'nested' levels.

On the most basic level, one of the two terminals on every drive-unit is marked in a distinguishable way. Typically, there is a dot of paint or ink, which could be red or blue or white (almost any colour). Or one terminal may be marked '+' or 'pos'.

If there is any doubt, it's easy to discover which

terminal is meant to be marked this way – the convention (for most drive units) is that when the '+' or marked terminal is in driven positive (with a small DC testing voltage), the cone or diaphragm should move outwards.

One of the two terminals on every drive-unit is marked in a distinguishable way

This test is readily accomplished with cone (chassis) drivers. An old style of 4.5V bicycle battery with screw terminals is commonly used. Flexible red (+) and black 'fly' leads are connected, usually with bared ends, to fit in the drive-unit's press-fit terminals.

For compression drivers, the diaphragm shouldn't strictly be driven so far that you can see which way it moves. A safer way is to fit the driver to a horn, and then use a 'phase checker' (polarity tester), which must be held at least 1m (3ft) back from the horn's mouth.

Even if the driver's polarity is known, you may not know which of the two wires you've removed fits on which terminal. Note: never take anything apart without making notes, unless you have a manual on it.

Failing this, you might be able to trace the 'high' side of the input (eg pin 2 of an XLR), which should go to the '+' or 'polarised' terminal in most cases. The trouble is, in some cabs, the polarity of some drivers is reversed. Often this is for good engineering reasons; at other times, the wiring may be wrong because of a manufacturing or maintenance mistake.

So it's worth getting definite confirmation from the maker on the correct polarisation of all drivers in a PA cabinet, each relative to the incoming signal.

Phase and polarity difference
Many people say 'phase' when they mean 'polarity'. The difference is that phase is a more 'catch-all' term – it covers other more gradual effects (phase shifts) that are nothing to do with polarity. With polarity, there are only two possibilities, which could be called forward and reversed, or suck and blow, or plus and minus. There are no possible intermediate states.

Checking cable polarity

PA's de facto standard mike and line pin-to-pin XLR-XLR leads can be checked by direct insertage. (In the process you may even come across phase-reversal leads you didn't know you had, which someone made in a hurry and forgot to mark up.)

Speaker leads would need two 'pin 1 to 3 converter leads' made up from hardwired male-to-female XLRs, each wired at their ends as follows: pin 1 to 3, pin 3 to 1, pin 2 to 2.

Other leads will require either their own conversion, or it may be easier to use a lead tester (see Chapter 9) which signals pin-to-pin connections, so reversals can be seen.

Microphone polarity

A small percentage of mikes may be wired with 'reverse polarity'. This will cause unexpected weird sound and cancellation (so a sound source's level strangely reduces when a fader is brought up above about halfway). If you suspect reverse polarity is the problem, try flicking the mixer's 'phase reverse' switch(es).

FX loops (FX send/return connections) polarity

As above, you may occasionally encounter unexpected behaviour when the faders are brought up. Solution: use correct leads, or polarity-flipping (phase-reversing) leads – whichever way sorts it.

Mixer output polarity

If one whole side of the PA has opposite polarity, this can have the effect of making the bass content of the sound seem unexpectedly light, and there may be a 'hole' in the centre of the FOH coverage. This problem needs to be traced and eliminated at source, rather than trying to fix these symptoms at later stages in a more piecemeal way (with EQ etc). The solution is to locate the wrongly-wired cable.

Or it's possible that the mixer has been serviced and a header altered so one side's output is re-programmed to 'pin 3 hot'.

Crossover output polarity

These are often fitted with polarity switches which enable particular bands to be polarity-reversed – which is required for some cabs, or just some of the drivers in some cabs. (See Chapters 4 and 6.)

But be wary when you use the crossover's polarity switches – it's possible that someone might have 'kindly' reversed the wiring in one or more of the cabs' frequency ranges. In which case, the switches aren't required – and will in fact cause the very trouble you're trying to fix (the root of which obviously lies elsewhere).

Either way, if an entire frequency range is wrongly polarised, the cabs' coverage and frequency response will be affected around the crossover point(s) surrounding the affected frequency band.

Amp connections polarity

Both inputs and outputs are balanced or floating, and so can get 'polarity flipped' without any obvious problem. Inputs can get flipped when apparently identical amps have been re-wired for 'pin 3 hot', or when they contain wrongly-set pin 2/3 header selectors. These may affect one channel only. If so, that should show up by a low or nil output when set to work in bridge-mode.

The key to thorough detection is never to assume anything, and to suspect everything

Outputs can readily get flipped when low-grade dual 'banana' plugs are used – a poorly thought-out standard that has nil polarisation. It can also happen when bridged outputs are derived from individual outlets. On the other hand, the use of Speakon and similar multi-pin speaker connectors greatly reduces or even eliminates the chances of polarity errors.

Polarity errors at this level will usually cause symptoms similar to reversals of individual drivers – which we're about to look at.

Cab and driver-connections polarity

When the polarity of just one or a few individual drivers in a PA is reversed by mistake, the effect can be hard to pin down, but is well worth detecting and fixing.

The key to thorough detection is never to

assume anything, and to suspect everything. For instance, driver wiring can get inadvertently wrongly polarised when cabs are serviced. Cab makers should really display information about wiring colour codes inside their cabs (it's not always the unequivocal red and black you might expect), but few if any do so.

When multi-pin connectors (Speakon, Socapex) are used, there should be at least a reduced risk of polarity mix-ups.

Polarity checkers

Drive-units can be safely checked with a low voltage battery (4.5V max). But for testing earlier in the signal path, at line-levels, this is not the best option – here a lower level (1.5V max) is sensible to avoid unnecessary risk of hard overdrive.

More important, randomly 'dabbing' wires on and off a battery, or switching, makes poor quality, lengthy and erratic pulses which signal-coupling and EQ may distort, possibly into the opposite of the intended pulse polarity, creating definitely unwanted ambiguities.

For this reason, a purpose-made polarity/phase checker should be used to verify polarity with all but 'raw' drive-units. Such checkers comprise a separate pulse generator and detector.

The pulse is shaped to work well, clean and sharp, and the detector is 'gated' straight after detecting the pulse's leading edge, so any 'error tails' and 'background noise' are ignored. Polarity is typically indicated by green and red LEDs.

The pulse generator is plugged into a line input at the stage in the PA signal path where you wish to start checking. Well-designed products are 'XLR-headed' so they can usually plug directly in without needing a cable. Otherwise any XLR-XLR cable can be used. The detector part can either be plugged into subsequent line (or speaker level) outputs, also XLR-headed.

A typical detector also has a built-in mike which takes precedence if the line input isn't in use, so that the ultimate, acoustic polarity/phase can be checked.

It usually doesn't matter which of the two polarity LEDs light – so long as the relative polarity is correct between channels, drive-units, frequency bands, etc, all will be well. To check if a unit or signal path flips polarity/phase, see whether the sending and receiving LEDs are simultaneously the same colour (which indicates 'in polarity') or not ('polarity is flipped').

Mike polarities can be tested using a speaker (driven at a suitable level by an amp being fed the polarity tester's clicks) that is reasonably distant (say four times the cab's width), and a fixed mike-test position (on a mike stand).

Problems with polarity testers

Polarity checkers can give wrong results in some conditions. Classic cases of wrong readings have been caused by the following:

- Noisy conditions – if it's line connected, are you sure all other sources have been muted? If using the mike, is the room quiet enough? Stop the test 'clicks' to see if the LEDs show there are other sounds present. If so, reduce these or wait until they subside. Or try moving the receiver position away from the noise source, if that's feasible.
- Overdriving a mike or line input. The test pulse is a narrow spike and so it does not sound loud. If turned up so it does seem loud, the signal path is probably being overloaded. Some common IC op-amps used in PA gear can 'flip polarity', or at least convey polarity-reversed spikes, when they are overdriven.
- Attempting to test acoustic polarity of several sources, usually by hovering the pulse detector midway between them. The best way to be sure is always to test one drive unit at one time. For specific frequency bands, this can be done by muting specific amp channels. The other frequency bands can be 'taken out' using the appropriate mute buttons on the crossover. Otherwise this may be tedious in a big system once it is rigged – but, then again, extensive checking shouldn't be required if polarity has been checked at earlier stages.

Alternative polarity check method

There is another way to check the relative polarity of any adjacent pair of enclosures. It requires use of the third-octave (or better) RTA (real-time analyser, or audio spectrum analyser) and a pink noise generator, which are standard equipment in most larger PA systems (there's more about RTAs and their use in Chapters 8 & 9).

Disconnect (or mute the amps feeding) all but

one of the cabs. Feed pink noise into the system and the one cab and adjust for a suitable level for measurement, one that's at least 10 to 15dB above the ambient noise level.

Now place the RTA's measurement mike (or any capacitor/condenser mike with a flat and wide response) midway between the cabs and a few feet in front of them – say 4ft (1.5m) for a bass or mid cab and 6ft (2m) for a sub-bass cab. While watching the display, reconnect the adjacent cab. The analyser's readout should jump up in level, showing acoustic summation (not cancellation) is taking place. If the level drops, the polarity – somewhere – is wrong.

If there are more than two cabs together, repeat the procedure until adjacent pairs of cabs (A to B, then B to C, then C to D etc) have been checked. This technique is particularly suitable for the rapid checking of bass and sub-bass enclosures.

When to check polarity

Polarity (particularly with the system back-end components) should be checked at all stages of system preparation, set-up and maintenance – for instance: when servicing individual drivers, cabs, cables, amps; after preparing new wiring looms; when first using new returns multis/snakes; and when first using new mixers, FX and back-end equipment – from house EQ through to new or different cabs. (For more on servicing and maintenance, see Chapter 9.)

Polarity should be checked at all stages of system preparation

---

Note that when polarity with equipment and cables is correct, output should be identical to input. But with speaker drive-units and cabs, reversed polarity may be correct in instances.

Checking arrayed speaker system polarity

The following procedure is typical for a multi-source, arrayed system:

● Disconnect all but one cab in the array.

● Drive the system with pink noise (a real-time analyser, RTA, is required) at a comfortable level.

● Boost the house graphic by (say) +6dB at the LF/mid crossover points (say 250Hz). This makes the pink noise predominant at the crossover point.

● Now mute the HF output on the crossover. Listen to the level. Then mute the LF and listen again. Which is louder of the two?

● Adjust, and listen to each alone again. As the LF and HF will sound different, some repetition will be required. An RTA may then be used (if desired) to check the levels.

● Now un-mute both the crossover's LF and mid outputs. Then adjust the crossover's 'LF-Mid' phase control for maximum addition of the two signals. If the level drops instead, flip the polarity switch on one output.

Take care not to disturb this switch setting when performing the same phase adjustment on the other pair(s) of bands. Sometimes, it is easier to set the polarity in reverse, as the phase is easier to tune by looking for the deepest cancellation.

The above procedure is then carried out for the other crossover points, un-muting just the frequency ranges concerned. If there are problems, check for wiring and, in particular, polarity problems – see preceding section.

● The graphic should now be reset to 'flat' and the LF & HF controls reset (if necessary) to give a flat response, using an RTA. Note that the measuring mike must be 'on-axis' to – directly in front of – the single cab being driven.

● Now perform a listening test and make any fine adjustments.

● Only reconnect the rest of the array when the solo cab sounds right. Trying to 'race ahead' and 'RTA' and listen to the whole array can readily lead to confusion.

● Lastly, house EQ can now be applied if required – see Chapter 4.

Note on phase alignment

Refer back to Chapter 4 – this sort of adjustment has nothing to do with polarity. But you must make sure the polarities of all the drive-units are correct before bothering with this. The system maker should offer optimum or 'starter' phase settings, in degrees.

Equipment organisation

Racks

PA gear racks (normally sturdily built and flightcased units holding outboard FX, EQ, crossovers or power amps) can be classified as either half or full-size racks. For PA use, where gear is portable, what's regarded as a 'full-size' rack is commonly about one metre (roughly 3ft) high. A half-rack is about 0.5m, or approximately 18in high.

FX racks and splitter racks are (relatively) light (compared with amp racks, for instance), so lesser-rated castors can be used. Half-racks may be used, so the FX can be placed side-by-side at one comfortable level. It's easier for the sound engineer's eyes to cope with scanning just 18inches-worth of vertical panels. (Spaces between units may also be fitted with blanking panels to improve visual appearance.)

FX racks and splitter racks are light compared with amp racks

---

1U equipment often has to be almost as deep as it is wide to accommodate the contents. Because this makes it very thin and vulnerable, consideration must be given as to whether it will fall apart when shocked. Unless the forward rack ears are extremely solidly coupled to the rack casing, and unless the casing has the rigidity of a rectangular tube, then it's a good idea to fit rear supports. These may be do-it-yourself jobs, or obtainable as accessories (as 'rear rack supports').

See Chapter 5 for details of power amp fan-cooling. Extra fan-assisted cooling within the rack isn't usually required unless the rack contains older or heavyweight digital equipment – with power supplies consuming above, say, 50 watts.

Aside from any other considerations, the size of your racks, cabs and trunks must also fit efficiently into the transport you will be using. For smaller set-ups this will just depend on the size of your van/mini-truck, though for larger systems requiring serious (heavy-goods-vehicle) trucking you may need to be aware of internationally standardised container sizes (see also Appendix B).

Full-sized amp racks can be extremely hefty, so heavyweight castors are essential, with brakes that can be clamped onto the wheels to stop them trundling off on an uneven or slippy floor.

Amps should be mounted on an inner frame that's joined to the outer casing at a limited number of points using suitably-rated shock-mounts. Rear rack-mounts are essential with most amps, unless units under about 8in (200mm) deep.

Flying & lifting terminology

(See Chapter 6 for more information on flying speakers.) The following terms are fairly specific to flying and hanging gear, so we'll mention them here as well as in the main Glossary – because any misunderstandings in this area can directly affect safety more than most words in PA.

Powering up & down

A PA can't just be carelessly turned on and off, because there's a high chance of creating some very loud and sharp (albeit brief) noises. This can even happen with some of the better-designed equipment fitted with power-up/down muting. Even if there is no damage to the PA drive-units, the sudden noise can be unpleasant, even alarming, for people in the vicinity. Chapter 5 deals with avoiding the worst of noises over the speakers. The rest of the system – the electronic path – is ostensibly immune to harm from the often large signal spikes that are applied when gear is powered up or down. Still, the sound of 16 swing-needle meters hitting their end-stops is a reminder that wear and tear is taking place.

To minimise this:
● Set all mixer faders down to their lowest level before either powering up or down.

● If the FX rack's signal loom/snake or multi-pin plug can be unplugged before the items are powered up or down, so much the better.

● Powering up the entire FX and drive racks at once is fine, but better still, the equipment in the drive rack should be turned off in order of the signal flow – for instance, first graphic, then crossover, then delays – and should be turned on in the reverse order.

Cable – this can apply to both wire rope or sheathed electric conductors.
Clevis – see shackle.
MBS – minimum break strength.
Shackle – a U-shaped fitting, closed with a pin.
Sling – an assembly of wire ropes/cables that connects the cab load to the lifting point.
Softener – sheathing materials used to protect the load or wire rope/cable from damage during lifting, or when in place. May be disposable.
SWL = safe working load.
Thimble – grooved metal fitting to protect the eye (fastening loop) of a wire rope.

Planning & PA system power

How much power do you need? There are two aspects to this question. One is 'system level' power, the other is gig or tour-specific power. A PA's overall 'power' is rated in watts ('W') or kilowatts ('kW' or just 'k'). This is a nominal, round figure.

If in doubt about venue capacity, it's better to play safe and have more power

---

Theories vary about the amount of power you need, but here's a very rough guide:

A 200-watt PA is really the smallest you should consider using, and this might just about be enough to cope with an audience of up to 50 people. If you wanted to feel more comfortable (sonically speaking) – and have a bit of 'headroom', and theoretically a less ragged sound – you could use anything from 500 to 1000 watts (1kW) for this size of gig.

Again, if you're being minimal, this same PA (500-1000W) could be used for audiences of up to 500 people (as long as the venue wasn't too large).

But if you were tempted to be extravagant, or just wanted to have some spare power 'in the bag' for dynamic effect, and you had the wherewithal

to get hold of the equipment, you'd be better off treating this size of crowd to a PA that could handle anything from 5kW to perhaps 20kW (20,000 watts).

The amount needed will obviously be influenced by the kind of artist and music involved, but can also be affected by the type, construction and size of venue, and even what the audience members are wearing – for example, on a hot day people will wear less clothing, which as a result reduces absorption of higher frequencies (so more HF horns and amplification might be required).

You'll learn by experience, using rigs in different venues. If in doubt about capacity, it's better to play safe and have more power (assuming you can afford it). Any excess automatically provides headroom, which helps sound quality.

Calculating 'k'

The power rating (or 'k') of a complete rig is never precise. The most honest way to calculate it is to add up the power that all the amps can deliver at clip (maximum) into the nominal impedance of the cabs they're driving.

This is as good a guide as you'll readily get, even though quite wide deviations can be caused by things like speaker impedance variations, mains voltage variations, and the fact that a PA should never be driven to these maximum levels for very long.

Horn-loading can make a huge difference. If the cabs aren't all this type (flared bins & horns), the effective amp power is ten times less – that's how inefficient non-horn-loaded cabs typically are.

Bear in mind too that the physical size of a rig may affect its usability. 'Creative' combinations of older PA cabs (known disparagingly as 'hotch-potches') with older power amp racks are likely to be bulky for the sound levels they give. More modern systems have been engineered to take up less space for the power they deliver.

145

soundcheck & showtime

PA voicing, EQing, mixing...

Quick view

Once the PA has been wired up, it's time to make sure the whole thing makes a decent noise. The soundcheck is the dress-rehearsal for the gig, a chance to confirm everything's working and sounding as it should. This chapter takes us through all those final preparations – including voicing the PA, then checking the performers can hear themselves through the monitors on-stage, and that the out-front sound is what's wanted – and then on to the show itself and the sound engineer's central role in that: delivering the best possible mix in any circumstances, indoors or out.

Voicing the PA

Before the full soundcheck itself can start there are one or two other tests to carry out. The first thing to do is 'voice' the front-of-house PA – to check it sounds OK on its own, before the band joins in and starts complicating matters. Using some favourite recordings and a good signal source (like a CD player) the PA can be tested and made to sound 'right' even before the performers have set up.

By doing this, you theoretically halve the possible sources of any bad sound, making it a lot easier to diagnose any problems that may occur. If the sound is wrong at the 'voicing' stage, you can't blame the performers or sound sources from the stage, because they haven't started using the PA yet – the fault obviously lies within the PA itself. Equally, once you're satisfied that

the PA is nicely voiced, any later problems are more likely to have been introduced by front-end signals – the part of the chain that runs from the performers to the mixer.

It needs someone with 'good ears' to voice a PA in this way – ideally this should be you, but if, for instance, you have a cold or are just feeling tired, it's best to get someone else to help with at least these preliminary sonic decisions. Also, if you've been working in a noisy environment for a time, relax your ears by going to a quiet(er) place for an hour before doing the listening.

The recordings you use for the voicing test should generally be of good quality (usually CDs or DVDs, but unscratched vinyl and a good turntable will do just as well), and should be very familiar, preferably from listening to them on a good hi-fi system. There should be no need to use EQ on the incoming mixer channels, except possibly small tweaks at the extremes.

If you're concerned by something you hear, a trusted pair of full-range hi-fi headphones (not budget 'midrange' ones) plugged into the mixer should be able to prove that the recording and mixer are not the cause.

If EQ is required, the reason should be established. It may be down to room acoustics, or it may be a set-up or wiring problem (with the crossover, for instance). With experience your ears will tell you which. If you're unsure, assume it's an acoustic problem, and try using the house graphic (see Chapter 4).

If a few tweaks fixes matters, fine. But if you have to push more than two or three adjacent sliders similarly deeply down, or far up, to get any remotely correct sound, it is time to turn off and start some checks.

Real-time analysis (RTA)

If sound problems cannot be easily diagnosed by listening alone, this kind of audio spectrum analyser may be used. A pink noise generator is used to feed a recognisable signal into a particular device, and the result observed on the RTA screen.

Pink noise (which consists of all sound frequencies at equal volumes simultaneously) has been used since the mid-1970s to 'analyse' PA systems and suggest corrective EQ. Like all audio measuring tools it may only be accurate under certain conditions, though it's very useful for quickly seeing what may be badly wrong (like crossover misconnections), and for getting a generally flat response as the starting point for by-ear tuning. Skilled listening is still usually the final arbiter.

It can also be helpful if you do a lot of tweaking and get 'lost'. And it's a godsend if everyone's ears are tired and you have to do your best using instruments only.

A classic analyser tool is something like Klark Teknik's rack-mounted DN60, though various similar but portable devices are acceptable. Unlike 'ordinary' spectrum analysers, which show thin lines on a screen, these 'first generation' audio analysers have bright LED bar displays at either half or third-octave intervals – like a

graphic where the sliders have mutated into LED bars. The LEDs can be set to indicate, for instance, steps of 1 or 3 or 10dB, varying the size of the viewable range.

Such analysers also generate their own pink noise signals for testing. For self-checking, first of all, the noise may be fed directly into the analyser – if all is well the response should be flat, except for a few LEDs bouncing either side on finer settings. Then the pink noise signal is fed into the system.

The test noise source may be fed into the system through a single mixer channel, initially panned midway. The system level is then brought up to a level that's above the venue's background noise – usually 10-15dB higher than the existing 'room noise' is enough.

The display is then adjusted so most of the LEDs are somewhere in the middle. Generally, averaging ('slow' or 'RMS') will provide a smoother and less jerky display. If the response is deviating by more than +/–2dB (except for falling off at the extremes) then you may try recovering a flat response using the graphic. Of course if a whole section of the frequency response is missing, leave the graphic alone for now and look for a wiring error or equipment problem instead.

On-stage

Now that we've voiced the PA, are we ready to start the full soundcheck yet? Not quite... Many inexperienced mix engineers might think the soundcheck starts with the front-of-house mixing desk. It actually starts well before this.

Set up and soundcheck the monitors before sorting the FOH sound

Making sure you will have good signals to work with at the mixing desk is crucial. This means setting up the microphones on-stage in the most appropriate way – not only using the right mikes for the job, but also placing each so you can get by with minimal EQ. Any EQ you have to apply to get the sound right – whether for the FOH PA or the monitors – has a cost in sound quality. 'Acoustic compensation' should always come first.

The second vital point learned from experience is to set up and soundcheck the monitors *before* sorting the FOH sound. If you set the monitors afterwards, any spill (leakage) from them – which you can't predict in advance – will mean you have to re-adjust the FOH sound again, channel by channel.

So, let's look at monitoring first...

Monitoring set-up

The purpose of monitoring is for the musicians to hear themselves and other key instruments and vocals on-stage. A good monitoring system allows everyone to hear exactly what they need to hear, and ideally nothing else. But what is really needed in a monitor mix may be different to what is imagined to be needed, especially by inexperienced performers.

Another view of monitoring is that the musicians should hear themselves and all the other players in a 'balanced' way – so they know exactly what the audience is hearing. Whatever the approach, the importance of good monitoring is first that it enables artists to perform confidently, and secondly (at least with louder types of music) it pumps them up with adrenaline.

(Note: in Britain on-stage monitoring is often referred to as 'foldback': this curious name dates back years before on-stage monitoring was invented (in the late 1960s) and is in fact an old BBC term for studio monitoring. It's also important not to confuse foldback with feedback – the dreaded howl – which we'll look at in a moment.)

Initial setting up of the monitor mixer, and EQ'ing of individual signals, is approached in a similar way (technically speaking) to that of front-of-house mixing – which we'll deal with shortly. But there are certain factors that a monitor mix engineer may be particularly exposed to – from feedback to personal abuse from stroppy artists...

The awkward thing about monitoring is that it has to make itself heard in a loud place. Backline instrument cabs and loud acoustic instruments will generally be filling the stage with over 105dB SPL – in other words levels where you can't properly hear yourself sing or talk. This means the monitors have to be pumping out a good level to cut through the other noise – and this in turn increases the likelihood of feedback.

If the sound from the monitor speakers is loud enough to be heard by a vocalist, say, it's also loud enough to be picked up by that performer's microphone, and fed back into the system, causing over-emphasis (and howl) at certain frequencies.

The main thing to bear in mind at all times is that a monitor mix is *not* the same as a front-of-house mix. A different discipline is required. A pertinent sign once spotted on a monitor desk of a UK PA company read: "You ain't out-front, so don't try to mix it, mate!" In other words, there's not a lot of scope for artistic subtlety and experimentation if you're mixing monitors.

Avoiding monitor feedback

Also called 'howlround' or 'shriekback', feedback can be defined as 'sound chasing its own tail' – in effect the PA becomes a (sine wave) oscillator at the frequency with the most gain.

Just before the feedback starts, sounds in the at-risk frequency range tend to take on a hollow, ringing nature, and you start to hear a steady tone – which may be in the form of a growl, rumble, boom, squeal, shriek or whistle, depending on the frequency.

Feedback is aggravated by having extra 'gain' in the system. Look for 'open' channels (fader up) that do not need to be, and either mute the channel or 'drop' the fader. Are all the noise gates working to keep all quiet channels closed, or are they triggering (opening) on others instruments' sounds?

The essential tools for 'tuning out' any feedback are either good output EQ, or a narrow-band graphic EQ – which usually means having 27 separate 'band' faders, each spaced one-third of

---

***Tips from the top* – monitoring**
The fewer instruments you put into each monitor mix the better – musicians may ask for more instruments than they really need to hear.

Avoid guitars as much as possible in monitors – they will block out the vocals (most guitarists should be able to monitor their playing successfully via their own guitar amps).

Use the input channel EQs on the monitor mixer to remove all low bass on each signal (the frequencies you can feel) – you can put it back later if the player really needs it.

an octave apart, though sixth and tenth-octave graphic EQs are also available. (See section 4 for more information on graphic EQ features.)

Note that a parametric EQ – whether outboard or in the mixer outputs – can remove troublesome 'spot' frequencies with even more finesse because the tuneable 'Q' control makes it easier to pinpoint narrow frequencies.

The usual idea is to get as much level as possible from a monitor without pushing it into feedback, and this normally involves some fine-tuning with the (graphic) EQ. In order to do this, it's customary for someone on-stage to chant 'One, two, test, test' (or perhaps something more imaginative) into a microphone while the monitor engineer increases the gain (if there are roadies on-hand, one of them is usually lumbered with the job, otherwise it may be one of the band).

When a howl/shriek/squeal begins, note its approximate frequency – that tells you which knobs to try on the graphic. Most 'squeals' are in the 1kHz to 5kHz area.

When you reduce the correct fader on the graphic (or sometimes two faders), the squeal will die away. Reduce it just enough to stop the feedback. Now increase the gain until feedback just starts again. This time it may well be at a different frequency – if so, repeat as above, finding and turning down a different slider. If it's the same frequency again, repeat the procedure, turning down the same slider(s) further.

The end result should be a louder signal than when you started the tuning, with enough faders remaining unchanged at 0dB to provide a balanced spread of 'pass' frequencies. You may then have to tweak (boost, in fact) other sliders – which don't cause new feedback howls – to recover the best monitor sound you can.

Tip: When tuning a 'moving' signal – for instance a vocal mike that gets carried around the stage – try to check all the different possible positions. It's better to discover in advance any lurking feedback 'hotspots', and fix them – or at least find out which sliders you might need to pull down quickly if required.

In general, large stages give room for the kind of separation – both between different instruments/amps and also between all monitors and mikes – that's vital for good sound quality.

When you hear monitors feedback, each channel's overload LEDs will light up to tell you which channel auxiliary send(s) to attend to. Only reduce the main send (the overall level) if you lose control of the situation (see Chapter 2).

As well as tuning by ear, you can use a pink noise generator and analyser to tune the EQ (as described earlier in this chapter) to ensure you're getting a flat response between each monitor wedge and each mike.

Monitoring psychology

Being a monitor engineer is sometimes held to be one of the most thankless jobs in all of PA. For a start the monitor engineer has to (diplomatically)

Tips from the top – **using sidefills**

As explained in Chapter 6, sidefills are often like mini-PA stacks positioned in the wings at either side of the stage, pointing in towards the performers.

There are loads of theories on sidefills – some people use them for vocals (some singers actually see them as a way of having their own 'backline' on-stage), or for drums, while others put through a full mix.

For rock'n'roll it's common to 'cross-fire' – that means the stage-left guitarist comes from the stage-right sidefill, and so on. Funk, blues, R&B and soul bands tend to play in a fashion which relies either on lead vocals or

what you might call 'rhythm backups' – kick drum and snare – so the sidefills end up with those elements, and maybe some bass too.

Some engineers feel the sound is much more controllable without sidefills, because the law of diminishing returns comes into operation. The louder the monitor vocals, the more difficult it is to get a natural out-front sound on the voice. And drum monitors can absolutely ruin the sound of the drums.

Often it seems that the actual sound of the monitor mix going down the mike is louder than the original sound itself. This happens when the sound from a wedge reflects off the vocalist's face and into

the mike, which puts a hard edge onto the sound that is actually very difficult to remove.

If you can take away the hard edges of the sounds on stage, then you can avoid the escalation of monitoring levels. There's really no point in getting a good FOH PA together until you've mastered a good sound on stage.

If you find you don't have enough wedges, and you need to handle, say, a brass section in a hurry, just spread out the wedges you have as much as possible. These sorts of musicians will invariably be skilled – and confident – enough to cope regardless of how good or lousy the monitoring equipment is.

soundcheck & showtime

get the musicians on-stage to distinguish between what they would like to hear and what they actually need to hear. Less professional bands, seeing an on-stage monitor mixer, feel it necessary to make maximum use of every facility in sight.

Musicians who airily say, "Just give me a bit of everything in my monitor" should be (politely) ignored. This is at the root of the 'everything-trying-to-be-louder-than-everything-else' syndrome which spirals into feedback and chaotic sound.

In all cases, edit down: fewer instruments means cleaner and louder. For instance, monitors can't really handle low bass – so you may as well get rid of it. Use a high-pass filter (HPF) to take out the low frequencies you don't need, which not only helps cleans up the monitor sound for the stuff that is needed, but also saves wasted power, so you should have more gain to play with.

Monitor volume/level

Sometimes apparent lack-of-monitor-volume problems can be solved in alternative ways: for instance, say a guitarist keeps asking for excessive levels of guitar in the monitor – the first thing to try is moving or pointing the relevant backline amp/cab more towards the player, so they can hear themselves better. Then you should be able to turn the monitor level right down.

Even if a mix engineer manages to set the on-stage sound at a reasonable volume to start with, it probably won't be long before several musicians are asking for something to be turned up in their monitors. If they get their way, the result will be monitor levels that spiral upwards, ending with an irate FOH engineer demanding the monitors be turned back down as they're interfering with the carefully balanced out-front sound.

Communication

The truth is that monitor engineers can't directly hear the results of their work (even if they have their own version of the mix in their own dedicated wedge or cab, they're not hearing it on-stage in the same way as the performers). So they do need some assistance from the artists/ musicians. This introduces more communication problems, especially if changes to the monitor levels are required in the middle of a performance...

Sign language is usually required – though unfortunately only certain elements of sign language are universal, and even ones you think you know can be misunderstood.

One case in point was a monitor engineer who was working on a particular name-act tour when he was alarmed to see the vocalist, on the first night, pointing and waving his microphone directly at him. In a panic the engineer switched in the PFL (pre-fade listen) to check the monitor send – but everything seemed OK. Then, a bit later, the singer waved the mike at him again. This carried on till the end of the gig, when the nervously-exhausted engineer asked the vocalist what the problem was. "Oh no problem," said the singer, "it's just the way I dance..."

A basic signal that should always be agreed upon is nodding the head to one side to alert the engineer that a lead has dropped out or a mike stand has fallen over, but there's a variety of others. If possible, use the soundcheck to confirm any such signs are mutually understood.

Use the soundcheck to make sure any signs are mutually understood

Generally speaking, it's the musician's responsibility to make sure that the engineer is told what they (think they) require – although some artists will get upset if they always have to explain what they want (especially the more precious 'stars' and 'wannabees').

Musicians' attitudes to monitoring tend to fall into two different camps. On one hand, disciplined and well-seasoned musicians (veteran US soul bands of the 1970s/80s have been prime examples) just get on with their job, regardless of whether their monitor mix is good, bad, indifferent – or even not turned on. On the other hand, more anarchic and high-energy players have grown to depend on the 'buzz' and adrenaline rush that loud monitors can stimulate.

Of course things can and do go wrong – and the monitor engineer has to be able to cope with abuse, whether or not it's deserved. Poor monitoring can certainly hinder a performance (causing a vocalist to lose their pitch or timing, perhaps) – but it's also an easy scapegoat, and is often used as an excuse for an inadequate performance. When a band plays badly, or the audience just doesn't like them, the monitor mix engineer – being the nearest 'outsider' – can expect to get the blame. It's human nature.

Adding injury to insult: being in a handy position at the side of the stage, monitor engineers sometimes have to help security personnel (or even deputise for them) in ejecting over-keen audience members from the stage…

Feeding monitors from the FOH mixer

As mentioned in Chapter 2, even if you don't have a separate monitor mixer, you can still supply monitoring of some sort to the stage. This will inevitably be a scaled-down version of what's available from a full-blown, specialised monitor desk, and depends on the size of the FOH desk. It's most likely to be a simple one-mix-fits-all signal sent to every monitor speaker on-stage via the auxiliary send output from the desk (using one or more separate monitor amps).

To set up a single monitor mix, first find out what the players want (or really need to hear). Get it down to the bare essentials. Then turn up 'Aux' knob on the appropriate channels.

To hear what's going on (in your headphones) you'll probably have to press the button marked PFL (pre-fader listen) on the relevant channel. The overall level sent to the monitor amps and cabs on the stage is controlled by the master aux knob rather than a fader (this is usually on the right-hand side of the mixer).

If there are other pre-fader aux sends on the mixer ('Aux 2' etc), a second mix can be created. A few mixers allow up to four pre-fader aux sends to be allocated – but only if you can do without FX on those channels (which is what the auxes are usually for – see Chapter 2).

To help control feedback, turn down any unused aux sends – for instance when the instrument concerned isn't being played. And if one instrument has to be turned up, others may have to be turned down. And watch the musicians closely for their hand or head signals.

FOH mixing

After the monitors are set up and soundchecked, attention can be shifted to sorting the out-front sound.

The first consideration is mix position – where the mixing desk and outboard racks will actually be sited. The idea is to occupy the best listening (and visual) location, while causing minimal obstruction and inconvenience to the audience.

If you're too near or too far from the stage/PA stacks, what you hear won't be representative of the average audience member's experience (different sound frequencies travel better than others through air, and some speakers have a longer 'throw' than others). The accepted, tried & tested norm is to have the desk positioned directly facing the centre of the stage, and between one-third and two-thirds of the way back into the audience. If the venue is small, you may have to work against the back wall.

Bear in mind too that the ideal listening position may be unsuitable for logistical or safety reasons in a public venue – you need to be flexible, as does your equipment (especially cabling: always carry more than enough).

You may find you have to run cables, snakes etc high up, secured to the walls, so they don't clutter the floor. This can also add significantly to the total length of cabling required. Cable-ties, rope and heavy duty hooks may be needed to help with this 'flying' of cables.

It's probably unavoidable that some cabling has to be run along the floor, but in case it has to cross public walkways or doorways, you should carry some pieces of carpeting, rubber matting or (preferably) heavy-duty custom-made 'channelling'

Tips from the top – **EQing**

Top sound engineers are well aware of the truth about EQ: it's a far more powerful tool when it's used to subtract what there's too much of, rather than add what you feel might be missing from the sound.

Here's one tried-and-tested EQ'ing technique, which is all about listening for the sound you want, and removing what you don't want:

First, turn the level up. Take out what you don't like. Then turn up again, and take out more of what you don't like.

Then reduce the level. This way you know you've got room to push the sound if and when you want to. Never try to use EQ to insert what you think is lacking in the sound. Novices add – that's your first impulse. Professionals take away. It's much more powerful.

to cover it up – to protect both the cables and the any passers-by.

Clearing the mixer

Before setting up the mixer for a new performance, always check all the channel controls are set to their 'defaults' – usually their neutral, off or zero positions. After a while this 'clearing' procedure becomes a habit and you barely have to think about it. Here's a rundown of the main items (most of these apply to monitor mixers too):

- Set master faders to the lowest position ('off', or -∞).
- Set all other faders at 0dB (usually about two-thirds of the way up).
- Ensure all 'solo' buttons are off.
- Set all input gains at 0dB (their minimum position).
- Set all EQ boost/cut knobs to their neutral or '0dB' position, usually the central (detent) position.

Temporary level markings and channel descriptions are normally written using a felt-tip marker pen on white gaffer tape, which can be laid in strips along the bottom of the mixer. Make sure any old markings are wiped off.

Setting the channel gain

(See Chapter 2, p34 if you need a reminder of mixer channel features and controls). For each instrument signal or vocal mike, turn the channel gain up until it's about loud enough over the PA. Then get the person singing or playing to give it their all – play or sing as loud as they can (bearing in mind that they may well perform more forcefully in the noise and excitement of the actual gig). Listen for distortion and check for clip/peak (red) LEDs lighting up on the channel or elsewhere on the desk.

If there is distortion (assuming it's not a deliberate, desirable effect – as on an overdriven guitar), or any clipping, back off the channel's gain control until it goes away. In the event of gross distortion, you might want to try using the mixer's pad (attenuator) switch.

Setting channel EQ

You can now begin to EQ the sound sources. As we've said before, bear in mind that EQ'ing introduces its own distortions – of signal timing ('phase') as well as coloration and damping (level dynamics). Always avoid using EQ if you can – the sound will be better for it. Just moving a microphone to an improved position, or tracing a wrong connection or misused switch could avoid the need altogether.

EQ should be applied gently and minimally, while listening carefully. Never 'over-EQ' – err on the cautious side. Always ask yourself first, "Why do I need this?". The bypass switch enables the effect of any EQ you apply to be tested at any time against the original. Do this until you're satisfied.

Large amounts of boost also raises signal levels, so the gain will need to be reset if you don't want to damage speaker drivers. Most EQ's have a lot more boost on offer than you should really bring into play: any more than +6dB (or at the very most +8dB) of boost can take headroom close to the edge (or beyond) in most systems. (Note that +6dB requires the PA to be rated at four times the amp power for the boosted frequency to maintain headroom).

In reality, if you over-boost, it's likely that system compressors and limiters (see Chapter 4) will dynamically defeat some of the intended effect anyway. (This is why you see some graphic EQs with asymmetric boost/cut ranges – for instance up to 20dB cut, but only up to 8dB boost.)

If you find you're boosting a lot of frequency knobs, try an alternative approach: reset all the boosted frequencies to 'flat' (0dB – the centre detent position), then cut all the others until you achieve the sound you want. You'll also have more gain to play with as a result, and help reduce feedback.

Tip from the top **– mystery noises**
A leading sound engineer was once mixing a dance act at an outdoor festival with multiple big-tops (hemispherical marquee stages). During the set he repeatedly heard a bass sound he didn't like, and spent quite a while trying to EQ it away. After about 20 frustrating minutes he realised it had to be caused by another band playing in an adjacent big-top – which was confirmed when the bass sound continued after he'd pulled down the master faders at the end of the set...

EQing & mixing tips

Some uses of EQ are not immediately obvious. With a bass drum or bass guitar you can use a low-pass filter (if available), or make a cut in the high frequencies, to filter away on-stage noise that's above the range of the instruments' highest useful harmonics (in this case that'll be above about 5kHz). This is often done by using a gate (as described in Chapters 3 & 4), but if you don't have one of these, using EQ cut is the next best thing – and better than nothing.

The same principle applies in reverse to cymbals and other high-frequency sounds, where un-needed low bass can be cut. How do you know if it's needed or not? Just listen to the sound as you make adjustments – if cutting certain frequencies has no discernible effect, it's probably not needed (at least for live sound, where high-fidelity is not such an issue as with studio work). Assuming you have a tunable high-pass filter, or use sweepable shelving EQ cut, listen to where the sound of the instrument changes then back off a notch or two.

Removing these unnecessary extremes on certain signals may only have a small effect on sound clarity, but lots of small improvements will add up to a nicer overall sound.

Also remember that a single signal on its own may be great, but when mixed it may mask other sounds, or clash or interfere with them in some other undesirable way. Then again, the opposite can equally apply – a signal may seem less than tonally perfect when isolated (or listened to solo'd on headphones), but when it's blended in with the rest of the mix, it can prove to be absolutely the best sound in context.

Don't expect to get EQ right straight away. It's an iterative process – something that needs constant repetition – because later settings will interact and change the total effect. 'Twiddle and listen', advance slowly, and if you can't hear anything obviously wrong, leave it till later.

Much of it comes down to trial and error with particular gear, performers and venues over many days, nights and years. But here are just a few brief pointers, gathered from accomplished and successful live engineers, which you may find useful additions to your own mixing experiences:

Vocals – in general these might benefit from a slight boost at 1kHz (for projection) and 10kHz (for sparkle). If there's a lot of popping on explosive consonants like 'p' and 'b', you could try trimming a little down at 50Hz. Female vocals, especially if they sound rather thin, might benefit from a slight mid boost at 500Hz.

As the vocal is the element most of the audience will be focusing on first and foremost, it's obviously vitally important that it sounds good. Ironically, vocals are also the most unpredictable of all signals – varying not only from singer to singer, and song to song, but sometimes from note to note: voices can change as they warm up over the course of a performance, or perhaps deteriorate towards the end of a long show or tour.

Singers' microphone techniques can vary dramatically too, which can affect both level and tone. Careful use of compression can help (see Chapter 3), but a sound engineer may often need to have a hand hovering within reach of the vocal fader to make emergency adjustments.

Lots of small improvements will add up to a nicer overall sound

Singers also obviously need to be especially aware of the dangers of triggering acoustic feedback – holding or grabbing the mike wrongly, pointing it in the wrong direction, or moving to a feedback 'hotspot' on-stage. They also need to be aware of what evasive action to take should they hear their mike starting to feed back.

Guitar – a touch of boost at 3-4kHz can add some extra bite; although, bearing in mind what we said earlier about avoiding too much boosting, a cut around 300–400Hz has a similar effect (you may then need to boost the channel gain/level).

Bass – it depends whether you want it to sit above or below the frequency of the kick drum. In most cases it'll be above – so a good frequency to boost would be between 120 and 200Hz. But deep dub-style bass that sits below the kick drum may have a boost around 50–60Hz. If the bass sound needs extra 'presence', or sounds a little muddy, you could add some boost at 1.5–2kHz, and/or make a slight cut at 350Hz.

If you're using a DI'd bass guitar signal, and it

sounds rather waffly or flabby in the low end, try rolling off some bass (which may seem odd on a bass instrument) and adding in a little low-mid, around 600Hz. This can revitalise it to some extent, making it sound more like a close-miked bass amp.

Drums – on the kick drum (also known as 'bass drum'), tastes vary, but one popular approach is the 'top and bottom with a hollowed-out middle': a reasonable 5dB cut around 300Hz scoops out some low mid (which might otherwise clash with the bass guitar); an 80Hz boost adds a bottom end thump; while a toppy 'click' can be added for penetration by pushing 2 & 6kHz a little.

Again, opinions differ about favourite snare drum sounds, but you can normally fairly safely lose the low frequencies below 150Hz (using a high-pass filter if you have one). For a snappier snare tone, add a touch of 4kHz; or for a fatter, fuller sound add in some 250Hz.

Much the same goes for the toms – although if they're sounding rather wooden and lifeless one idea is to 'stagger' the EQ over them, boosting them slightly at 'stepped' frequencies – one at 800Hz, the next at 1.5kHz, the next at 2kHz...

With cymbals, a cut at 1kHz takes out some of the clanky mid, and a boost at 8-10kHz adds a shimmery top (as well as hiss if you're not careful). Cymbals and hi-hats don't need much below 250Hz, so an HPF can lose that.

The better you are at using PA, the less you move any EQ controls

---

Synths/keyboards – beware of extra low and high frequencies at loud levels. These can damage or at least stress tweeters and bass drivers. Suggested precaution: set low-pass filter at between 16 and 18kHz. Set high pass filter at about 40Hz – or get yourself some better FOH bass/sub boxes.

Brass/horns – all members of these families can be amazingly loud at their mouths. Watch the overload LED. Pad switches may need to be used – compression as well – to prevent over-loud blasts.

Hiss & hum

Hiss can be a problem, but with most professional

gear it's acceptable within limits. Lower-frequency noise (buzz and hum) may be heard if audio wires or susceptible gear is placed near any transformers, particularly of larger gear such as power amps or mixer power supplies – and any AC frequency hum (50-60Hz) is likely to be increased when you boost the graphic at those wavebands.

To eliminate hums, try cutting the harmonic frequencies related to AC power lines – for instance in the UK/Europe this means taking out 50, 150, 250, 350Hz; in North America, make cuts at 60, 180, 300, 420Hz. For PC monitor noise, drop the sliders at 8kHz, 16kHz, and (if available) 32kHz.

Finally, remember you don't need to use every slider/knob, or have wild settings. The better you are at using PA, the less you move any EQ controls – most will be left resting in the centre.

FOH sound levels & quality

Our ears acclimatise to loud sound quite quickly. The most powerful PA soon becomes undynamic, even unimpressive if it stays very loud throughout the gig.

For subtler acts, sound levels (measured in dBs SPL) should start at the quiet end of the range, and levels can be gradually worked upwards towards the end of the set, to add a lift for the finale.

For more 'dramatic' performers, you may want sound levels to kick in at the PA's maximum level straight away – but this leaves you nowhere to go, and if you stayed at this level the impact would soon be lost.

Fortunately, audiences won't notice a sound level being reduced if it is done slowly enough. So you can crash in with levels high at the start, then gradually (imperceptibly) pull them back down slightly – for maybe 20 or 30 minutes, to give the audience's ears a chance to readjust. Then you have the leeway to increase the level when needed for extra impact.

Some sound engineers are addicted to the adrenaline they get from loud sound – but these individuals shouldn't be inflicted on an audience by allowing them to mix FOH. If you need an adrenaline fix, work as the monitor mix-person up beside the stage, close to the performers. Monitor-mix people are able to listen to their mixes at very high levels (if they want to) without upsetting the audience's ears.

Sound level restrictions

At many licensed venues and events noise levels may be limited to a particular dB SPL. If sensibly applied, this will be a particular kind of sound level metering, called 'Leq', which is shorthand for 'equivalent level'. It calculates average exposures to high sound levels over a period of time.

If this kind of metering is used it means you can (legally) have loud music in bursts, provided you compensate for it with a balancing period of quieter passages. (In some countries/states if you exceed the maximum SPL level, the promoter or venue owner may be fined by the local council/government authority.) In some venues, the power is cut – sometimes automatically – and/or a red light flashes.

Outdoors, sound level limits are often employed mainly to limit the SPLs that stray off the site. Typically a level of 60dB SPL Leq will be set at the periphery of the site – at this level you should be able to comfortably speak to someone a few feet away. This level will be averaged over 15 minutes (so if a noisy truck passes in a measurement period, the reading has to be scrapped).

It's important to work with any officials who are monitoring the levels. When there's cooperation, they can (for example) warn you if the level in any 15-minute period looks like it's going to exceed the limit. A radio link from the monitoring team at the site's periphery is essential, since the sooner you're told the level is heading over the limits, the less sound level you'll have to 'ditch' to stay within the restrictions.

Sound level monitoring isn't all negative. It can help sound quality by preventing FOH engineers from just endlessly turning up the level – sometimes bit-by-bit to 'get out of trouble', almost without fully realising how loud things are getting.

Sound quality

As the volume gets cranked up there will come a point (particularly indoors) where the sound quality 'hardens'. This may be a sign that the PA system is being overdriven – probably because it's underpowered or the gain structure is wrong somewhere (see Chapter 2, p46 for a reminder about gain structure). But it can also happen with a large PA that's peaking several dB below clip.

Again, this may be traced to a limiter operating, or the cumulative sonic effect of particularly high levels at various points in the PA system that individually still sound OK.

Alternatively, the cause may well be the venue (outdoor stages, like big-tops and polygonic stretched skin types, are partly included here). The laws of acoustics are not fully written, but we can be sure that they break down when musical energy is being piled in faster than the room can respond to it, or dissipate it. In effect, powerful PA systems can create real acoustic chaos.

Either the sound levels have to be kept below the trigger point, or the room has to be made more able to accept continuous high energy input. Opening all available windows and doors would help a bit, though is probably illegal in licensed venues.

(Note that triggering this kind of 'sonic chaos' will cause 'backlash' or hysteresis – a kind of delayed reaction. To stop the effect, the sound level will need to be reduced to quite a bit lower than the level that initially triggered it.)

Curing front-of-house feedback

The section on monitoring earlier in this chapter covered how best to avoid feedback generated through the monitor system, usually by testing and re-EQing one mike or instrument at a time. Some of the same procedures can be applied to detecting and eliminating sources of feedback through the main FOH PA, but (as you'd expect) there are complications – not least because the mix itself is fuller and more complicated than the generally simplified individual monitor mixes.

It's also not as simple as, let's say, moving one performer or one speaker cab slightly. If there seems to be a feedback problem with more than one mike, for instance, you have to consider the possibility that the FOH PA cabs might be too far back, and *all* need to be moved forward...

Alternatively they may be far enough ahead of (or wide of) the stage mike positions, but one or two cabs may be projecting sound directly at something highly reflective – such as a metal barrier. Maybe this leans up a little, on a ramp, so sound is being bounced straight back to the stage.

In this case you might need to negotiate with the venue management and security to get the barrier covered with a heavy layer of soft material, or changed for a hollow 'weldmesh' type. Failing that, you may need to do an ad-hoc tune-up:

155

reduce the power amp gain settings on the two or three cabs/boxes causing the reflected sound. In extreme cases you might have to take the cabs down and re-organise and re-angle them.

In instances like these, the last thing to do is use EQ. Why? Because that would corrupt the whole FOH sound when there is no need – all because of a stupidly placed piece of metal.

Some instruments (mainly big brassy ones) set off howling simply because they are hard, reflecting surfaces being waved around near microphones. If you're working with a band with fairly inexperienced brass/horn players (particularly for more than one show) it may be worth running over with them the sort of positions that can trigger howls, as well as surging levels, and what to do about it – namely move away from the mike, quick.

When feedback happens in the middle of a gig, make a mental note of the approximate frequency. Look at the channel clip or peak LEDs – these will tell you which channel is howling (sometimes more than one).

'Hit' the fader of the affected channel. What's needed is a quick but small downward movement – pulling back just 3dB to 6dB is often enough, though slightly more is sometimes needed.

This will inevitably have an audible effect on signal level, so ideally you want to get this fader back up again as soon as possible. But you need to take steps to avoid the same problem as before. So, before recovering the gain/mix on the affected channel(s), do this:

If you can recall what frequency the feedback was at, set some cut around that area on the channel EQ (use the 'dip' or band-reject response if available). If lots of channels are affected in one frequency area, you might set the dip on the house graphic instead (see Chapter 4).

Slowly recover the gain setting. If feedback starts again, adjust the frequency of the EQ, then the amount of cut, until it stops. Then push the gain up again. When the feedback starts again (a ringing effect may be heard at first), tune the EQ again.

If you have a powerful-enough EQ, this can be repeated many times, with many dBs of level being gained as a result. Some of the better FOH consoles have highly tunable parametric EQs to assist in this.

If there's a persistent problem, consider changing the mike for another model – one that might be less prone to feedback – between numbers (this requires good crew logistics).

If there are re-current problems outside of your control, consider inserting an outboard parametric EQ into the affected channel.

Mixing outdoors

The FOH mix position (usually shared with the lighting crew, who generally do their stuff on two decks above) needs to be agreed with the PA operator before it is sited. The mix position will ideally be aligned dead centre and some way back from the stage – normally between one-third and two-thirds of the way back into the audience (which may be between 100 feet and 100 yards), though this varies according to scale, practicalities and personal preferences.

An outdoor mix/lighting position will be on a raised scaffold, and must have a canopy – a plastic tarpaulin or canvas sheet that protects the equipment (and perhaps even the operators, if you're lucky) from rain/snow/wind, etc. This canopy should be able to fold right back, when the weather is fine, so the mixing person's ears are in relatively free space (except for the sheet at the back, and minimal scaffold at the sides and above).

Strictly speaking, the sound mixing position should be thrust forwards of the lighting decks, like the nose of an aircraft. This allows much clearer perception of the sound that the audience is hearing.

Outdoor sound

Wind, temperature, and changing humidity can all play havoc with what you hear at an outdoor gig FOH mix position. In some areas (the British Isles is a notable example), the weather can change abruptly and/or repeatedly throughout a show or festival day. Some sonic/acoustic effects you can expect are as follows:

● Wind literally blows the sound away and breaks it up. There's not a lot you can do about this at an outdoor gig. The best you can do is avoid side and rear winds affecting your own hearing – so even if it's not raining, this is one time you need the sheeting or canvas to be pulled in at the sides, as well as at the rear.

It might be best to leave most EQ untouched,

since if you can't hear the sound clearly, or for long enough between the wind bursts, it may only end up sounding even worse, and you wouldn't discover this until the wind drops away.

● Sun warms up the air, causing it to rise. Rising hot air can 'lift' the high frequencies away. This can be overcome to some extent by over-angling the HF cabs downwards, pointing more into the audience. This is going to be easiest to adjust if the HF cabs are flown – otherwise angled wooden blocks/wedges (or similar) are placed under the rear of each HF cab (being careful about stability).

The best humidity level for good sound transmission at all audio frequencies is either very, very humid (basically a mist) or else very very dry, as in a desert. Middling humidity – say 50 to 80 per cent RH (relative humidity) causes increasing high frequency losses over increasing distance. The FOH mix position at a large-ish outdoor event is usually far enough away (say over 50 feet) to hear the effect of the lost 'highs' – and it will be even more marked for those further away.

A Baxandall-type HF boost (which may be set with the house graphic), shelving off by the time it reaches 20kHz, will provide fair compensation for the effect. But make sure you don't apply it to any near-throw HF 'fills', which won't need the boost.

This is a good place to mention that when you use a PA outdoors, it is possible (depending on the wind and weather) to get complaints from locations up to 30 miles away – even though the sound level seems low enough at the site boundary, and is even inaudible inside houses two miles away. The sounds which roam furthest tend to be in the mid-range, so keeping these down a bit may help limit the excess unwanted travel.

Unexplained sonic changes

It's long been recognised by engineers that the sound of any complex and/or finely-tuned system alters each day (and this applies to both indoor and outdoor set-ups). Personal mood shifts and physiological changes can account for some of this, but not all. Individuals of different temperament can note the same effects.

Small, random variations in many influences have a cumulative effect on sound or audio signals. Firstly, nearly all PAs are affected by weather – and not just at the more obvious outdoor gigs discussed above. The frequency response (and other sound quality features) of most speakers and mikes is affected by local temperature and humidity – variations of which can be just as wide indoors as out.

Mixers can sound different from one day to the next: their metal frames expand when they've been on for a while and warm up, and then they contract again when off and cold at night. After they've warmed back up the next day, the plug-in connections may have slightly shifted. This can result in either worse or sometimes better connections. (Use of a 'contact enhancer' helps.)

Mixers can sound different from one day to the next

A second blatant cause of sonic variations is the amount of RF (radio frequencies) in the air. If a radio transmitter near the venue is off-the-air one day but starts up the next, you can expect the PA's audio quality to alter. Thorough or curious sound engineers may want to scan the local airwaves for strong transmissions, or see them on a scanning spectrum analyser, or detect the nett effect with a wideband field strength meter – all using wideband aerials (it's a new toy).

Who should mix FOH sound?

A recurring mistake made by endless bands, or their genuinely well-meaning management or record company, is to employ the engineer who made their studio recordings to mix their live sound.

Two things happen. First, the soundcheck takes several hours. Studio engineers are not always in the habit of working quickly – especially those who work on big-budget productions (where the longer they take the longer the job lasts, and the more they'll probably get paid). With live sound, the end-result of this is some pissed-off production people, late or missed meals, and a band that's tired-out before the show even starts.

Secondly, the 'perfect sound' that took hours to get at the soundcheck will be virtually pointless when the gig starts. The audience warms up the venue, making it humid and probably filled with smoke, and completely altering the acoustics. (Elements of the performance will vary too.) An experienced live engineer knows and expects all this. A recording engineer will often go into panic mode when it happens, as they don't have another three hours to recover that 'perfect' sound.

The ironic outcome is bad sound – the opposite of the reason the studio engineer was hired in the first place – and the risk of a bad reputation for all concerned. So the wrong sort of sound engineer (however well-meaning) can end up being an artist's worst enemy.

Experienced live sound engineers work quite differently. They will need only 30 minutes to an hour to do a soundcheck. They know that perfection is pointless, particularly before an audience has warmed up the venue. And they know how to fix a less-than-ideal sound during the opening number, calmly yet quickly, by progressive steps – adjusting first this, then that, in strict order of priority.

Something that *is* a good idea is for a dedicated FOH engineer to be invited (and paid) to spend time with a band while they're recording in the studio, alongside the full-time recording engineer. Then the live engineer gets to know the band, their songs, and their sonic and musical requirements in an intimate space – before they get separated on tour by 100 yards of cable.

DJ sound

Soundchecking a DJ set-up can be much simpler than checking a full band – there are fewer sound sources and fewer egos to deal with.

First make sure the turntables and any standalone CD players are stable, and there won't be a problem with stylus-skipping or vibration-induced rumble or other problems when people dance – see Chapters 1 & 9 for more advice on checking and curing this.

DJs, of course, normally 'mix' their own set – that's part of what they do and what makes different DJs unique. But while some DJs (usually the more traditional 'mobile' type) may carry

enough PA gear to be heard by a small crowd, most larger club DJs will need to plug into a substantial sound system. This will either be an installed 'house' system, or one supplied by a PA hire company, and perhaps shared with a live band on certain nights.

When a DJ's mixer is plugged into a PA supplied by another organisation, the usual approach is to set all the gain controls on the DJ's mixer at maximum (assuming the DJ mixer has enough headroom for this).

The higher-than-normal line signal should be handled successfully by any good PA mixer. The gain is then backed off here for a 'normal' level, and the maximum sound level controlled from the PA mixer by the channel faders (L & R) or possibly a 'DJ' sub-group.

The upshot is that the DJs (whoever takes the controls throughout a possibly long night) can't overdrive the system. Most DJs do like a good bit of volume – which may not always be in the best interest of their audiences or, for that matter, the gear that's being played through.

DJ monitoring

Monitoring is needed by DJs on big stages, and particularly outdoors. DJs can lose any connection with what they're playing if they can't experience the same 'soundfield' and the same sort of sound levels that the crowd are enjoying.

DJ monitors (2x15in wedges or 'DJ fill' cabs) can be pointed away from the audience and adjusted so DJs can bathe in exactly the sound they need to perform, without inflicting it on everyone else. Typically, DJs who badly need a rest from bad sound systems will want to apply lots of treble boost to create the hard sound they've become addicted to.

If not supplied from a monitor desk, with a monitor engineer in charge, such DJ monitoring should be 100 per cent separately adjustable, with a self-op (self-operated) mixer of its own to provide all the required EQ and gain control. It may also have a graphic and compressor/limiter. The PA's monitor engineer could help to set up the controls.

CHAPTER 9

troubleshooting & maintenance
diagnosing & solving problems

Quick view

Occasionally when you have wired up a PA, bits of it don't work, or they work unexpectedly, or all of it is pretty silent. The cause could be something very simple but vital (like you've forgotten to plug in a cable), or something far more subtle and hard to pin down. This chapter takes you through some of the common (and some less common) reasons why PAs don't work properly, and how you can diagnose and remedy these difficulties as quickly is possible. The second half of the chapter is all about regular and effective maintenance – because prevention, as we know, is better than cure.

Spotting and fixing problems
Electronics, like computer programming, is fussy about 'syntax' (the order in which things are structured). The slightest thing out of place, or working differently from usual, might cause the PA to malfunction unexpectedly. In fact the amount of circuitry employed in even quite a small PA system makes it almost miraculous that it ever works smoothly at all.

Almost any problem that manifests itself can have more than one possible cause. And often, as a sound engineer, you will be under pressure to diagnose and cure the fault in a hurry. You may have to trace and scrutinise a perplexing chain of events, systematically and methodically, like a detective solving a complex case – and probably against the clock.

Following the signal path
The majority of problems will be caused by a fault with the PA's signal path. But how do you track down the fault? The main point to keep in mind here is that this is a path – musical information from instruments or pickup cartridges simply flows in one direction only.

It travels from the signal sources, through various bits of gear, sometimes branching off then rejoining at places (via auxiliaries, sub-groups etc), and onwards to the speakers. And it gets modified in quite predictable ways as it goes along.

Compared to some systems, it is relatively simple. For instance if only the mid drivers are silent or distorted, you know the fault is most likely at or after the crossover, and in the mid-range section only.

Another useful feature is having 'stereo': the PA's two (or more) channels provide not only 'parallel redundancy' (meaning one can take over both in an emergency), but a point of comparison and sometimes 'sanity check' at each stage.

We'll look in detail at some specific problems in a moment, but we'll start with a troubleshooting overview. Many PA signal path problems can be traced by answering one or more of the following questions:

● *What is stopping the signal getting through?* To answer this, the first step may be to find where exactly the signal *is* getting through to? Whether it's quickest to work backwards from the speakers, or forwards from the beginning source, is something only instinct (or experience) can tell you.

After that point is found, you can begin to figure out causes – such as a half-mated connector, broken wire, blown fuse, wrong connection etc. Where tests says it has to be a cable or unit, prove this by temporarily substituting the gear's other channel, or other channel's cabling, as feasible – or, if not, use fresh cabling and/or an equivalent unit.

If you can't find the fault, you must have skipped testing something

● *Where and how is the signal changing unexpectedly?* Let's say it's getting distorted, or losing treble at some point. As above, this requires tracing back (or forwards from the beginning) to a point where the problem isn't occurring. After checking for obvious causes (which we'll look at shortly), don't let anything escape suspicion. Many times the thing you decide in advance cannot be the cause turns out, hours later, to be the precise cause – possibly in some highly unusual yet explicable way.

● *Why does the PA sound different on one side?* As above, this requires tracing back (or forwards), to a point where the difference isn't occurring. While the stereo signal lines are run side-by-side – that's up to the returns multicore/snake terminating into the feeds to each of the stage wings – it is easy to troubleshoot by 'flipping'

the sides over, and noting what happens.

These different techniques may be flexibly applied to get the quickest results. Let's take the example of a simple audio path, comprising a sound source (let's say a turntable) DJ mixer, crossover, amps and cabs. The problem? No sound at all in the left channel of PA.

Put on a record and observe the following (it's helpful if you use an oscilloscope, if possible – see later in this chapter for more on 'scopes – or at the very least a good pair of headphones):

● See if the metering on the mixer registers activity – check all controls are up, and routing is correct. Use 'PFL' (pre-fade-listen) switch to check if source can be heard. If not: check the leads, and cartridge connections, or try another turntable. Also try another (stereo RIAA EQ'd) mixer input, or an outboard disc input pre-amp, etc.

If the mixer's metering does show the signal is reaching that far, now connect test headphones or scope directly to mixer output. Can you see a signal of at least 1V peak. Into medium impedance (75 to 150 ohm) headphones, that's a fairly loud signal, but it will be easier to hear if other sound sources can be kept down while you are looking.

If there's no signal here, you need to ask what control on the mixer is stopping it. If you can't find one, does the mixer have another output that does work? If so, why? Maybe that helps guide you to the problem. Failing that, try another mixer.

● If the signal gets this far, now move on to the crossover input. The crossover may have metering which shows if the signal has reached it, but if not, connect the 'scope or test cans to the cable coming in to the crossover.

If there's no signal here, check the cable for damage and faults, and fix – or trace the length and find out where connecting leads have been joined and come apart (or were never plugged together when the system was wired up, late last night). Failing all else, simply try different cables.

● If the signal is getting to the crossover OK, next look for it at the crossover outputs. Plug the scope into the low (bass), mid and high (top) outputs in turn. If they are not passing signals (or not at the appropriate frequencies), check the crossover is correctly set up. Are you sure the controls are set right (for instance mute or polarity invert)? Panel markings can be ambiguous. Try the opposite switch settings, just in case. If this fails, is the crossover really powered up?

If no meters or LEDs show, maybe a fuse has spontaneously died (of old age). Check any fuses you can, by removing them and using a fuse tester. This procedure goes for any other gear that shows no evidence of being powered. Some crossovers have plug-in cards which may have fallen out, or may be part dislodged, or may have been removed or 'borrowed'…

At any stage up to this point we have the option of 'cross-patching' the L & R ('stereo') channels. This acts at the very least, as a sanity check – to make sure you're not just going mad.
● Next, check the signal is reaching the amps. Good power amps have 'signal present' LEDs – if these are lit up, you know the signal is getting this far. (In fact, if the amps have such indicators, and they're lit, you could start your tests at this point at the outset, to save time.)

If the signal is reaching the amps (and the amps are turned up, and producing a signal themselves) and there's still no sound coming from any speakers, then the cause must be something in the last link affecting all channels – such as a multi-way speaker connector that's undone.

Obviously if only certain speakers are not working at this stage, you can be fairly sure there's a problem with those drivers and/or connectors.

Never assume anything. If you can't find the fault, you must have made an assumption and skipped testing something. Be cynical and test everything to your satisfaction before moving on.

Tracing faults by symptom

Noise – covers a multitude of sins. Try to be more specific about the kind of noise. Is it a buzz/hum/tone (which is a 'non-random' noise, or a hiss/sshh (more random noise). This can help determine what you should be looking for, as follows…

Buzzing or humming – check wiring. Work back from the amps, checking the signal (as described on the previous pages) until where unplugging no longer stops the hum or buzz. There's your cause (or at least the final link or co-cause).

Loud, steady 'raspy buzz' – if it's just on one or two drive units fed by the same amp, the amp has 'gone DC' (see Chapter 5). If feasible, pull out the speaker cables (or power plug) to save drive unit from burning out. Do not attempt to restart. Use another amp for now – if the cab still works.

Fuzz/distortion – when heard on decaying notes and in quieter passages, may be crossover distortion caused by an amp fault; or a rubbing voice coil (see later in this chapter); or a stylus encrusted with gunk…

Hiss – there are many different types, and causes, of hiss. The five most common are as follows:

● A fairly clean, smooth hiss – means the mixer has too many unused channel faders opened up.
● A loud, rough 'surgey' hiss (like a seashore) – can indicate damaged IC chips.
● A more synthetic, machine-like hiss, like a

Top ten causes of PA problems
● Connection failure. This includes: lead/cable failure – caused by broken or open wire, wrong wiring, corroded/very dirty contacts, or two or more shorted wires; plus plug or socket problem – bent, retracted or broken pin or receptacle, corroded/ very dirty contacts, general disintegration (quickened by use of flaky, low-budget connectors); and human error – leads or connectors not properly mated, or wrong connector forced into place, causing damage to pins or receptacle.
● HF diaphragm fractured, or voice-coil 'blown'.
● Bass driver blown, or rubbing.
● Amplifier RF burnout, or amp protection cuts in.
● Compression-limiters badly set, including excess compression.
● 'Open' mikes being dropped.
● Unsafe electrical condition or 'automatic re-closure' repeatedly trips-out and resets the power when PA is running. Likewise, turning a PA on-and-off all at once.
● Inadequate or incorrect ratings of power cabling or supply, or speakers, or amps.
● Spilling liquids over the mix console's faders.
● Generator (if used) running out of fuel or stalling before the gear's power is turned off.

'noise modem' – probably a digital problem.

● A swirly or 'birdie' hiss – RF oscillation (interference).

● A hiss that sounds like someone hoarsely whispering – TV or radio pickup (possibly from a mast nearby). Either that or a poltergeist.

To trace which part of the system is introducing the hiss, break connections temporarily – starting at the mixer outputs, moving on to the next item in the chain until noise reduces substantially. When it does, the gear you have just 'passed over', or its connections, are implicated.

Odd or 'coloured' sound – are drivers reversed (for instance is bass sent to mid, mid to top, etc)? If so, mute system immediately (to minimise damage to high-frequency speakers fed with LF signal. Errors could be at cabs or crossover.

One sharp shock can effect miraculous cures

Intermittent sound – look for a mechanically loose connection. Try new cables. If not obvious, the fault may be hidden inside gear. Where all else fails, turn each item off, remove from rack (where necessary) and shake about. If loose parts are heard, get the amp serviced. If not, get physical... One sharp shock can effect miraculous cures. Lift the gear's front up about two inches and slam it back down. (It's been known to work, but can't be guaranteed – and we can't be held liable if you cause more damage...)

Weak sound – are all the frequency ranges connected to the right drivers? Are the drivers still alive? There may be shorted inputs or speaker cables – or a short inside a cab. If amp 'signal present' isn't showing, check for low gain control settings, amp or drive rack or mixer mutes, mixer solo... Or look for unpowered (switched-off) equipment in the signal path (some of which may pass clean but weak signals).

Fixing painful sound

It's easy to have gross distortion in an otherwise healthy PA – only one little thing need be amiss.

Compared to the effect on the ears, the causes can be surprisingly subtle or trivial-seeming. If the PA causes discomfort or actual ear pain (more of which in Appendix C), check for the following immediately:

● Rubbing cones in the cabs. Usually caused by past (thermal) excess. These sound horrid at lower levels, though OK when driven harder. Try turning off different sections of the PA at a time. Walk across the front of the cabs to detect possible sources.

● Ripped cones or diaphragms. Sound horrid at all levels. Trace as above.

● Loose cones. Drive units are held together with glue, and once in a blue moon the adhesive fails or is somewhat absent. Sometimes an emergency 'quick-fix' can be done using 'superglue'.

● Dead or dying amplifier. This creates gross distortion. In any ideal set-up, all amp outputs would be 'scopeable' (see later in this chapter). The better computer control systems may permit this. Failing that, check by listening to each cab and driver in turn, after narrowing down by driving different bands, etc.

● Dead IC (integrated circuit, or chip), transistor, power supply, or other key part, anywhere else in the main signal path. All cause gross distortion. Try putting items you can temporarily do without – like house EQ – into 'bypass' mode. Be aware, though, that if the distortion is caused by a bad relay (it can happen), that's the one bit the signal may still be passing through. Follow signal-tracing procedures described earlier in this chapter.

Mixer/FX area problems

Problems around mixers are relatively easy to solve, as all mixers and many FX have meters. One approach is to find something that a meter says is working, and work 'forwards' from there.

Missing signals or no sound

First of all, don't start pushing fader levels up to try to hear the missing sound. It won't help, will simply mess up your mixer settings, and can even be harmful (to equipment and ears) should the signal suddenly reappear.

Work out what is missing. If some signals are

still present, what makes those ones different?

The most common three things that cause lost signals are: mute switches activated without realising; solo functions activated without realising; or polarity (phase) cancellation – in fluke circumstances (see Chapter 7 for more on polarity). Check all of these first, before you start tracing the signal path as described earlier in this chapter.

If there are no signals at all, are there at least any indicators showing? These may show, for instance, whether the power is on, or whether there's a 'signal-present' at any points.

If the mixer is definitely not receiving any power, check the power supply. A well-designed stage supply outlet should have an indicator light. Or else plug a desk lamp, or some other gear, into the same outlet. If the power supply seems OK from the outlet, and the plugs are properly fused and rated for the job, try another suitable power cable (always carry spares). Or try a new fuse. Whichever is quickest.

If this still doesn't restore the power, check any externally accessible fuses in the mixer's rear, or on the panels of the outboard power supply. Be aware that fuses sometimes look OK even when they have 'opened' or blown. Use a fuse tester to be sure, or simply fit ones known to be fresh.

If these fuses are all OK, the 'lack of power' problem may require a major repair in a workshop – either by yourself or by a qualified service engineer.

Power amp troubleshooting

If a general 'fault/error' LED lights – but the amp still appears to be working, check whether there is some obvious cause, such as a gross cooling blockage (is air being sucked one end and blown the other?), or an excess number of speakers connected, a shorted speaker cable, or maybe too much drive level or a higher gain setting than other amplifiers doing the same job. If a cause can't be found so easily, switch off if that won't stop the show. But if it will affect the show's sound badly, leave 'as is' but reduce level by say 6dB.

If an amplifier shuts down, try to find and bypass a suspected cause or fault, before attempting to reset.

Before undoing or making or otherwise touching the output connections (speaker cabling), turn down the gain controls (note where they are first, for correct re-setting), or disconnect input drive, or mute the signal drive – whichever is easiest.

Tips from the top
Practical production solutions

● Say the direct feed from the guitar amp is horribly buzzy and noisy – it sounds like an earthing problem, but there's no time to trace and fix it. And the precious VIP guitarist won't allow any backline cabs to be miked up... So what do you do? One answer is to use a spare backline amp, fed with the guitar signal, placed under or beside the stage, and mike this one. A noise gate, properly set up, can also quieten the guitar buzzes between notes.

● A safety official (with crinkly yellow jacket and clipboard) has condemned the tall stack of out-front PA cabs as unsafe. Solution: ask the venue management where the rigging points are in the ceiling. Rigging straps or suitably rated ropes are then used to secure the stack to the rigging points.

● To avoid the clutter and visual obstruction caused by bulky floor monitors, one (high-budget) solution – as used by Pink Floyd, among others – is to use under-stage monitoring, with the monitor cabs pointing up from beneath open grids fitted flush into the stage floor.

● When the PA is flown, it's possible that the front rows of the audience might miss out on some of the signal – the sound can travel over their heads and they only hear the monitors and backline. This can be overcome by using 'groundfills' – full-range PA cabs placed under or beside the stage.

● Miking an orchestra that's seated underneath a flown PA can be a problem – strings are fairly quiet, so the mike level needs to be high, increasing the risk of picking up spill from the PA, and even feedback. If it's not feasible either to move the players or re-position the PA away from them, one solution is to alternate the polarity of neighbouring players' mikes, to reduce (partly cancel out) the ambient soundfield. Alternatively you could use lapel (tie-clip) omni mikes taped to the rear of the string instruments' bridges, which helps reduce spill into individual mikes.

● When using a revolving stage (not common, but used in some big-name productions) it is normal to reverse the stage's direction after every two acts, to avoid twisting the multicore cable/snake. Multicore lines have been lost in this way before – effectively by strangulation.

troubleshooting & maintenance

Also, before powering down any preceding gear in the chain, either first mute or power down the immediately preceding item if known to be 'clean', or disconnect the amplifier inputs (by unplugging the returns multicore/snake). The amplifiers and all preceding gear can then all be powered down safely and in any sequence.

Set up a programme to check and clean the amp-cooling airflow's pathway regularly. Replace or rinse filters and use an airline to clean the tunnel or heatsinks inside – see maintenance section later in this chapter.

In general, avoid 'adverse' loads (see Chapter 5).

RF troubleshooting hints

If an RF problem is suspected, there are two initial approaches. The 'blind' approach is used ad-hoc in the field, and assumes the RF problem produces some sort of symptom – as described above. The 'sighted' approach requires an oscilloscope, fitted with an RF 'sniffer head' or coil, or else a stand-alone RF 'sniffer'. These are specialist items that a qualified practitioner can set you up with, and which should be carried as part of a large PA's troubleshooting kit (see Tools section later for more on these and how they're used).

RF problems can also be caused by wiring faults

The RF is usually either being injected by the input cables, or it is being provoked by something to do with the output cabling or connections and parts (drivers, DC protection capacitors, possibly passive crossovers) inside the connected cabs. Unplug one at a time to see if the symptom then changes or goes away.

If the input signal appears to be responsible, trace back down the line. First try bypassing the crossover or delay – or whatever else is preceding the amplifier. Of course, this may only be feasible outside of showtime. Also, for the crossover, this is safe with bass, sub-bass and low-mid signals, as the drive units for these bands can be safely fed higher and lower frequencies, as applicable. But for amplifiers handling HM (high mid) and HF,

you'll either have to try swapping to the other channel's output (L or R), or else connecting to another crossover's appropriately set HM or HF band outputs.

If the problem isn't caused by a particular output on the crossover, then it's easy enough (at least while setting up) to bypass the graphic, input delays, and to mute the signal from the mixer. If the problem seems to be coming from the mixer or an EQ unit (whether the house graphic or an inserted PEQ, say), check the settings, and try turning down any HF EQ back towards 0dB (less boost or cut).

Once the cause (or aggravator) is found, a way to recover the required EQ can be looked into. In other words, the HF EQ (say) that appears to cause the problem may be boosting RF signals picked up elsewhere, or it may be making them itself. Further experiments will be needed to determine which.

If any equipment generates RF by itself, and once you're sure it's not caused by a wrong connection (which we'll come to in a second), it's a good idea to put gaffer tape over the main controls and outputs, with a label like 'Danger, RF problem – do not use', just until that gear can be serviced.

RF can be caused not just by sick or damaged equipment, but also by wiring faults. For example, if one side of a balanced output is shorted (perhaps because a balanced-to-unbalanced converter lead has been inadvertently plugged in-line), the output side that has been shorted may sometimes produce RF oscillation.

This sort of misconnection may well remain invisible without astute cable and connection checking, or noticing that one output strangely needs a little more gain – because an unbalanced output will be 6dB lower in level than a balanced one, in ordinary conditions. Loops (dual path ways) in cable shields can also aggravate RF pick-up (see 'pin 1 lift' mentioned elsewhere).

If the RF problem re-occurs with nothing connected to the amplifier's input, but just after re-connecting the speaker cable(s), first check the cables alone without the cabs – ie by unplugging at the cab end. If this solves the problem, the cab concerned may be wired wrongly.

Most often the RF is either being provoked by the cabling, or is being picked-up by it. To eliminate the latter, it's useful to know that RF pickup is more efficient in some 'planes' than others – meaning the direction and positioning of the cables can make a difference. This suggests trying another route to the speakers, with different lengths of horizontal and vertical runs, and with some of the horizontal runs changed to lie at 90 degrees to the previous cable run – if possible.

RF that's picked up can also be magnified by some cable lengths which happen to cause resonance. So shortening or (surprisingly) lengthening the cable may overcome the RF. This can also work if the cable itself is making the amplifier unstable.

If the cable does appear to be the cause, a more practical solution is to fit an 'external inductor' in line with the '+' output of the affected amplifier. This stands between the amplifier and the cable, so the cable is no longer able to upset the amplifier. It can also reduce RF entering the amplifier, in case that's the problem. A suitable inductor also has no effect on the wanted audible signal, at any power level.

DJ troubleshooting

As well as the kind of buzzes, hums, unwanted silences and other faults already explored in this chapter, there are some problems more specifically of concern to DJs, and any sound engineers working with them.

For a start, all DJ turntables (and any standalone CD players) must be stable, meaning all four feet are touching the mounting surface. If that's a problem because the mounting surface is overly warped or bumpy (for instance the turntables have been optimistically put on an axe-hewn bar table), consider using another surface. Check it with a spirit/bubble level if possible, while adjusting the unit's feet as required. If the feet aren't adjustable, or there aren't any, use time-honoured methods: beer mats, flyers or other thin card or thick paper, folded as required to get the right depth of wedge or shim – but large enough so the decks don't slip off the edge of one during the show.

Serious vibration of a CD player can cause jitter, and force automatic error correction – which is cleverly designed, but still doesn't always work successfully.

If there's a problem with stylus-skipping on the turntable when people dance, one common solution is to add weight to the 'tone arm'. This is not the best first approach, for two reasons: it not only ruins the sound, particularly the treble, it also ruins the records.

One way to reduce bouncing problems is to add weight and solidity to the decks or the console stand – so there's less chance of accidental movement – perhaps by using bricks or concrete 'breeze' blocks, or anything available.

A less weighty and more organised approach is to use an isolation platform, as described in Chapter 1. In extreme circumstances the stage-boards may need to be 'tensioned' from underneath – this is a rigger's job. Another good idea is to keep the decks away from bass and sub-bass cabs, which can cause a lot of vibration.

Sometimes the sound just isn't right – too 'hard', too quiet, muffled, distorted, or generally dreadful. The cause could be one (or more) of many things...

First of all, before going through the whole PA looking for faults, check the records/discs themselves – dirt, scratches, sticky fingermarks, even condensation (if the DJ has just come in from the cold) can have detrimental effects on the performance of both vinyl and digital discs.

A turntable's tone-arm has to be balanced and weighted properly; a diamond stylus is pretty tough, and in fact rarely wears away, but can become encrusted with grime, or get chipped or break off; or the cartridge can become loose or damaged – all possible sources of sound problems. Work with the DJ to check these; DJs should also always carry spares. (See Chapter 1 for more on DJ gear.)

All DJ turntables must be stable, meaning all four feet are touching the mounting surface

Check connections too – look for loose or dirty plugs and sockets. Spray with a contact enhancer/de-oxidiser like 'ProGold' (if gold-plated) or 'Deoxit' (if not) – assuming this sort of treatment hasn't been done recently. This will reduce radio noise interference (in case there is a lot of it about), which can clean up sound remarkably.

Once you're sure the sound sources aren't the problem (or part of it), start to work through the rest of the PA, asking the kind of questions we've covered elsewhere in the book. Is the gain structure OK? Are any overload LEDs on? Why is this lead plugged in here? Should that button/switch be in or out...?

Be prepared

A good general rule about troubleshooting is to always expect the unexpected. Gear is vulnerable to all kinds of abuse at live events – not always from likely sources.

To suggest some of the more extreme examples: speakers have been known to rip after being drenched in all manner of fluids by audience-members; the mix of chemicals that PA gear is exposed to during a life on the road (from mouse excreta to stashed drugs) can rot away circuit boards, and can itself lead to weird, intermittent distortion.

It's not unknown to find cats, and even people (perhaps somewhat the worse for wear) snoozing inside active bass bins – which can deaden the sound somewhat...

Vocal microphones get an especially rough deal. As well as physical damage from dropping and knocking, they can develop distorted and lowered sensitivity thanks to blocking of the capsule's sound entrances by saliva, tobacco residues... and whatever else comes out of the mouths of various singers.

All this abuse makes it obvious that good maintenance plays an important role in the working life of any PA system – and hence any sound engineer.

The section coming up now lists and explains many of the tools that can be used for diagnosing and repairing PA faults, or for counteracting wear-and-tear.

This is followed by a section on maintenance, offering practical advice on keeping various parts of the PA in clean and efficient working order, effectively extending its useful life – and making yours a lot easier.

Tools & testers

'Scope (cathode ray oscilloscope)

A 'scope (as it's commonly known) is by far the most useful, basic test instrument for spotting blatant problems with audio signals. It allows you to literally see signals and their waveshapes – it shows you whether or not a signal is present on a particular path or line, as well as the frequency and level of that signal, and whether any unwanted signals are competing with it.

To make absolutely full use of a 'scope, a knowledge of electronics and signals analysis is required. But as the data is presented as a pattern, to some extent it can be used instinctively – particularly when there is a good (healthy) signal alongside for comparison. You don't need any special mathematical or even verbal skills – you can just 'point and grunt' at what looks different.

Users of computer sequencers or samplers which display music as waveforms (or students of physics or electronics) will already be familiar with this kind of visual representation, and you can soon learn what certain sounds and problems look like on the 'scope screen.

For example: a pure, constant tone produces a sinewave, which displays the simplest and smoothest undulating waveform on the 'scope screen; the higher the pitch, the closer together are the undulations (higher frequency of cycles = more Hertz).

To some extent a 'scope can be used instinctively

RF noise on the line shows up as a thickening of the wave trace itself, while hum on the line is seen as a secondary 'ghost' wave that drifts above, below, behind and in front of the main signal. A distorted signal with hard clipping will appear as a wave with flattened peaks.

A pure, mid-frequency squarewave (in its pure form seen as an entirely right-angled waveform) can be used to test general frequency response through parts of a PA system 'at a glance'. But note that a squarewave can damage some power amps and drive-units if applied at medium to high levels, so it's best to disconnect the power or mute all amps, and confine injecting and reading square-waves to all the preceding parts of the PA.

If there's a high-frequency emphasis to the signal, this puts a visible spike on the leading edge of the squarewave on the scope screen; while an HF cut (whether caused by EQ'ing or cable deficiency) means the leading edge of the squarewave image is rounded off.

With an LF boost, the horizontal sections are seen to be sloping (resembling a shark-fin pattern); while an LF cut or early roll-off is shown by the trailing edge 'fading away'.

Choosing a 'scope

To be useful for sound engineers, a 'scope should have a 'storage facility' – where the display has a 'slow-fade trace' option that keeps the waveform on the screen long enough to examine even quick-changing music signals.

Repetitive signals, on the other hand – whether these are ones you intend, such as steady test tones, or ones you are tracking down, such as hum and RF instability/oscillation – are 'triggerable'.

'Scopes come covered with knobs – almost as many as on a mixer

This procedure is something similar to using a strobe lamp to make moving items appear stationary – the idea is that a stable display can be obtained of an object (or wave) that's changing maybe thousands of times per second.

The trouble is that few if any good 'scopes are readily available in compact, portable formats. Decent quality 'scopes use cathode ray tubes (relatives of those used in TV and video monitors), which offer superb brightness, are viewable in pitch black or bright sunlight, and allow distance viewing, wide viewing angles, good sharpness and fast response. But they are also bulky and delicate, and because they use enough energy to make battery powering largely impractical, they need to be AC-powered.

The best 'no-nonsense' 'scopes are also very expensive, though they can sometimes be picked up second-hand (look first for American or European brands such as Tektronix, Hewlett-Packard, Philips).

'Scopes with liquid crystal and other flat displays have been attempted but have yet to be widely accepted for general use. The rather poor 'consumer grade' design (as opposed to 'professional grade') of many modern test instruments means even expensive and so-called 'high impact' plastic units can be wrecked under real-world on-the-road conditions.

Modern 'digitising' 'scopes, claiming huge bandwidths and many fancy features, are less than ideal for audio type (varying) signals, except in digital signal format. Of course, makers may not tell you that.

'Scope features

'Scopes come covered with knobs – almost as many as on a mixer – but fortunately many of these are not needed for audio testing. For PA work you should look for the following features:

- Dual inputs – two channels allow comparisons. In fact it's rare for 'scopes to have only one channel; some have extra plug-in receptacles. A minority of top-quality 'scopes have differential or 'subtractive' inputs, which allows 'no hassle' connection to balanced lines.
- At least 30MHz bandwidth – well above audio, this allows you to see some RF noises or oscillations. 100MHz is better. In fact, few 'scopes have bandwidths below 10MHz.
- All decent 'scopes have BNC sockets, for coaxial cables. To connect into a PA, you will need to have suitable BNC-XLR cables for line or speaker connections.

Fitting BNC plugs requires some skill or, if crimp-on types, special tools. But pre-made cables of a suitable diameter can be bought and chopped up and the open end(s) fitted with suitable XLRs etc.

Radio frequency (RF) detection

For reliable detection, a 'fast' 'scope with a wide bandwidth of at least 100MHz is needed – otherwise you might just see nothing on the screen, even when there's a lot of RF around.

Specialised RF 'sniffers' are simpler. They either have LEDs that light to show the presence of RF over a wide range of frequencies (from 100kHz or so up to 1GHz is ideal, though up to

250MHz will be almost as useful), or else a meter needle moves. Radio amateurs and CB'ers use a similar device with an aerial/antenna, called a 'field strength meter', but you will need the 'cable sniffing' type with a detector coil.

Using an RF sniffer, or a 'scope with an inductive 'RF sniffing' sensor, doesn't require any electrical connections, so it can't cause grounding problems or trigger instability itself. The sensor coil or head is just held against the speaker cable, preferably the amplifier's '+' output wire. Once the problem's location has been established, the sensor may be temporarily taped or tie-wrapped (using a disposable plastic tie, which has ratchets to keep it tightly closed) to the cable or wire, so that there can be consistent monitoring.

Cable/lead testing

There are two types of 'continuity tester' – audible and illuminating. Both are useful for odd jobs where speedier dedicated testers won't easily reach. Use low-voltage types with limited current, say 4.5v and 50mA. Take care to use only on free cables, wires and connectors, not on equipment inputs and outputs or raw electronic circuitry, which might be damaged.

Industry-standard units have sockets organised so that common cables can be plugged in, with LEDs indicating 'OK' or faults. Buttons and switches are used to do different tests, for instance core-to-core, end-to-end, etc. For multicore/snake cables more specialist testers are needed. Typically these will speed up the job by checking all the connections at once.

Multimeters (DVM, DMM)

For field work it's best to use a portable type. It's best to go for professional grade 'originals', rather than cheaper copies, despite the higher prices. The difference may well include years of life and survival in real working environments – cheap ones go 'phut' easily.

At the very least, a 'rock'n'roll' grade meter should be rubber-covered – they need to withstand being dropped regularly.

Conventional V-A-Ω (volt-ampere-ohm) ranges will be OK. A diode test range is also essential. Frequency and capacitance are possible extra test ranges, and may be more useful than a transistor tester option.

LEDs (light-emitting diodes) or other bright, self-illuminated displays are excellent for readability and can be used in dark places, but are too heavy on power to be found in portable instruments, other than serious military kit. Instead users are forced to accept the hassle of an LCD (technically 'liquid crystal display', though sometimes disparagingly known in professional circles as 'liquid crap display'). If you have to use an LCD-fitted device, try finding one with high contrast for better visibility, and a screen designed to be viewed from a variety of angles, with a back-light you can leave on (ideally auto-timed to turn off after a few minutes).

Test meters should have analog as well as digital readout, to help tune-ups. Types that can read DC+AC as well as pure AC can be useful. Types that have true RMS (TRMS) reading are more useful for AC and audio work.

Fast reading can be essential. Budget types tend to be slower. A faster response time is a lot less stressful if you are working under pressure, and possibly also in an awkward position – like dangling from a scaffold – when you're taking the readings.

A rock'n'roll grade meter should be rubber-covered, to withstand dropping

Leads should include a thickly insulated set with shrouded high-voltage plugs and in-line fusing, to be safe for use on high-current power supplies (more of which in Chapter 10), and a choice of sharp prods and various croc-clip ends, all insulated.

A pluggable, modular 'ends' system can be made with 4mm plugs and line sockets. Plugs that have spring-loaded shrouds are best – they can't short on anything when unplugged, or if they drop out.

Signal generators

For bench use, there are many suitable models – even if some are almost museum pieces. Few cover the audio range you will need (at least

20Hz–20kHz) in one sweep. Expect to switch through several ranges of tuning, like the short wavebands on an old radio set. Still, the less switching the better.

Cheaper signal generators have just a single output level knob. Professional grade models have switched 'decade' (10dB-per-position) ranges as well, so a level limit can be set. The frequency dial calibration will usually be mechanical. Accuracy is not so important if you have a frequency meter useful down to 10Hz – though even today surprisingly few types have this built-in.

For field use you need small, battery-powered and lightweight signal generators that are also rugged, though these are rare.

Signal tracers

Most are 'bespoke' (custom-made), though a few makers have supplied them as accessories from time to time. The simplest DIY 'tracer' (and lightest to carry about) is a set of high impedance headphones – which should be 500 ohms or more.

With both sides wired in series (to give 1000 ohms) into a 'mono' quarter-inch ('phone/jack') plug, these can be driven by most line level outputs (feeds) without loading them down excessively. This means that lines you probe across shouldn't 'know about it'.

To get easy access to lines (where they can be unplugged for just a moment), several male/female tails will be required – for instance an XLR line F to M (which inserts into any XLR line connections), or quarter-inch line socket to line plug (mono and stereo types).

Bench Tools

For your regular workshop. Basic equipment is listed first, while 'extra' suggests more specialised additions you might acquire later.

Screwdrivers ('drivers): flat-bladed – wide selection from 1-15mm blade widths; cross-head (x-head), Pozidrive ('Pozi' or 'Supadrive') – sizes 00 up to 3; and Philips (an older variation on the cross-head screw).

Choose handles that feel comfortable, not what looks pretty – bear in mind you may be sweating to tighten up a screw, and soft handles aren't very clever after they've got soaked in sweat and grease. Cabinet maker's handles are used for

high-torque jobs (wood); engineer's handles are made for ease of precision turning and 'feeling' fasteners.

Extra: a hand ratchet may be useful, and a battery-operated (cordless) drill/screwdriver. Plus special screwdrivers – stubby ones, with tips as above, or 'bendy' ones that access round corners, and types that grip screws until started. Also a plastic trim tool with a fine metal blade (for mini and micro presets).

Choose handles that feel comfortable, not what looks pretty

Cutters, shears & drills: a selection of tools rated to cope with assorted materials of varied sizes and toughness, from cutting 1mm to 50mm copper, and up to 5mm steel wire. Should include mini and standard hacksaws and selected blades. And drill bits – buy as a metal-cased metric set in half-millimetre steps from (say) 1.5mm up to 12.5mm (half-inch).

Extra: set of drift punches. Set of 'non-metric' drills (for older gear). Taps from M2.5 to M12, and several tap-wrenches.

Pliers, grippers & clampers: range of long-nosed, smooth and gripping types, from three to ten-inch length. Decent types have jaws that meet solidly and don't bend or get loose-hinged with use. They'll cost at least £30 ($45). Long-nosed and ordinary mole grips in various sizes. Crocodile clips, selection of clamps (plus bits of wood to make clamping jigs), portable bench vice with jaw softeners.

Extra: types with bent and offset jaws, and/or round jaws.

Hammers: a few different hammers, wooden and rubber mallets; centre punch.

Spanners: open and socket types, also nutspinners, with precision engineering sizes (8BA and M2.5 upwards) up to general engineering sizes (British BSF, BSW; American UNC, UNF).

Extra: sockets, hex to square drive-converters; torque setting gauge.

Soldering iron: most jobs can be covered by a

troubleshooting & maintenance

temperature-controlled soldering iron (about 50 watts) with a range of tip (or 'bit') sizes. Usually AC-powered, but with its own low-voltage power supply. Compared to full-AC powered irons, this is a lot safer for PA gear and users.

Solder itself is usually a mix of tin and lead in a 60/40 blend. For removing old or excess solder, you should have two de-soldering 'guns' (or 'solder suckers') with spare nozzles. Also de-solder braid, for the times when vacuum sucking doesn't work.

Extra: use low-melting-point (LMP) solder for repairs. If you have to modify a printed circuit board (PCB), track slicing and scraping tools are useful. It may also be worth obtaining an extra-long, low-voltage extension connecting cable for the iron.

For soldering at outdoor gigs, you may need a butane gas soldering iron (in case the generator is unavailable for any reason), and some kind of wind-break – even slight breezes stop soldering irons getting hot enough to solder properly.

Even slight breezes stop soldering irons getting hot enough to solder properly

Field tools

A selection of the above will also be taken on the road, as well as some specific gig-related utility tools – for example:

- Clamp meter, to read up to 300A
- Sound level meter capable of LEQ readings. With rechargeable batteries, charger and spare battery packs.
- Unbreakable stainless steel-cased torch (ex-military) – may be rebuilt with 75W, 12v halogen flood and spot lamps. Seriously powerful if short-lived portable lighting. Needs charger and spare 12v battery packs.
- Portable bench vice.
- Hair-drier, or two, to help dry-out any PA gear that gets wet.

Sundries

- Stanley knife and lots of spare blades.
- Felt-tip pens, various colours, fine and chisel tips, which can write on metal, plastics, glass, etc.
- Set of crimp terminals, various sizes, matching types.
- Several 'clip-on' desk lamps, angle-poise type, etc – must be robust enough to withstand knocks and shocks. White bulbs for general illumination, red or orange lights for showtime illumination of mixer, FX rack, etc. Always carry plenty of spare, assorted light bulbs.
- Heat gun and heat shrink tubing in various diameters and colours.

Consumable spares – replacings things that wear

- Spare fuses – a kit of the commonest values (various A ratings), sizes (mostly 20mm and 1.25in, but some odd US sizes occur), and types (F or T, and glass or HRC). A good starting point is to list all the fuses in the gear you use, define categories as above, then bring a pack of ten of each sort.
- Spare lamps – for instance for mix console switches and meters.
- Spare drivers – assorted bass & mid chassis drivers.
- Spare diaphragms – for HF drivers.
- Possible addition: assorted recone kits.
- Newspaper or a rubber 'service mat' – a safe surface for working over when changing diaphragms.
- Vinyl tape – for cleaning out gaps in voice-coils (see under Maintenance, p173), among other uses.
- Bottle of Ferrofluid (see p173).
- Blotting paper (ditto).

'Get out of shit' kit (or 'punt' kit)

A box packed with goodies used to sort out problems in the field. Well-equipped hire companies will station two such units, typically at the monitor console and FOH mix location. Here is a typical list of contents, listed in order of what they're intended to fix – though the kit will vary in light of individual experience:

Microphone fixes – Outboard phantom power supply – cables from capacitor mikes are plugged 'through' it. Useful if the mixer's phantom supply goes down.

Passive mike combiner – enables two mikes to

share one cable, where appropriate, and where levels don't need to be adjusted separately. Could be useful if more mikes are required than there are mixer channels, or if some vital input channels are lost.

Line level fixes – A collection of cable connectors and adapters, in as many combinations as you can think of – mainly needed to solve equipment incompatibilities, and when musos don't bring their own kit... For example, you could have:

- Polarity reversal XLR-XLR leads (2 of these)
- XLR 'Y' leads – female to two males (x2)
- XLR 'Y' leads – male to two females (x6)
- XLR male to male leads (x2)
- XLR female to female (x2)
- Male to female XLR with pin 1 lifted (un-connected) (x2)
- TT (mini RTS jack) to male XLR (x2)
- TT to female XLR (x2)
- Mono quarter-inch (jack/phone) plug to male XLR (x2)
- Stereo quarter-inch (jack/phone) plug to male XLR (x2)
- Mono quarter-inch (jack/phone) plug to female XLR (x2)
- Stereo quarter-inch (jack/phone) plug to female XLR (x2)
- Phono to male XLR (x2)
- Phono to female XLR (x2)

Mixer fixes – Box of spare lamps for meters, etc (as appropriate); set of spare power supply fuses; spare power supply cable; spare input channel.

Headphones
- One pair high-Z headphones (200–400 ohms)
- One pair low-Z headphones (15–30 ohms)
- One converter, from 3.5mm to quarter-inch jack/phone plug

There are two types of maintenance: fixative and preventative

Maintenance

There are two types of maintenance – fixative and preventative. Some of the 'fixative' stuff has been touched on in the 'Troubleshooting' section earlier, and takes the form of emergency repairs.

Ideally you should spot fledgling problems before they fully develop – and this is where regular, on-going maintenance checks can be invaluable, however time-consuming they may seem. It's worth putting aside a couple of days on a regular basis for these kinds of jobs – something to fill a rainy/boring afternoon, perhaps.

You'll find a number of hints here on recognising and averting trouble. We've also included examples of some of the more common repairs and fixes you're likely to have to make to an average pro PA system.

You may be lucky and not need to deal with the more serious ones very often – on the other hand they may become regularities, and you might find a few different ones cropping up as well. All part of the job.

Microphone maintenance
Every month, or just before a tour:

- Check overall condition. Look for cracks in the body, loose screws, or loose grille.
- Remove grille, and clean (first ask the maker about the best procedures), then dry out, then replace grille.
- Store under cover in a dry place.

Speaker maintenance
In general, to avoid speaker damage:

- Be on guard at all times to avoid clipping – anywhere in the system but particularly in the amps.
- Use amps with DC and RF protection– preferably some form of 'anti-clip' (see Chapter 5).
- Suitably set-up compressor-limiters can help (see Chapter 4) but are not the whole answer. While they reduce peak power, they increase average power, so in extreme cases their use is self-defeating.

The 'wrong' sort of signals passed into the speakers (see Chapter 6 for more on speaker

171

troubleshooting & maintenance

signals) can cause breakdown in one or more of the following ways:

- Ripped cone, or surround. Caused by large, sudden excursion (movement), produced by something like a dropped mike, or hard-clipped kick-drum sound. Problems can be exacerbated by an ageing surround, so it's more likely to happen if drive units have been exposed to sun's UV light, or are a few years old.
- Burnt voice-coil. Usually it's just the plastic insulating coating on the wire that chars. The surface effectively expands, causing the voice-coil to rub on the magnet, whereas the gap between them must be kept small for the best

Rubbing voice-coils sound horrible at low levels, but OK worked hard

efficiency. Some speakers that appear to be rugged have wider gaps. These allow the speaker to keep working even if the voice-coil enamel is slightly charred or 'bubbled'.
- Rubbing coil. Rubbing may be caused by burning (as above), but it can also be caused by magnetic (usually steel) 'swarf' or metal hairs that have been attracted into and then trapped in the magnetic gap. This is inevitable if a drive unit's inner surround or dust cap are slit, ripped or otherwise open, but it can also be caused during a repair (recone) operation, or if the cab has received a severe shock. Re-alignment may be all that's needed, but usually a drive unit maker has to do that.

(Note: The sound of rubbing voice-coils is usually horrible at low levels, like a portable radio running on a flat battery, but it often sounds fairly OK when the driver is worked hard.)

Inspecting and testing chassis drive-units
First, carry out listening tests. Sweep (using a test signal) at a modest level, each over rated frequency range. This is best done through the usual crossover. Sweep slow then fast, listening (with ear close to each speaker in turn) for 'scraping' sounds, or other anomalies. Then set-

up the crossover for music at a similar level, and listen to different sorts of material through all the drivers (ie 'full-range') to check for any sonic anomalies. Then re-check at several higher levels, as appropriate.

Next, perform some visual and hands-on tests. Check cone-drivers' front and inner surrounds, and cone inner and outer edges – look for signs of brittleness, perishing, slimy liquid, cracks, tears or other dubious-looking characteristics. Check airtightness of overall cab (see Chapter 6).

Check the cone moves both ways without scraping. Support evenly. Move gently. Check the surround (edging) isn't brittle, or coming loose, or decaying in any way. Similarly check the inner surround (though this generally gives less trouble).

Check that the dust cap (the central dome) isn't loose. If slightly indented, use a pin to 'hoick' it outwards. If considerably dented, obtain spares from the maker (of the cab or of the drive unit). The old dust cap can be gently prised/cut away from the cone. Follow the drive-unit maker's instructions. The new dust cap will usually be glued in place.

Repair or replace?
Emergency repairs can sometimes be made to ripped cones using papier-maché techniques, with several layers of newspaper and paper glue, or a rubber-based adhesive (like Copydex or Evostick). Similarly, surrounds that have come loose can be glued back into place (with due care over alignment), using 'superglue' (cyano-acrylate), and this can save the driver, assuming no 'scraping' damage has already been caused to the voice-coil by misalignment.

A few speakers are 'field reconable'. Unfortunately, this admirable trait comes hand-in-hand with wider voice-coil gaps, hence lower efficiency. There isn't much choice in driver, either. In any event, this type haven't greatly caught on.

Recone kits can be ordered from the drive-unit maker, or sometimes from the cab makers. Get instructions at the same time. The procedure involves some mechanical skills, adhesives, heat and time. Some experimentation on old drive units is advisable before reconing front-line units. Alternatively, seek professional help with repair.

Replacing drive-units

Damaged drivers are usually removed and swiftly replaced. Exact replacements must be used – a wrong model (even one with only slight apparent variations) may be more readily blown-up, or greatly reduce a horn's performance. For example, some versions of drive-units are made for horn use, with slight excursions, while other superficially similar units are optimised for use in direct-radiating, vented cabs (see Chapter 6).

Damaged drivers are usually swiftly replaced

When replacing drive-units, correct tightening down is important. As with an engine's cylinder head, the bolts must be tightened gradually, and in stages.

Never tighten one fixing fully then move to the next – that could strain the chassis and even cause the voice-coil to rub, either now or later. Tighten one slightly, then the one opposite it, then the others, moving like tying shoelaces (or tuning drum skins), from side to side. Then tighten all in turn, to a second level, and so on.

Air tightness is essential. Many replaced drivers should have a rubber seal on their rims. If this is absent, obtain and fit, before replacing the driver.

Fitting/replacing compression drivers

Compression-drivers' innards are best left alone until the driver sounds wrong. Air tightness doesn't matter in cabs that surround these, but it does matter – a lot – where the horn throat is coupled. This means checking that screw-fitted drivers and adaptors are tight, as well as flange fixing bolts.

Compression drivers can be very heavy, and the larger types (3in and 4in) hard to handle. Arrange your work so the driver can't drop far, or will fall onto something soft, or at least back inside the cab. Shocks caused by dropping such a heavy (yet delicate) object onto hard surfaces (like concrete floors) can reduce the magnet's potency, or crack and destroy it, or shatter other fragile parts inside.

It will be easier to mount a heavy driver (whether screw-threaded or bolted) onto a horn (and cab) that's face down. In the field that will require the cab to be 'de-stacked'.

Diaphragm replacement

The complexity of this operation varies widely. With older designs it's generally more likely to be an intricate, precision engineer's job.

Note: if you're going to attempt this, read the whole of this section before starting – and practise on an old driver before doing it for real.

The first thing to ascertain is whether or not the driver uses/contains 'Ferrofluid'. If so, see the panel below on 'Handling magnetic fluids'.

Tips from the top
Diagnosing mismatched drivers

If you find a particular driver has a lower sensitivity than another, a likely cause is that the magnet has been 'smacked'. This may be evident from chip marks (if ceramic) or dents. The shock of such an impact can easily 'discharge' the magnet, effectively ruining the performance. The magnet can be recharged but will have to be dis-assembled from the rest of the driver. If it's worth doing at all, it's a sending-away job.

Handling magnetic fluids

Ferrofluid is used in some drivers to help increase power handling. When a diaphragm is removed, some of the 'Ferrofluid' (which isn't, incidentally, a significant health hazard – it's just messy) remains on the sides of the voice-coil. If the part is re-inserted without much fluid being lost (onto fingers, bench, etc) that's OK, but if the amount of residue left in the magnetic gap is uncertain, the unit may fail again prematurely from voice-coil burnout, as its power handling will have been reduced.

But the shortfall in Ferrofluid can't be measured accurately by inspection. The only sure way is for all the fluid left in the gap to be removed, and to start again. Ferrofluid, which sticks in the gap like a 'liquid limpet', is removed using an ingenious tool – namely fresh, clean blotting paper (designed for soaking up ink, and usually obtainable from stationery suppliers).

Before you fit the new diaphragm you have to recharge the gap with Ferrofluid. This requires a precise amount to be measured and injected (a process you can find out about from the makers of either Ferrofluid itself (see *www.fero.com*), or the drive unit(s) that contain it.

The final part, getting it to flow into where it's wanted, is surprisingly easy – the magnetic gap sucks it up.

troubleshooting & maintenance

When replacing a diaphragm, it's essential you have a clean workspace, free of any (particularly iron or steel) swarf or filings. Newspaper is a good temporary work-surface. Plenty of light is strongly recommended. Also, a soldering iron, adhesives (rubber-based 'contact' adhesive, and cyano-acrylate 'superglue'). You will also need self-adhesive plastic parcel tape, or at least ordinary transparent sticky tape (eg Sellotape).

Now, dis-assemble the driver and remove the old diaphragm. Keep it, don't bin it – for two reasons: firstly, in case the new diaphragm you have doesn't fit and you need to know some detail about the original; and secondly, you could perform a post-mortem – maybe learn something to help prevent more failures, or spot a manufacturing defect (in which case you might get a free replacement).

Dead diaphragms have been used as ashtrays or paper-clip holders

(Even once you're sure you don't need it any more, it can still come in handy – traditionally, 'dead' diaphragms have been used as ashtrays or paper-clip holders, or just arrayed on workshop walls as trophies.)

The next stage is to clean out the gap. This should already be clean, but if anything has crept in, like a tiny hair of swarf, it could create horrible distortion later, or even damage – regardless of the particle's size. You may be able to see particles or 'grot' in the gap using a strong light and/or magnifier, but whether you can see any or not, this is the chance to remove it.

This is what you need the adhesive tape for. The first step is to remove any grot on the outside of the magnet (most magnets have picked up something) – it's these particles that are most likely to 'infect' your driver's gap in future. The sticky tape is used to pick up dirt, dust and tiny flecks of debris.

Then a short length of tape can be slid into the gap, and swept around or back and forth. Do this methodically, one area at a time, using fresh tape for each sweep – it's cheap enough and

avoids recontamination. Tape needs to be applied in all directions, so any unwanted debris lodged on the inner and outer sides of the gap should be captured.

Fitting & testing

The final step is inserting the new diaphragm. Follow the maker's instructions, and always try a test fit before going ahead with adhesives or other irreversible steps.

A very important feature is getting the polarity (or 'phasing') of the diaphragm or its lead-outs correct. There are usually two possible ways, and only one is right. To help, there are various systems of coding and keying, and the makers' instructions will guide you. Often the sole coded terminal is positive, whatever the colour used.

If in doubt, perform a test fitting and compare with a working diaphragm, using a test battery, and watching which way the diaphragm moves.

When re-assembling, any multiple fixings around the sides should be done-up a bit at a time. Final tightening of a purely cosmetic piece like the lid is not worthwhile until you've tested that the new diaphragm works.

First, you should couple the unit to a horn, then 'sweep' the unit with a low level of about 1V from a signal generator, from the lowest frequency of use, or try 3kHz and up to 15kHz (say). Go up and down. The sound may be piercing, but you shouldn't hear any crackles, buzzings, distorted bits or the like. (If you do, make sure it's not caused by a crackly carbon control pot on the signal generator.)

Then try music – which is fairly important – obtained through a crossover's HF output, and with a suitable frequency setting for the driver being tested. If bad symptoms continue, they are likely to be caused by the driver. This suggests there is still dirt (or something) in the gap, and the only thing to do is to start over again.

Planning ahead

As soon as you start to use a particular cab, it's a good idea to obtain:

● several spare diaphragms for each of the (compression) drivers inside it.
● a copy of the fitting/replacement instructions.
● any special tools or jigs (such as shims) that

might be needed – or will at least help the job along when it inevitably needs to be done backstage, under pressure and in poor lighting.

If you want to increase your chances of getting the right parts, you'll need to know the exact model of the cab and, if possible, any serial number on both cab and driver. That's because, over time, parts used in manufacturing the cabs or drivers are likely to vary. Ask the maker, importer or distributor, or a specialist speaker repairer.

After doing this, it's a good idea to actually try making a replacement. If this isn't possible without wrecking a good diaphragm, go through the motions as much as you can. The idea is to check that the parts are the correct ones and that any tools and jigs also work OK. Then when an emergency occurs, there won't be any unpleasant surprises.

Inspecting & testing PA enclosures
Things to look for:

- Loosened screws – must be tightened (note that 'sweeping' the cab with a signal generator from about 70 to 900Hz, at modest levels, can help you find loose screws or split grilles – listen for rattles, odd sounds and vibrations.)
- Loosened components – inside or out. Tighten.
- Evidence of corrosion, especially on metals, inside. Consult the maker. (This can be exacerbated by smoke machines. Avoid them where possible.
- Evidence of dampness – if so, dry out thoroughly before use.
- Shrinkage of plastic or wooden parts. Consult the maker.
- Broken leads or loose connections to crossover or protection parts. Resolder, and use silicone paste to withstand vibration more resiliently.
- Bad or loose 'press fit' connections (to drive unit terminals). Remake with cleanly stripped and twisted wire ends.
- Dry, decayed or simply bad solder joints. Tug sharply to test.

Painted cabs need occasional repainting (or respraying) or touch-ups. Check and tighten up or replace any loose fastenings or connectors. Straighten or replace dented grilles.

General electronic maintenance & PA protection

- Fuses deteriorate with age, and the chances of them blowing increases, especially if they're marginally rated. Don't re-use old (pre-used) fuses.

If possible, make a note of the replacement date of all gear fuses (with a rugged laminated label) – if they're still in use after three years (or about 2000 hours of use, if that comes sooner), change them anyway.

To be sure correct types are fitted, it's easiest to obtain a spares kit from the gear maker. Never substitute any non-fuse wire or foil in place of a properly rated fuse.

- Keep spillable drinks and 'mixer-unfriendly' food away from the mixer, and indeed all electronics. 'Mixer unfriendly' means anything wet, crumbly, greasy, or where there's a chance that less than 100 per cent of it will make it into your mouth – and inevitably drop into fader slots, vents, or vulnerable areas. It not only causes long-term maintenance hassles, but can have drastic, immediate, and even dangerous consequences.
- As soon as knobs show the first signs of 'scraping', get them serviced. Don't leave them to get worse. Similarly if anything is coming loose, it's best to fix it sooner rather than later.
- If any gear hasn't been used for a while, you can help clean the contacts by turning all toggle and rotary switches rapidly back and forth, say 30 to 100 times. This can seem harder work than spraying the switches, but at

If anything is loose, fix it sooner rather than later

least the gear doesn't have to be taken apart first. And it works just as well.

- Use a mesh to protect 'outdoor' gear from wind-blown grit.
- If pluggable modules (on a mixer or other gear) have to be removed, spray contacts (with appropriate contact cleaning fluid) or otherwise clean them, as a matter of course.
- Guard PA against unauthorised tampering or

troubleshooting & maintenance

unsupervised use. A representative of the PA's owner should always be around to supervise, even if there's a 'guest' FOH or monitor mixing person.

- Supervise or (preferably) be involved yourself in the packing-up of at least the most vulnerable parts of the system – the mixer, FX and drive racks, mikes, cabling, splitters, DIs, and amp racks.
- At outdoor events, be prepared to cover the FOH mix position promptly, on all sides, particularly from the front, with clear plastic sheeting, in case of rain, hail, sleet, snow, windborne dust and grit, smoke, or (the bane of some gigs) 'flying' beer or other liquids.

Cleaning

Keeping control surfaces clean

Knobs and controls on PA gear soon get grimy and 'grotty' with a mixture of finger oils, greases, dust and airborne particles. Control lubrication also leaches onto panel surfaces, attracting more dust.

Before you even think about cleaning, it's a good idea to turn off the power (and/or unplug) if the equipment contains AC voltages. (Sometimes you can clean a large mixing console when it's still switched on if it contains only low DC voltages – but in general, to be safe, switch off the power.)

Many everyday cleaning products are unsuitable for use on control surfaces and knobs

---

Many everyday cleaning products are unsuitable for control surfaces and knobs.

Strong solvents like petrol or acetone should never be used – they can strip off markings/ legends and melt plastic parts.

Even a weaker, generally benign solvent like isopropyl alcohol (a purer form of 'surgical spirit', often used for cleaning tape heads) can

strip or thin printed legends. It can also leave a dull surface. Using hot water with lots of detergent may work best on the panel, but if any gets into the electronics, a major rescue mission may have to be mounted to prevent damage occurring.

One substance that appears to do most of what you want with the least hazard is a proprietary cleaner called 'Armour-all'.

Sold by car accessory shops, it's a milky liquid that is pump-sprayed (really splattered) over the panel and knobs you want to clean.

All that's then required is to brush it over all the offending areas and crevices, lightly, with a soft paintbrush. Then rub away or 'pull off' with tissue paper or other absorber. This process leaves largely polished surfaces that are also anti-static treated.

Cleaning faders

All faders can get noisy when moved if they have collected too much dirt. The elements inside 'plastic film' types can be removed and washed in soapy water when they get dirty, but with cheap carbon types, aerosol cleaning sprays have to be used, and some of these can eventually dissolve away the track material – leaving you faderless

Cleaning connectors

Start by vigorously wiping with a soft cloth partly wetted in isopropyl alcohol (available from chemists/ drugstores). This may be all that's required. The alcohol may be mixed with some household ammonia to help degreasing and dirt dissolving.

If not, a special micro-abrasive is best. Failing that, a mild abrasive-based bathroom tile cleaner may be used. Never use wire wool, brushes, abrasive paper, or metal-cleaning chemicals, which may strip the plating.

The contacts of (cleaned) connectors may be enhanced by wiping them with a soft carrier soaked in some contact cleaner/enhancer. This will enhance conductivity when the contacts are next mated. (Gold-plated contacts need a specialised contact cleaner.)

If there are a lot of very tarnished or heavily soiled connectors to clean, hire an ultrasonic cleaning bath and some solvent (alcohol can be used). As alcohol is harmless to connectors and cables, cable ends or whole (short) cables, can be dipped into the bath.

Cleaning and re-forming PA cables

PA cables can get disgustingly dirty by any standards. In a worse-case situation they'll have been left marinading in a gungey mixture of spilled drinks (alcoholic and/or sugary), condensation, mud, grease, plus all kind of unsavoury by-products of an evening's entertainment, congealed on the venue floor… Until it's time to coil them in.

When the time comes, wear industrial gloves. Then use lots of toilet/kitchen tissue or other absorbent materials as a 'filtering' system, drawing the cables through and wiping them clean as they are reeled in.

PA cables can get disgustingly dirty by any standards

To clean less obviously soiled, dry leads and cables, draw the cable through a rugged cleaning cloth wetted with diluted detergent. Hot water is best. Periodically wring out, recharge and reapply the cloth. Change the water and detergent regularly. Repeat back and forth until no more 'gunge' comes off.

Important: don't dip the connector ends in the water, or even get near them. Clean the last 12 inches at each end with a slightly damp cloth. Dry before re-using.

To 'reform' and neaten PVC cables that don't lie or coil correctly, feed them through a bath of hot (recently boiled) water – again apart from the plugs. Then coil into the desired shape/size. Use a weight or other restraint techniques to quickly hold the cable in its new format until it has solidified and 'learnt' the new shape. One hour is the minimum, overnight is best.

Fan-filter cleaning

Fan-cooling of power amps and other gear concentrates air-borne dirt inside the equipment, so fan intakes should be fitted with filters. Good equipment comes equipped with these, but if the air is dirty these get clogged and the airflow can be impeded.

Even if the equipment survives this reduced airflow without over-heating, the bearings of the common propeller-type fan are overworked by this condition, which reduces the fan's lifespan.

Cleaning a filter is easy enough, but the conventional (and eco-friendly) type of filter – made of reticulated foam – requires something that doesn't mix very well with electricity or audio gear: a washing-up bowl of hot water and detergent. Make sure filters are dry before you re-fit them.

(Note: filters with black coatings must not be washed – see panel below).

The trouble with this procedure is that, depending on touring conditions, it needs to be done every few days or weeks. Another option is to buy a year's supply of filters. Filter pads are then changed every week and dirty ones collected up; after six months you take a job-lot of filters to a laundry and wash them inside a net-bag.

It's worth organising this, one way or other, since using a filter is, alas, the best approach to keeping your amp clean and in working order.

One way to reduce the problem of numerous filters getting clogged is to pre-filter the incoming air with one very large filter pad, which is easier to change – daily if necessary. Hence some racks have a common air intake.

The other approach is simply to remove all the fan filters. This often happens in the real world, because PA people are 'real men' (even if

---

Cleaning electret filters

If an amp uses a black-coated filter, it's probably impregnated with carbon (charcoal) or an 'electret' substance. These need special handling.

It's best to turn off the amp before touching or handling these filters – as they're conductive, you might get a shock if the pad touches any live parts while being manipulated. The electret type of filters may be washed and squeezed dry – but will lose their properties in the process. So this is only worth doing in an emergency. Other sorts of cleaning are difficult, so this type is best replaced and binned. At the typical rate of one replacement per month, this is not an eco-friendly process.

they're female) and gear designers haven't really thought how practical it is to rig up washing lines and have the monitor engineer in a frilly apron every Monday morning washing fan filters.

The inevitable result of having no filters, of course, is that the inside of the equipment gets clogged. This causes greater and quicker increases in operating temperature, as the heatsink's dissipation capacity is directly impeded by an insulating layer of dust.

To counter this problem it's common practice to 'blast out' PA gear's airflow paths using 'airlines', or compressed air blowers. If used, these must be moisture-free, and pressure must be limited when blasting around electronic parts to avoid detaching or breaking them.

Periodic 'blasting out' can work to an extent, but it is not ideal, particularly in the many items of gear where the air flows past electronic components, circuit boards and solder joints.

A filter is still the best protection from airborne pollution – and it's not just a matter of dust. The main problem is one of corrosion.

A filter is still the best protection from airborne pollution

For example, smoke machines pump highly corrosive chemicals into the air, and the sweat from 1000 clubbers dancing for eight or more hours (not to mention any other chemicals that may be in the air) also contributes to a highly acidic atmospheric cocktail.

If these corrosive liquids settle on electronic parts – even just power transistor cases – serious damage can occur, whether in a few days or spread over months, to a point where repair may well become uneconomic.

The problem of air-borne corrosion would be helped if PA equipment makers either kept airflow fully away from parts, or else covered all the delicate electrical and electronic parts with 'conformal coatings' – a smooth, shiny surface that makes air blasting safer and more effective.

With due caveats regarding the solvents used

– which might just soften some parts, or dissolve some board lettering etc – users who want trouble-free parts can perform DIY conformal-coating from aerosol cans (although this can only satisfactorily be achieved on new or at least completely clean gear).

Maintenance sundries
Useful bits and pieces that can help with the organisational side of regular maintenance:

- Laminated labelling machine. Most are made for Japanese schoolchildren – other than the lack of professional symbols and limited colours, excellent adhesion on suitably smooth and clean surfaces.
- Laminated labels (vastly superior to crude 1960s 'vinyl' type that are still peddled): 'Tested, by/date', 'Tested for electrical safety, by/date', 'Do Not Use!' etc.
- Fine and medium pens that write on metal, plastics, glass.
- Silicone paste – holds notices onto some rougher surfaces yet can also be removed easily.

CHAPTER 10

electrical power & safety

voltages, fuses, precautions...

Quick view

It would be a much quieter, duller world without it, but electricity is not without its quirks and dangers. In this chapter we look at AC and DC power supplies, comparing some of the main international voltages and frequencies, and what these mean for you and your PA gear; plus wiring conventions, correct fusing, circuit breakers, and how to dodge dubious gear and survive scary situations. There's also a crash course in electricity – an idiot's guide to volts, watts, amperes etc. (If you think Coulomb was the guy who discovered America, you really shouldn't miss this.)

t he first electrical networks – in 19th century Britain, Europe and the USA – used DC (direct current) power supply systems. These were hopeless, as the power could only be transmitted at best a few thousand yards without very serious losses, despite using massive amounts of copper wire.

The vastly more efficient AC power (alternating current) system was haphazardly invented and developed just over a century ago by various individuals, mainly in the USA, Britain, Italy, and Germany. (One of the most famous was Tesla, a émigré genius from the former Yugoslavia.)

Using AC enables electrical energy to be generated centrally at a small number of efficient generating stations, then distributed over hundreds or thousands of square miles, with low losses, by stepping up the voltage. In the 1920s, the idea of 'national grids', interconnecting all the power stations and supplies in a country made electrical power increasingly reliable. In turn the devices that were needed – alternators, high-voltage cables, glass insulators and transformers – all had to be perfected.

After experimenting with many variations, worldwide power systems have come to be fairly similar in design and operation. But there are two major differences that prevent equipment being used interchangeably across world without some thought and/or adjustment – namely the voltages and the AC frequencies used for power delivery in different countries.

Voltages

World voltages split into two principal levels, often called simply high and low. High voltage is

electrical power & safety

approximately 240 volts, and is used, among other places, in the UK, Australia, New Zealand, Malaysia, Thailand. This voltage is theoretically the most efficient, using a fraction of the copper that lower voltages need. It is also more likely to kill by contact than lower voltages. But then again it does at least demand greater care and respect, which is no bad thing for safety, as well as for reducing fire risks.

Voltages can often be converted using adjustable transformers

In western 'mainland' Europe the voltage used is 220 volts – so between the UK and the rest of Europe there is a hypothetical 'Eurovoltage' of 230V, and some gear can be set to this rating.

Low voltage is roughly 115 volts. USA, Canada, parts of central America, and Taiwan are among the areas that use this voltage, with local and regional variations from 110–120 volts. Japan uses the slightly lower 100 volts, with regional variations from 95–110V.

(If you want to check particular supply details in other countries – plus view pictures of 13 different types of international plug – try the useful resource of website *www.kropla.com*).

These lower voltages are roughly half those of 'high voltage' supplies. They are less likely to cause fatal electrocution, because there's less actual 'charge', but current is approximately doubled, meaning there's a higher 'flow'. (On an economic note, about four times as much copper has to be used to keep losses as low as with high voltage supplies.)

If equipment set for these low voltages is connected to a 'high' voltage, it will hopefully only blow a fuse – if you're lucky. But it can do damage, sometimes severe.

Voltages can be converted using transformers, which are often built into equipment, usually adjustable by either switches or 'taps' (which may be internal or external). On the better professional-grade PA equipment there will be settings for all or most of the main world voltages.

Power variation and operable range
All AC power supplies vary, typically by as much as ±10 per cent over some periods. The variation will be biggest in non-urban areas and third-world countries, as well as in any area where there's a big movements in population – for instance office and shopping areas which are deserted at night.

Better-designed PA gear should be able to handle such power variations – even up to ±15 per cent, to give a good safety margin. For equipment used in some areas of the world, where the power grid is 'flaky' or non-existent, an even wider safety margin, up to 20 per cent, is advisable. Less well-designed equipment may well fail under these circumstances.

Changing the rated voltage using a transformer
A large, overall step-up or step-down transformer (converting voltages to higher or lower levels) can be used to supply all the amps, with the transformer rated according to the amps' total power consumption – a VA (volt-ampere) rating about four to five times the maximum wattage delivered, to get an unwavering, 'stiff' supply.

For amp power ratings totalling several kilowatts, this will be a large and heavy (but reliable) item. (It will sometimes be called a 'pig' transformer because of its bulk and awkwardness to handle.)

If the AC power voltage is fluctuating, you will

220 versus 240 volts
Most PA 'control gear' equipment made for 220V will work fine on 240V, and vice-versa. Sometimes, marginally-rated power fuses may need slight uprating (+20 per cent) so as to not blow occasionally with the switch-on inrush. (See Chapter 9, and later in this chapter for more on fuses.) The partial exception is that most power amps set for 240V will deliver a slightly lower maximum voltage 'swing' (explained in Chapter 5 & the Glossary) and so less power. This will have hardly any audible effect on volume, but it will reduce the level at which the amps clip by a few per cent – exactly in line with the change in the power voltage peaks.

Of more concern, some amps made (and set) for 220V will be overstressed by use on 240V, under worst-case conditions.

Many amps have 'taps' that can be used to switch between voltages.

know about it if you use ordinary light bulbs ('incandescent' as opposed to fluorescent or energy-saving types) for lighting the mixer location(s). These are superb detectors of voltage variations – it's seen as flicker, or dimming and brightening, an also changes the colour of the light itself.

If the voltage is varying widely enough, some equipment may be at risk of malfunction or damage. The supply voltage can be regulated, or at least the variations kept within acceptable bounds, by using a motor-driven Variac (an adjustable transformer, available from specialised electronics suppliers).

This is a benign method, and well proven. Other, more electronic kinds of 'power conditioner' are less suited to use with audio gear, typically introducing interference or limitations of their own.

Three-phase (mains) power

Most AC power is generated and transmitted in this form – it's not only the most efficient way to do so, it also suits motors, as the three-phase sequence provides a fairly smooth spinning effect.

Three-phase power means there are three live terminals: it has the usual national voltage between each of the three phases, but between any two phases the voltage is about 73 per cent higher – hence around 415V in UK, 380V in Europe or about 200V in North America. The UK and European voltages are again more efficient – but also more likely to kill if abused.

Nowadays a lot of equipment is designed for dual frequency

Three-phase power can't be used directly by most PA gear (except some motor hoists and specially-equipped amp racks). But nearly all power used is three-phase – it's just not recognised as such, as the branch distribution is between one of the phases and neutral. Small venues may effectively only receive a single-phase voltage.

When power is accessed from high-current power sources, the three phases will be met, and a decision may need to be taken as to which to use to get the cleanest, least disturbed power. For example, where only one or two phases are used by the lighting rig, these should be avoided. In some venues, a particular phase may be known from past events (or from testing) to be cleaner or provide a more stable voltage.

The higher voltage between phases can create a hidden, increased danger in the event of a fault condition if you connect different parts of the PA to different phases. This can happen just by using different, widely-spaced power outlets in a large building.

Two-phase power (two-wire single-phase)

In the US and Canada, AC power is also supplied as 'balanced' two-phase, also (inaccurately) called 'two-wire one-phase'. This is a development of the old three-wire DC system, where two voltage levels suited different loads.

This system provides the national phase voltage (nominal 115V) on two legs, with neutral in the middle. Equipment connected leg-to-leg receives 230V, or whatever double the phase-to-neutral voltage is. Assuming it's available with a suitable current rating, this arrangement can be useful for operating 220, 230 and 240V rated equipment from outlets in North America.

World power frequencies

There are only really two frequencies used for power distribution – 50Hz and 60Hz. Using higher frequencies than this would make a more efficient power system, though it isn't physically feasible. If the frequency was any lower, you'd notice light bulbs start to flicker.

These frequencies are also the primary cause of most hums and buzzes in sound systems – though they were not chosen to cause grief to the PA and audio world. They are set by the maximum rotational speeds of giant turbo-alternators at power plants: 3000rpm = 50Hz (50 revolutions per second). (It seems rather ironic that most PA systems and a 'high tech' world ultimately depend on the revolutions of large physical machines.)

Frequencies apply across entire countries or regions because grid systems need this to be

electrical power & safety

Electricity (volts, amperes, watts...)

Electricity is a hard word to define. It's sometimes used to mean power, charge, current, voltage – all of which are really separate elements that combine to make the 'electricity' we use every day.

The first thing to grasp is that electricity is not 'created' by power stations, or batteries. These both just act like pumps, producing the drive required to push chunks of electric 'charge' along wires at various energy levels. In fact, the raw material of electricity – and this might come as a shock (sorry) if you're not aware of it already – the basic 'charge' is already in the wire itself. No, really.

As you know, all matter in the universe is made up of atoms, and there are zillions of atoms in something the size of a pin head. Atoms contain three particles called protons, electrons and neutrons. Both protons and electrons are 'charged' particles – they are essentially microscopic magnets, attracting and repelling each other – but only electrons can move around freely (jumping from atom to atom).

It just so happens that metals (like copper) are really good at letting their electrons move around from atom to atom whereas plastic, rubber, wood, ceramics etc are bad at it (any further explanation of this goes into mind-boggling quantum theory which can show there's a good chance you don't exist – so let's skip that). Materials which let electrons move around freely are called conductors.

Charge is measured in 'coulombs' – a term which sadly doesn't get the coverage of its famous relations, volts, amps and watts, but this is where the story starts. The amount of charge carried by a single electron is very small, and a quite staggeringly large number of electrons have to move around in a copper wire before any measurable 'flow' of charge is observed. As a matter of fact, electrons are continuously flowing around (almost imperceptibly) inside all metals in a kind of electron sea, which is initially directionless and fairly random (varying with the type of metal and its temperature).

When you connect a piece of metal (such as copper wire) to a battery (an electrical pump, as we said) then the electron sea will be forced to flow in a uniform 'current' – a current of electrical charge. This process is known as 'electro-magnetic induction'. The actual flow of charge (the current) is measured in 'coulombs per second', which is also known as amperes, or amps.

The thing that dictates the crucial extra ingredient of 'energy' carried by the charge in the wires is the output of the battery, or generator, or power station (or even, in the case of an audio signal – in a different but comparable way – a voice singing into a microphone, or note played on an electronic keyboard...). The number of units of energy carried by each unit of charge is described as the 'voltage'. The unit of energy is the 'joule' – so 1 volt equals 1 joule of energy per coulomb of charge.

Voltage, then, is the amount of energy that the charge carries (on its back, if you like). Major power cables – the type you see suspended high up on pylons – can carry electrical charge at an energy level of 400,000 volts. This is why a sizeable herd of cattle can be toasted to a crisp in seconds if a cable snaps and hits the ground.

The reason it takes several minutes to brown a slice of bread in a toaster is that our toaster is only fed the normal 110-240 volts provided at power outlets in the average home. The fact that this low voltage can still damage you (as in an electrocution on-stage) is to do with the combination of the energy level and the amount of charge that's flowing at the time.

If you've ever seen or felt static electricity (a spark off newly washed clothing or a car door, for example) it may surprise you to know that this static charge was probably at an energy level of 50,000 volts (or more). The reason you didn't end up like well-done steak is that the actual charge carried by the clothing or the car door is so tiny. So when you multiply the 50,000 volts by this teeny weeny charge you don't get much 'power' – although it can still sting.

In electrical circuits it's the 'charge' that carries the energy – so the amount of charge that passes a point in a wire per second will determine the amount of power (or joules per second) that the particular circuit can deliver.

Electricity's 'power' – its 'ability to do work' (ie toast something, pull a train, boil a kettle, pop your eardrums at a gig) – is measured in watts. As power is defined as units of energy delivered in a given period of time (joules per second), then a 100-volt circuit with a 1-amp current flowing through it can deliver 100 joules of energy per second – which is 100 watts. Put another way, a 100-watt light bulb uses 100 joules-per-second of electrical energy – so in one minute a 100-watt bulb will use 6000 joules (60 X 100).

If volts are the number of joules of energy on each coulomb of charge, and the current is the number of coulombs per second, then electrical power (or energy per second) is calculated as volts multiplied by amperes. To spell this out, watts = energy delivered per second (joules per second) = energy per coulomb of charge (volts) x number of coulombs per second (amps) = voltage x current. This can be abbreviated to $W = V{\times}I$ (where I is current).

In short, the **power** *at any point in a circuit is calculated by multiplying the* **volts** *at that point by the* **amps** *at that point.*

In the real world you also have to take account of resistance (measured in ohms). This introduces some additional formulae, such as Ohm's law, which states that:

current = voltage ÷ resistance ($I = V/R$); and therefore voltage = current multiplied by resistance (or $V = I{\times}R$).

Since $W = V{\times}I$, and $I = V/R$, you can work out that the power calculation now reads: $W = V{\times}V/R$, in other words watts = voltage squared ÷ resistance ($W = V^2/R$).

And all of this is just in simple DC circuits. With AC systems it all gets much more complicated: everything has to be averaged out (using root-mean-squares and logarithmic values) because of the 'back & forth' nature of the current flow; and there's also the fact that resistance to AC fluctuates depending on the frequency – a process known as reactance – which changes resistance into our old friend 'impedance' (see Chapters 1, 5 & 6).

synchronised. Power frequency is held stable for this reason – though it can run fast or slow for limited periods.

Here are the main geographical areas that adopt one or other of the two frequencies:

50Hz: UK, Eire, Europe (except former states of USSR), Caribbean/West Indies (part), Argentina, Africa (most), India, Pakistan, Malaysia, Singapore, Australia, New Zealand, Japan (part)
60Hz: USA, Canada, Central America, South America (most), West Indies (part), Russia, former states of USSR, Saudi Arabia, Korea, Japan (part)

Nowadays a lot of equipment is designed for dual frequency – this will appear on a label at the back of the gear as: '~50-60Hz'.

Occasionally issues can arise when touring abroad or using directly-imported gear. With some older equipment, using the 'wrong' frequency might affect the speed at which mechanical parts move, for example.

Equipment made for 60Hz use will usually run hotter and less efficiently on 50Hz, while parts made for use in 50Hz countries are over-rated when used in 60Hz territories. This is a concern for power amps and power supplies particularly.

Anyone using US and Japanese-made gear in Europe, or particularly in the UK, Australia and NZ (where the voltage is also higher than double the US) – and if the gear isn't marked as OK for dual-frequency use – should be checking transformer temperatures in use, noting unusual mechanical buzzing, and quizzing the maker or importer.

Makers in 60Hz territories can use less copper wire in their transformers and less core material, which cuts weight and boosts their profits. But when this gear is used on 50Hz supplies, a tightly specified unit (designed down to the nearest penny) runs hot, can vibrate heavily, delivers less power, and will probably fail or need servicing sooner.

Equipment suspected of being in this category can be helped by forced-air cooling (by powerful fan) of the transformer. Fuses may also require uprating – but much more than 50 per cent is a sign that something is seriously amiss. (There's more on fuses shortly.)

Electrical wiring colour codes

This is essential information for safe electrical wiring – and also useful when wiring up recently acquired equipment (new or second-hand) found to have unrecognised colour-coded wiring.

There are two separate areas where colour-coding of electrical/mains wiring occurs – flexible cabling (power cords etc) and permanent/fixed wiring (in wall outlets and the like). There are also variations in these colour schemes among different countries.

Flexible cables
In 1970, many countries (except, significantly, the US, Canada and Japan) agreed to change their colour codes to the following:

Live, phase, line	Brown
Neutral	Blue
Earth/ground	Green with yellow stripes

Note: in Britain this replaced the older system of colour-coding, which was:

Live, phase, line	Red
Neutral	Black
Earth/ground	Green

In the US & Canada, the coding in flexible cables is:

Live, phase, line	Black
Neutral	White or pale gray
Ground/earth	Green

Permanent/fixed electrical/mains wiring
With portable PA, or even installs, this is any wiring you encounter when wiring into a main power feed (which must be done by a qualified electrician). Here you will often meet three live phases.

In the UK and Eire:

Live phase 1	Red
Live Phase 2	Yellow
Live Phase 3	Blue
Neutral	Black
Earth/ground	Green & yellow stripe

In the USA & Canada:

Live phase 1	Black
Live Phase 2	Red
Live Phase 3	Blue
Neutral	White or pale gray
Ground/earth	Green or green & yellow stripe

Power connectors

There are three principal classes of connector: standard 'national' power plugs for 'everyday' use; lightweight connectors for use inside racks; and heavy-duty 'building-site-proof' connectors for external use.

National mains/power plugs and sockets

Many countries have their own unique domestic format. Britain's rectangular-pinned '13 Amp' plugs (designed back in 1943, nowadays made to BS 1363) with their integral HRC (high rupture capacity) fuse, are also used in many Commonwealth countries, and are widely regarded as the world's most sturdy, workmanlike domestic power connectors. They are also probably the safest (as long as you use the correct rating of fuse), helped by the fact that the plug cannot even be inserted into the socket if the large, strong earth/ground pin is bent or absent. In recent years safety has been improved further with sleeved pins.

Only the 'premium' types should be used for PA – either Duraplug's rugged, nylon-based plastic type, or MK's top line model 655. This and other MK models have superior ring-compression terminals (where a metal ring squeezes the wire into place, so the conductor isn't bitten as it is with a simple terminal screw), and a superior, fast-fit cable clamp, with no screws to do up or come undone.

(Note: before the 1950s the norm in Britain was a round-pin electrical plug – often brown or cream in colour: if one of these is encountered nowadays it should be changed.)

In the US, most gear uses two-pin plugs, many of which are nowadays 'polarised', meaning one pin or 'blade' is bigger than the other, so they can only be inserted one way. Three-pronged 'grounded' plugs do exist, and are considered to be safer, but are not the norm – largely because of the lower voltage supply, which, as we've said, reduces the likelihood of serious electric shocks. Most US plugs don't contain fuses either – that's all dealt with inside the gear itself.

Among the good, bad and ugly diversity of national connectors, Australian plugs are considered to be pretty solid, with generous contacts – not unlike some UK circular three-phase plugs. Among the least impressive is the substandard European device with two primitive 'scraping' spring blades operating as an earth/ground – there are two of them in case one snaps off or makes a bad contact.

Lightweight connectors for use inside racks

Examples include the rectangular, all-plastic IEC type (like kettle plugs), available in three ratings: 6A, 10A & 16/20A. These variants look similar but are different sizes.

'Moulded' types prefitted with power cables are usually fine to use, but you should look for safety approval marks on the plugs. IEC connectors aren't usually latchable, and are not fully-formed 'professional' connectors. Even where fitted with a latch it's a skimpy add-on which would fail to mate with plugs that are flared. But the soft-plastic, moulded IEC plug can be surprisingly resilient, with enough give to protect the less rugged socket as well. To keep IEC plugs in place, a bead of silicone paste ('liquid gasket') is often used around them. This may be unsightly, but will tear off neatly later.

The soft-plastic moulded IEC plug can be surprisingly resilient

A premium grade type of IEC connector, re-engineered from scratch for solid electrical performance, can be obtained from US company Wattgate (www.wattgate.com).

Handy IEC accessories include splitter leads which comprise two (usually 6A) power cables (cords) fitted with IEC plugs, wired into one 13A plug (or equivalent national plug, rated above 10A, in other territories). These can be bought or made. An IEC splitter block is a plastic block with two outlets emerging from one intake. It needs male-to-female IEC cables to be usable. Male-to-female cables also enable power cords to be easily extended. Not latching, but some plug/socket pairs fit quite snugly, and it's easy to tape-up the mated pair.

Less often you may find XLR power plugs. Originally made by Cannon, but Neutrik have since made a superior range at lower cost. These look like their respective makers' standard XLR signal cables, but altered and insulated for safe electrical connections. Used for FX and other low-power gear only. Maximum cable rating is about 3A (0.5mm2). Latching.

One Neutrik version is a plug with a socket in the rear (type MRCT) – this allows easy daisy-chaining of power, greatly reducing the amount of cabling used in FX and drive racks. Made for PA but more fiddly to wire than some, and rugged enough to be used outside of racks.

'Construction-site-proof' connectors

The Neutrik Powercon is a development of the rugged Speakon (speaker connector) plug design (see Chapter 7), adapted to house a rugged, 20A-rated power connector for PA. Accepts quite large, 15mm diameter cable, and is available with both male and female case and cable (line) mounting types.

To help identify these, the power inlet types (female line, male chassis) are blue, outlet types (male cable plug, female chassis socket) are gray. A male-to-female in-line cable-coupling accessory is also available, so suitable leads can be extended safely. These connectors are particularly cost-effective.

Users should also bear in mind that as Neutrik specialise in serving real pro-audio needs they will listen to the gripes and other feedback you may have as a PA user. This is not true with most connectors, which are made for much bigger markets, and to international specifications 'cast in stone' – so no improvement is possible. The following is a good example of such, but is luckily a very good design already.

C-form power connectors are used for large PAs

CEE ('C-form') connectors (British Standard BS 4343, International Standard IEC-EN 60309-2 – these are world-standard connectors for serious portable operations. They're used on building-sites by contractors, and are also the power connectors used for large PA systems' main power distribution. They're a rugged development of the British 5A and 15A round-pin plugs (which date back to 1930) – UK mains/power connectors have

always used pins that have hugely more current capacity than the rated amount. The 63A CEE form uses pins the same size as the old 15A plug that are still generously rated.

CEE connectors are available in ratings of 16A, 32A, 63A and 125A, and in both male (pins) and female versions for both unit and cable-mounting. Both line (cable-mounting) and trunk/chassis-mounted parts can have flip-top covers to protect the terminals when out of use. Various levels of waterproofness are also available.

Different, incompatible types are made in distinct, rather garish colours, for 115V, 230V, 375V/415V three-phase, balanced mains, etc. This makes safety relatively foolproof where mixed electrical systems (like 240V and 115V) are in simultaneous use:

Yellow	110V, single-phase
Blue	230V, single-phase
Red	400v, three-phase

Prices can be surprisingly high, particularly with the higher current and higher-spec models.

Some useful Cee-form connector accessories include 'Plug guard' – which locks on a plug to stop use; and 'Interlock' – a series-connected switch that can be padlocked, and where the connectors cannot be withdrawn while the switch is on.

Hubbell twist-lock connectors are high quality US/North American industrial-grade power connectors. Rated for 20A at 240V (as well as 115v), they have long been used in the UK by professional operators, from the days when the only other option was EP (see next paragraph). They can also cope with chunky cables up to 18.3mm diameter.

EP3 (Cannon)

These were used as ad-hoc power connectors in the PA systems of the 1970s and 1980s, before more suitable types were made or discovered. They are quite rugged, 20A-rated and compact, but can't handle really thick cables, and are no longer acceptable in today's safety-conscious climate because their socket contacts aren't shrouded very deeply. They're also metal-bodied, which increases handling risk.

Equipment using them should really be updated to use CEE (C-form) or Neutrik Powercon connectors. Another concern is that as EP connectors with four or more pins are used for PA speakers, there's obviously a risk of danger in the event of any of mix-up, if the higher pinned parts are used for any reason.

Electrical safety

Verifying a power outlet is safe

It's not always advisable to plug a PA system (or any part of it) into other people's power outlets without checking them first. In some countries electrical wiring inspection is quite strict – but even then, electricians make mistakes, and may fail to check every circuit with every permutation in a big wiring job when working under tight deadlines.

So errors can persist, which in specific circumstances can be dangerous. Examples includes transposed neutral and live wires, neutral and earth/ground touching behind an outlet, unconnected earth/ground wires etc.

Technically dangerous wiring is more prevalent in some southern European, far eastern and many third-world countries, where bribery is rife and inspections may be cursory. Electrical safety is generally better in most parts of northern Europe, Scandinavia, Australia and north America.

Where the outlet is a 13A or C-form (or some similar types), plug-in testers (for instance by Martindale) can be used quickly show the status of the connections. The visual pattern of an array of indicator lamps fairly instantly indicates if the outlet is safe to use. And if not, it should reveal what's wrong with the wiring. This is a simple yet valuable tool as it provides objective evidence of risk to life, can demonstrate it is not to do with the PA's wiring, and can help persuade owners to call out an electrician to fix the safety problem, whatever day or hour it is.

Before you use any cable, it's wise to inspect it

Verifying a cable is safe to use

Before you use any power cable it's wise to get into the habit of inspecting it – this can save bother,

injury or death. Running a hand down it (before it's plugged in) quickly divulges if the cable is damaged. If it's not smooth all the way along, something is wrong.

A cable which is not tightly clamped to the plug at the end, or which has a grazed sheath showing the inner cores, or where the sheath is slit or has been partly melted or severely crushed, should be put to one side and not used until rectified. It's not worth taking chances with electricity.

A power cable can be verified for basic safety and correct connection using an ohmmeter. There should be low resistance connections between conductor ends. Conductors should also connect in the right order or places, and not to each other. For proper testing of insulation, a 500V ohmmeter – called an insulation tester – is required. But an ordinary ohmmeter should show insulation between conductors of at least 10 Megohms (10MΩ). For mass testing, lead testers will speed up the job (see Chapter 9).

Earthing/grounding

The connecting together (bonding) of all exposed metalwork in AC power systems, and connection to earth/ground (literally the soil or other substrate, somewhere under your feet) provides essential, primary protection against shock and fire.

In electrical wiring, the earth/ground conductor is called the CPC (circuit protective conductor) in the British 'IEE Regulations' (BS 7671), or the 'earth-ground' in the US.

As we've already mentioned, safety earthing/grounding is also a major cause of buzzes and hums in PA systems using unbalanced (and sometimes even balanced) audio connections – but only by interaction, when earthed/grounded equipment is connected together. In spite of this, the safety earth-ground should never be removed. In all cases the problem can be overcome (or at least ameliorated) by other methods – see Chapter 9.

Safety earthing/grounding is rarely 'clean'. Other than the frequency hum and harmonic currents caused by loops in the wiring system, the CPC (earth-ground) wires also carry leakage currents and all manner of noise signals from filters. There can also be high levels of induced harmonics when a venue has huge numbers of transformer-supplied quartz-halogen lights, or other adverse loads.

The extent to which these signals will affect a

PA system is highly variable. Rather than take chances, it's wise to assume that a PA system will benefit from a clean(er) earth-ground connection.

When a larger PA system's power is wired into a switch-fuse with tails (individual conductors), this can sometimes be achieved (with the help of a qualified electrician) by taking the earth wire a few more feet across a switchboard, to a more central point – nearer to the actual earth (whether that's a metal stake in the earth or the bonding to the incoming cable's outer metal sheath or armour).

When a smaller PA's power is plugged-in, a cleaner earth-ground may be obtained by trying different circuits. Largely unused higher-current circuits will tend to be cleaner.

Fuses & fusing

Fuses (and/or circuit breakers, which we'll come to in a moment) are essential to stop fire, serious and costly equipment damage, some shock risks, and general tragedy. For electrical supplies, sturdy white ceramic-cased 'HRC' (high rupture capacity) types should be used, rather than standard glass-cased fuses. Only these types (filled with sand to deaden any explosion) are rated to be able to break the 'arcing' caused by the worst-case level of fault current likely on the power supply.

Use ceramic-cased high rupture capacity fuses

Simply put, a short circuit in AC equipment or wiring can cause currents of several thousand amperes to flow. This can quickly cause a fire, generate toxic fumes from smouldering components, or cause expensive damage to the electrical system.

Amazing as it may seem, a fuse wire that melts and breaks the circuit is not guaranteed to stop an arc at this level of current. Such a fuse is useless here, and a fire could start before the next fuse down the line breaks the fault.

Fuses do 'wear' and can fail in time, apparently randomly. Spares of all types used need to be kept. If fuses blow repeatedly, the problem needs investigating. There may be a real fault. If so, a fresh fuse will usually fail immediately and

violently. Or there may just be a high inrush (surge) current, requiring a slightly (50 per cent say) higher value of fuse, or a fuse with a 'delay' ('T') alias 'anti-surge' characteristic.

Fuses must be rated according to the current rating of the conductors they serve, and also graded, as the electrical power is divided into a series of lower current branches. This is important, otherwise a fault in some minor, low-current unit could blow a whole chain of fuses, causing needless expense and a lengthier fault-finding exercise.

The high inrush current drawn by amp racks will require either oversized wiring to permit larger fuses, or the use of industrial fuse families (like BS-88 type) which can withstand surges.

Circuit breakers

Breakers are resettable fuses. Most breakers give clear visual indication of which circuit has opened. In Europe and the UK, use of German DIN-rail systems enables arrays of breakers to be quickly constructed and housed alongside terminals.

These factors make breakers the preferred choice in the AC distribution racks that larger PA systems require. Some circuit breakers have 'aux' contacts that can be wired to sound alarms or light big red lamps etc.

Circuit breakers used for the mains 'front-end' must be permanent wiring/install types rated to handle high fault currents of at least 6000A (6kA), if not 10kA (10,000 amperes). 'Equipment-type' circuit breakers have a lower fault-breaking capacity.

On some power amps and amp racks such breakers are seen as ones which double as power on/off switches. They are fine in this position so long as the incoming supply is protected by HRC fuses, or 6 or 10kA-rated breakers.

Avoiding risk of electrocution or burns

Electricity usually kills by upsetting heart rhythms. This only takes anything above 30mA (30 thousandths of an ampere) passing for a fraction of a second. The higher the current (in amperes) and the greater the duration, the more likely it is to prove fatal. It takes a certain voltage to pass 30mA (or any higher) current. But as the human body's resistance varies with individuals, emotion, sweat, etc, the voltage needed to kill varies widely, from just above 50 volts, up to many thousands if you're very, very lucky.

electrical power & safety

You can largely avoid any risk of getting a shock that disrupts your heart by always making a habit of keeping your left hand in your back pocket when you handle, examine, or test-probe any potentially dangerous or unknown electrical situation. The idea is that the current, looking for the quickest route to earth, is more likely to travel down your arm and bypass the heart.

Certainly think twice before ever using both hands – this is a common way to be zapped, considering that most feet (in Western countries at least) are well insulated by footwear, so the current stays in the body longer rather than passing straight to earth.

Don't hesitate to wear insulated gauntlets. Even if it flows nowhere near the heart, electricity can still kill by causing internal burns.

If you see someone you suspect is receiving a shock, the urgent step is to separate them from the electricity source fast – but don't touch them (or any equipment they're holding) with your bare hands. Use insulating material (a dry wooden stick, thick dry cloth). If possible, try to switch off or disconnect the power supply first.

Safety trips (RCD, GFI)

RCD stands for residual current device (also called 'RCB', residual current breaker, and 'RCCB', residual current circuit breaker). In North America it's called a GFI, ground fault indicator, or GFT, ground fault trip.

Circuit breakers trip when any small part of the total current is detected as being diverted somewhere (or into someone). RCDs provide fairly comprehensive protection from both electrocution (if rated at 30mA or less) and also fire.

Note that they do not trip when there is excess current. Either a separate circuit breaker or fuse must be used as normal, or else use an RCBO – which combines both kinds of breaker in a compact package.

Water & electricity

Water (usually at outdoor gigs) greatly increases the likelihood of electrocution. At outdoor sites – or gigs held on floating barges, near fishponds etc – safety can be achieved with a combination of localised 30mA and even 10mA RCDs, waterproof connectors, well-insulated floors, balanced low-voltage power (as used on building sites). Consult and employ a qualified electrician.

Organising AC distribution

Small PA power distribution

A small PA system can be powered from flexible cables leading directly to multiple socket outlets, usual national outlets (eg 13A sockets).

Low-cost plastic 'plugboards' (power strips) are built to inferior standards and durability. Avoid them. Much time, grief and frustration will be averted by using PA-grade socket strips. These may take the form of either metal-cased flush national outlets, or by mounting ordinary metal-clad surface sockets (13A dual-switched type in the UK) in a recessed enclosure, eg a wooden case. Then the switches won't get smashed if the case is turned over and jumped on, or hauled over the stage upside-down (by tired helpers at the end of the night).

It's often useful to make the extension cable pluggable, so different sizes of 'socket strip' can be mated with different lengths of extension cable. If so, Neutrik Powercon or CEE-form 16 or 32A connectors are probably the best choice. Obviously due care needs to be taken to fit the power cable end with a female line plug (receptacle).

As DJs and musicians are generally incapable of getting decent electrical hardware together (and who can blame them?) it can pay dividends to carry extra 'PA grade' socket strips. These will often come in handy, saving time and trouble if, for instance, you discover that DJ or keyboardist is using an unsuitable plugboard/power strip, with a forest of loose contacts, cable almost falling off, and fuse holders rigged with silver paper. (Just make sure your loaned-out plugboards/power strips don't vanish – unintentionally or otherwise – when the artist does).

Large PA distribution

The short-term 'peak' current draw for a large PA can be surprisingly high. Very roughly, if the music is averaging (on a VU-type reading) about 15dB below clip, and clipping on the odd peak – ie the PA is being driven to its undistorted limits – then the average current draw will be about the same as the maximum power rating suggests.

For a 10kW rig that's about 40A at 240V (= 10,000/240), and for a 100kW, 400A. In low mains/AC power territories these figures are more than doubled. And these are not the highest peak currents, which can be quite a bit higher – at least double, assuming the AC power system permits it.

A medium-to-large PA system will typically have a distribution ('distro') system as follows: a 125A (or greater) rated flex cable leads to a 'distro trunk' – a flightcase which carries the large flexible mains cables. It also has a control panel, typically containing breakers which lead to high current outlets (typically 63 or 125A) for say the left and right PA wings, and also the stage – including the monitor PA, and a lower current outlet (typically 20 or 30A) for the FOH gear.

Power cables

Power cables for PA should be flexible types – sometimes also known as flexes or cords. PA power cables that will be exposed (ie not wholly inside racks) should be sheathed with one of the various kinds of synthetic rubber or special plastics made for industrial use, instead of the PVC used for the cables of most ordinary domestic products. Usually, industrial-type power cables are only available in black.

PA power cables that are wholly encased in racks can be any of the common materials – one of several grades of PVC or rubber. Cables sheathed in everyday PVC can come in a far wider range of colours, but may be unsuited to very hot racks, where freshly heated air meets cables running up the rear of the rack and softens the protective insulation.

Heavy flexible power cables (over $^5/_8$in or 15mm diameter) should be treated like PA multicores/snakes: don't coil them too tightly, or twist or force. Coil and store neatly.

'Arctic' grade PVC is beautifully flexible at room temperatures but it will soften and melt at even lower temperatures than ordinary PVC cable.

On the other hand, ordinary PVC will split if handled a bit below its minimum temperature, which is near to freezing.

Although few gigs are held in frosty air, PA truck innards can get very cold, and cables may well need be handled and unraveled before they have fully 'de-frosted'. Cables made with rubber and the more advanced plastics do not have this problem so much.

Cable sizing

Working out minimum cable sizes for a given loading is a complex calculation, set by the average current passing through them, the amount of cool air or number of other cables around them, and also the sheath used.

The most important factor, especially when the cables are long, is the voltage drop or loss caused by the cable's resistance. On long runs the outcome is that cables of much higher ampere capacity will be required than the current being passed. This is why competent extension cables are over-sized.

The amount of oversizing can be calculated by taking the cable length, doubling it, finding the resistance (in ohms per metre), then multiplying this factor by the total length to find the total resistance in ohms. This amount is then multiplied by the expected highest current draw to find the voltage loss. Keeping this loss below five per cent of the system voltage is a good rule – that's below 12V on a 240v system.

Failing this, as a rule of thumb:

- use the largest cable that will fit your connectors.
- use oversized (higher-current-than-you-need) connectors on main cables requiring long runs.
- keep all large cables as short as possible. They are heavy and will add to trucking costs.

Unexpected electrical dangers

Electrical dangers are not always obvious, even if you know electrical theory. For example: high-current wiring inside main feeder boxes (rated above 45A, say) can kill you from burns with just one small slip. If a slipped screwdriver causes an arc, it's not always anticipated that a sheet of flame can grow to envelope the face and neck, let alone arms, causing burns that can be fatal, if not just horribly disfiguring. This kind of risk is reduced by using plastic-coated '6000V' tools, HRC-fused testing probes, and by isolating circuits you are working on, by finding the next switch back down the line, and simply opening it (switching it off, or detaching from the power supply).

As a general rule, it's a good idea if at least one member of a stage/touring crew is trained in first aid – you never know when these skills might come in useful, perhaps in a life-and-death situation when there's no time to call professional medical help.

For guidelines and advice on electrical safety, contact a body like the Health & Safety Executive (www.hse.gov.uk) or the US Occupational Safety & Health Administration (www.osha.gov).

Special power cables/cords

Shielded mains/power cables have been available for some time, but they only shield against electrostatically-induced hum, not the more common and harder-to-get-rid-of magnetic type.

Where electrostatic shielding is useful – for example when mains/power wiring has to be run near sensitive 'high-impedance' signal cabling from electric pianos and some other older instruments – great care is required with conventional shielded power cables. These use a braid like the heavier shielded audio cables, and if this isn't earthed (grounded) and sleeved carefully at each end it may easily touch the live 'phase' and either blow fuses or, by making the connected equipment casing live, risk killing someone.

Shielded power cables are rarely used in PA

For this reason, shielded power cables are rarely used in many areas, including PA, since there is no 'secure use' or guaranteed system of regular inspection.

More recently, Swedish brand Jenving has overcome this problem with 'Suprasafe' – a patented flexible cable (suitable for low current equipment) which uses a conducting nylon tape that's a good shield but isn't a good enough conductor to cause any harm if it should touch the live conductor. It also copes with severe handling, unlike ordinary shielded cables. (There's more on cables in Chapter 7.)

Power cable accessories

- **Fault tracers** – electric field detectors. These are pen-shaped, with a red LED in the tip that lights brightly when within half-an-inch or so of a live wire. Used to quickly trace where volts are getting to in electrical wiring networks, to locate blown fuses, loose wires, hidden switches, etc.
- **Cable 'ramps'** – solid rubber ramps with protected channels for cables, allowing people or vehicles to pass over without snagging or squashing the cable.

Voltmeters (voltage monitoring)

Distro boards and/or other equipment racks may be fitted with voltmeters. If these are of the ordinary swing-needle type, they are almost useless in areas of the world with reasonably stable and adequately rated power supplies (eg the UK, much of Western Europe, and the better supplied areas of North America, Australia, and South Africa). They either read roughly the voltage you expect, or nothing at all – and a panel lamp could tell you the same at a fraction of the cost and mental effort.

But because these basic meters enable a fast diagnosis when things are not well with the power supply, they're useful when touring 'wilder' areas (such as third-world countries where the voltage variation can be very wide), or where power comes from on-site generators ('gennies').

Also, if you only have one type of meter, the swing needle will tell you more. If the electricity varies at the 'wrong' (fast) rate, then digital meters tend to scramble and are hard to read, whereas swing-needle meters waver to tell you what's going on.

On the downside, swing-needle meters are delicate and more vulnerable to damage, or accuracy degradation, when trunks are shoved about – or roll back down ramps. To 'ruggedise' these sorts of readout, consider mounting on shock-mounts, with maybe up to 1in (25.4mm) of recessing, and bolting some polycarbonate 'riot-shield' plastic in front of the display.

RMS and true peak reading

Most swing-needle meters (except laboratory instruments) read average values, but this only works for pure 'sine' waves. The electrical supply is only like this in academic textbooks. Because real AC supplies are almost always distorted, such meters read slightly amiss by a few volts.

It's often stated that making accurate mains/power readings depends on using metering that can read the 'true RMS' (TRMS) value. This is true for measuring power delivery for heaters and lamps, but it's not very useful for power amplifiers, whose delivery capability depends solely on the electrical supply's peak voltage value. Whether or not the amp uses an ordinary or light(er)-weight switching power supply, unless the supply is regulated, the power supply's peak voltage is directly reflected in the maximum voltage swing the amp can deliver to the cabs – meaning the exact level at which it clips is AC-peak dependent.

More sophisticated metering reads both the

true RMS and the peak level. The difference between the two shows the peak-to-average or peak-to-mean ratio (PMR), also called 'crest factor' (see Chapter 6) – which is a good gauge of AC power quality: if the PMR is 1.4x (or above), the power supply is clean; if the PMR is below 1.25x, the supply is very polluted.

Ammeters

These are optional fittings that can monitor the total current draw. This can be useful if some part of the system is capable, through its design, of being overloaded in some circumstances that may just arise with peak usage. A swing-needle ammeter allows the situation to be monitored at a glance, at any time. It also indicates, at a glance, that various parts of the PA are switched on and doing something.

Again, traditional swing-needle meters give the most useful sort of readout. But PA systems need ammeter types made for industrial use – these have an overload zone, so a meter made for, say, 80A may be scaled up to 400A (five times higher) over a short 'override' section. This may be scaled red. If not, suitable marking can be easily added (with the better quality meters) by taking the meter's front off, gently removing the scale, marking it up, and gently replacing it. The over-ride scale allows inrush currents to be metered, with less 'pinging' of the needle against the end-stop.

Connecting to mains/AC supply in buildings

This should usually be done with the assistance or guidance of the house electrician. And see colour code details earlier in this chapter.

Single-phase (1-phase) – select a switch-fuse or fuse or breaker-protected supply, which can be isolated and wired into, and can accommodate the current you will require. Isolate the supply – use a voltmeter, set to read over 450V AC, to prove this. Dress (twist and tidy, if necessary fold-over) the conductor ends, test fit, and adjust. When fitting snugly, tighten down. Wait a few minutes. Then tighten again, a tad more. After the cable has been physically restrained, and also after checking that any crew at the other end are aware, the isolator can be turned back on, to energise the PA's supply.

Three-phase (3-phase) – this is the same except you need to be sure to connect to ('between') one phase and neutral only. Test gear should be used to verify the correct voltage is present if working with any unfamiliar national system or unfamiliar colour codes, or in case of any doubt whatsoever.

Electrical safety with used gear

In many countries, legislation concerning electrical safety has become stricter and its coverage wider, and some older gear you see on sale may not meet the new standards. In many cases this can be rectified quite simply in second-hand/used PA gear by upgrading.

The following work should be carried out (or at least guided and inspected) by an electrically qualified person.

- First, make a visual check of all main power cables and wires. Look for damaged, decayed, cut or worn insulation, and corroded, broken or loose wire strands. Tidy and make good.
- Do a mechanical check (test for physical movement) to make sure all terminal and cable clamping screws are tight.
- Check that the connection between the earth/ground wire (where there is such a wire) and the gear's casing is good, using a 'low' ohm-meter capable of reading below 1 ohm, and looking for a reading of this sort of value. This helps to prove that the earth connection will work as intended, and will blow the fuse in the unlikely event of a fault.
- Check the gear's insulation resistance at 500V DC (called 'meggering' in UK) using an insulation tester (aka 'megger'), connected between the power plug's live and earth/ground pins. The resistance should read well above 10 mega-ohms (10 million ohms), showing that any substantial AC current won't leak somewhere it shouldn't.

Safety upgrades that may be carried out include covering exposed power wires on fuseholders and related parts. (Bare power connections used to be more acceptable inside gear in older times, but not in today's more safety-conscious workplace...) These can be covered either with moulded plastic 'boots' (like heavy condoms but with wire outlets), or using heatshrink sleeve or so-called 'self-amalgamating' tape. This is a heavy electrical insulation tape that sets into the wrapped shape, which makes it good for covering parts where there's no other option.

buying PA gear

Since you're unlikely to find many handy PA superstores (no 'PAs-R-Us') in the Yellow Pages, and as practically no single manufacturer makes a satisfactory all-in-one PA system of any size, putting together a live sound set-up is quite a tricky, specialised and time-consuming activity. This section gives you some tips on where to look, what to look for, and what to avoid.

Buying PA gear

PA systems seem simple enough in principle, but as we've seen there are lots of complex and subtle elements that have to be considered when you're putting a system together. Each item of gear not only has to do its own job well but also interact successfully and smoothly with the other parts – not to mention the interaction with the users, and the highly-pressured environment of a live gig.

In other words, the gear may fit the bill technically, but will it suit you, the music, and life 'on the road'?

It's easiest to start with what you shouldn't do. It's not sensible to buy a PA just by selecting the prettiest or 'coolest-looking' boxes. Nor is it generally a good idea to buy all your gear from one manufacturer, without at least first making an independent evaluation or some separate assessment of each unit. That's because it's highly unlikely that any one maker can master every area of the field, and there are few (if any) makers today that can offer every PA component under one brand-name.

So be prepared to do a fair amount of detailed investigation and shopping around.

There's plenty to choose from too – though it must be said that most of the PA gear you'll find in an average 'high-street'/shopping mall music store may not be up to full pro standards, even if it's suitable for small and semi-pro gigs and rehearsals.

Professional PA products are generally distributed through trade channels, and there are several routes to finding these. You'll probably have to leave the main shopping streets and malls for a start.

Pro PA products are generally distributed through trade channels

First, there are trade magazines. These contain adverts and contact details for the makers and distributors of many pro products. These titles aren't found in the majority of news-stands/bookstores – ask some PA manufacturers or sound engineers for names or back copies, and try to get on mailing lists. If this isn't possible, look for a copy of *The White Book*, or another live-industry trade directory – try reference libraries first, then if you want to buy a copy, relevant details will be in the book. (These directories are not cheap, but can be very useful for finding not just suppliers but lots of other pertinent contacts. Think of it as a career investment.)

Then there are mail-order suppliers. Again, these will mostly be advertised in trade magazines. (The usual warnings apply about buying anything by mail-order from a company you don't know.)

There are actually a number of 'trade counters' – the equivalent of a full-on retail store – except for PA gear they'll rarely be in busy shopping areas but hidden away in some relatively drab industrial, commercial or other unlikely-looking buildings.

Visit trade shows if you can (they're generally free, but you usually need to be 'in the industry' to be admitted – which of course you are, as a sound engineer). The main world shows for PA gear are the PLASA (Professional Lighting & Sound Association) show held in London, and the AES (Audio Engineering Society) events, held several times a year in diverse exhibition venues around the world. The Frankfurt Musik Messe and US NAAM shows are more wide-ranging musical equipment shows, but

incorporate a fair amount of PA-related gear (often shown in dedicated halls).

If all else fails, you could try visiting distributors or even manufacturers directly – but only if they'll give you an appointment. Never just turn up. (They're most likely to say yes if you're planning to spend a sizeable sum of money.)

Which manufacturer?
The question is a thorny one. No makers are always 100 per cent perfect. And it's not enough just to design and create a good product – it has to be assembled to a reliably high standard, and have consistent sound quality, time after time, product after product.

What you're looking for are manufacturers with a track record of producing dependably well-made and useful gear. A reasonably competent manufacturer may meet (say) 80 per cent of your needs perfectly. The trick is being sure that the other 20 per cent of the features and functions that aren't quite right – or at least right for you – won't severely impair the sound, or your ability to use the gear effectively.

The ideal situation would be that on a particular piece of gear (a mixer, for example) all the features you're not so happy with happen to be ones you know you won't use much, or you have alternatives for. Or let's say a certain piece of gear is known to have a 'hard'-sounding top-end,

but you're planning to use it on bass only, so that's obviously acceptable. Equally, try to think ahead – one particular choice might seem OK at the time, but leave you with less flexibility in future.

It's not enough just to design a good product – it has to be built reliably time after time

A small number of perfectionist 'blue chip' makers can be trusted to be globally discerning in the quality of what they produce. In every case, these are small, specialist makers who are likely to have backgrounds steeped in the PA industry and be in close personal touch, at all levels, with the end-users and their requirements. If they don't dependably come up with the goods, they're unlikely to stay in business for long. The same is not automatically true with the larger, corporate makers, who can survive unsuccessful 'dabbles' into certain area if other sectors of their business are doing well…

This is why experienced PA equipment users won't be heard saying, 'This is made by ZXY (a

Tip from the top
If you're looking for gear for a specific gig or event, it's important (though rarely mentioned) that the gear should be 'rider-friendly'. This means you need to make sure the equipment you provide is acceptable to the artist/band, or their technical advisors, and is either what appears specified on their written contract with you (their 'rider'), or is an acceptable equivalent or alternative.

buying PA gear

globally-sold 'brand' of PA gear) therefore it has to be good.' For a start, teams of designers differ and budgets fluctuate with every product.

Innovative product design is also risky – as with musicians and acts, history is filled with clever products that failed, in one way or another.

And as with all companies, big or small, circumstances can change, and standards can slip, particularly under the financial pressures and rapid pace of change in modern business.

Some of what were once small companies have, by their efforts and successes, gradually evolved or have been merged into large, multi-national corporations – which may or may not retain the insights and ideals that first inspired their founders.

Some non-specialist broad-based corporations have also moved into the PA market. Some have good intentions, and have recruited PA system experts and top designers, while others may just wish to muscle into what has become a lucrative part of the music equipment industry. Then if the enterprise doesn't work out, they can always retreat to their more successful areas of production.

(Cynics might at this point quote the English social commentator and early product designer John Ruskin, who, as far back as the late 19th century, remarked: "There is hardly anything in the world that some man can't make just a little worse and sell just a little cheaper, and the people who buy on price alone are this man's lawful prey.")

Here's a cautionary tale to show what can happen: a few years back, a new manufacturer started to get a good name in semi-pro circles, by selling a (shameless) copy of a well-researched US-made processor unit, produced at low cost in a large far-eastern country. But once they started getting positive reviews and public interest and market share, the product was re-designed to use a cheaper, lower-quality key part (the VCA chip, in this case). So purchasers blindly following the trend were soon, without realising it, buying quite a different unit to the one that had received the good reviews.

This is not, thankfully, common practice, but it has been known now and again.

Alternatively, you might have received a glowing recommendation about a certain piece of gear, but it still may not be right for you: what suits one application may be less useful for another. In particular, what works in studios or broadcast may be quite unsuited to PA. On the other hand, if the occasional user with money to gamble or burn doesn't try things out, nothing new would be discovered.

When assessing gear you also need to be wise to the fact that, occasionally, a perfectly reputable maker may have produced and supplied a one-off dud (or dud batch of units) in what's otherwise a seemingly faultless range. If other users say they get good results, and you don't, this should ring alarm bells. If you then explain to the maker or distributor that the product you're trying has problems, they may either help by figuring out and telling you what you're doing wrong, or else they might take it back – and having found the problem, reward you with a unit that works far better.

All consumers should try to be proactive – not only voting with your wallet or purse, but also offering feedback, positive and negative, to the maker of the product. After all, if everyone bought the cheapest and shoddiest-made things – and, more importantly, didn't complain – makers would say, 'No one's objecting, so we must be getting it right'. Without any comments – bad or good – makers are working in the dark, and progress is slower.

It's up to every buyer to appraise each piece of gear on its own merits

To summarise, it's up to every buyer to appraise each piece of gear on its own merits. Magazine reviews, word-of-mouth testimonials, evidence of previous brand quality, and of course price, are all part of the decision-making process, but in the end it has to be right for you and your set-up at the time.

Some features of decent PA gear
Over the next few pages we offer some general pointers to the kind of attributes you should be looking for, and some things to be wary of, in various categories of PA gear.

These tips are not exhaustive, but do hit some essential nails on the head. They should also help you develop the right sort of thinking and expectations about professional PA products. (You'll find additional tips on good equipment are scattered throughout the book's main chapters.)

A good PA mixer:

- should have knobs and buttons that are an appropriate size, don't feel that they're about to snap, and which can be turned quickly without knocking adjacent knobs – when used by people with 'real-sized' hands and fingers.
- shouldn't have important buttons that are both tiny and in compact rows/groupings, which makes them all-but impossible to operate easily and speedily.
- should have knobs and buttons where you expect them (on one otherwise reputable Japanese mixer the pan control was placed where the gain knob normally is, and vice-versa – presumably for some spurious 'style' reasons).
- should have colour-coding and legend (text) that's readable in dim lighting, particularly in red light.
- should have LEDs placed where they are visible – not hidden behind tall knobs.
- should have switches and buttons that don't cause unwanted noises such as 'zhits', 'zaps', 'plonks' or clicks, when they're operated.
- should have headphone outputs that go loud enough to be heard over monitors or FOH sound with pro-standard medium and high impedance headphones, and without causing distortion or ear pain.
- should have switch-on delay –

where outputs are muted for a few seconds after switch-on to prevent unexpected thumps and bangs. Vital, for example, when the power suddenly fails but then comes back on before the sound engineer has time to react.
- should have high-frequency EQ that doesn't boost frequencies way above 20kHz when supposed to be only boosting audible treble frequencies. A boost of one or two dB is OK, but many boost by 10 or 15dB and this can certainly prove stressful (or fatal) to the PA's HF drivers, in one way or another. (Ask the maker to supply plots (graphs) of the HF boost curves, and see for yourself.)
- should have mechanically flexible busses – the common connections running under all the channel and group strips. Some mixers have had rigid busses which can either fracture or develop dicey connections when the mixer is transported about in trucks etc, often making the console practically unusable.
- should (even if it's a small mixer) preferably have modular, pluggable channel sections, and groups, etc. This means the mixer is highly repairable, on the fly. With the power switched off, a damaged channel can be simply pulled out and another one plugged in to replace it.

A good power amp:

- should not produce any substantial noises over the speakers when turned on and off.
- should have clip LEDs that remain accurate and slightly

'pessimistic' (in other words reading clip slightly 'early') under all conditions of use.
- should offer good visibility of all important LEDs from all angles.
- should be supplied with spare fuses, recognising that these do age and fail.
- should not have fiddly and/or inaccessible or hard to tighten-up output connectors.
- should not stop working completely if there's a fault on just one channel.
- should not have to be manually reset after a momentary power-down or power failure.
- should not shut down when overheated such that there are no indications or cooling fan action.
- should not have cardboard 'safety insulation' (found between circuit boards and the metal casing) which acts as a major moisture trap if the amplifier ever gets rain blown into it, or beer thrown at it.

Knobs and buttons should be where you expect them and easy to operate

- should have minimal or no airflow over any small electronic parts or circuit boards. The dirt deposited by the air can be corrosive, and if so, will start to eat away the

buying PA gear

delicate circuit board tracks and thinner component legs/leads – which action is hidden under a layer of dust. Dust build-up over electronic parts also causes them to run hotter than they should.

- should offer easy access to internal parts that are known to wear or get broken easily and need replaced. For example, power transistors shouldn't be hidden under one-hour's dis-assembly work.

Good PA cabs:

- should have adequate power handling capability and efficiency for present (and future) jobs, and some flexibility in how they can be deployed.
- should offer good sound quality with a variety of amplifiers.
- should have acoustically dead construction. No rattles or buzzes when driven hard or surrounded by loud bass sounds.
- should be designed so it's easy to get inside and replace or service drive units or panel-mounted connectors.
- should have comfortable handles placed sensibly to minimise stress and strain when carrying and stacking.
- should be easy to stack (up) or fly in useful combinations.
- should be just narrow enough to fit through a standard-width single door (roughly 30 inches, or 75cm).

With **monitor cabs**, if can be helpful to realise that there are many imitations – watered-down look-alikes – but only a few professionally dedicated PA monitor makers. To play safe,

discover which monitors are the few that are widely used and liked by musicians and monitor engineers, and evaluate these models first.

A good microphone:

- should have suitable frequency response for the job – for instance if you want a bass drum mike, it must handle low frequencies.
- should survive being dropped – not something you can really test before buying, but can learn by word-of-mouth (especially if they don't survive very well) from experienced users.
- should have low handling noise and sensitivity to 'blasts' (see Chapter 1).
- should have low sensitivity to performers' grasping, clasping, even putting the mike in their mouths, which with some mikes triggers acoustic feedback (howlround).
- should be easily accessible, so you can replace the most breakable parts.
- should be easy to clean the internal parts that will get 'gakky' (clogged with breath deposits).

Good PA electronics gear:

- should have no flimsy controls on exterior bodywork.
- should not have power switches that can be operated by accident.
- should not have sharp, un-radiused (non-rounded) metal case corners – which can rip your hands, particularly when heavy units are being manhandled into racks.
- shouldn't have unlocked screws, which will vibrate loose in real use. (Trucking does it,

even if stage vibrations don't.)

- should not have unsealed pots or switches – controls with open insides – in a case that is openly fan-cooled (compared

Electronic gear should have no flimsy controls on exterior bodywork

to air passing through an isolated tunnel only); or that has any holes in it, which will inevitably let in dust and grime over time.

Good battery-powered electronic devices for PA (radio mikes, beltpack transmitters, DI boxes, testers, etc):

- should have knockproof switches, so the unit won't be accidentally turned on (and battery drained) when packed and moved about.
- should have battery metering on demand ('push to check'), and preferably flashing low-battery alert LED (eg 'only 20 per cent left').
- should have auto turn-off (or low current sleep-mode) if there's no signal or activity, or if key connectors are withdrawn – for instance, if nothing is plugged into a DI box, there is no point in the power being on (except for lab testing).
- should, if it uses rechargeable batteries, have clear instructions on the unit about how to treat them for longest life and best charge

buying PA gear

capacity – not often seen explained on much equipment. Once you've been let down a few times, you'll understand why it matters.

Buying used gear

Professional PA gear lasts much longer than mass-consumer products (cars, for instance), in part because it's made to be sturdy, and also because it tends to be relatively well treated – and also because, like other electronics-based products, there are few, if any, fast-wearing mechanical components.

A great deal of PA gear from the 1960s is still going strong, somewhere – you may spot it semi-retired, in seaside pubs… Like older cars, this gear will generally continue to be mendable longer than most modern equipment, some of which is increasingly composed of specialised and micro-miniature parts that aren't 'repair-friendly'.

On top of the advice given already for new equipment, here are some tips relevant to buyers of used or second-hand gear:

Buying used microphones
Don't judge by looks alone: a mike that's badly scratched and looks well-battered may actually work just fine – though it should ring a few alarm bells; equally, a shiny new-looking mike might be damaged internally through misuse. In both cases, plugging in and testing is the best way to tell.

Other than testing with voices, try miking an instrument, and comparing it with your usual (or any other) mike.

Shake the mike too – is anything loose? If so, that's a sign that it's seen some rough handling, and that some level of repair may be required. If you can find out how to take the top off the mike, do this (when it's unplugged) to see how clean it is inside – which will show well it's been looked after.

If a capacitor/condenser mike sounds crackly, it's probably just got a bit damp. If you're allowed to take it away for a trial, try drying it out slowly in an airing cupboard for a few days before using. Or suggest the seller does so, then try it again.

If the mike has actual or suspected defects that you may have to fix, this cost will need to be ascertained (by getting a repair estimate) before you can come to a fair price reduction with the seller. Mike repairs can cost anything from a few pounds/dollars to hundreds.

Buying used mixers & auxiliary gear
First, check the feel of the faders and switches – are they free of 'grittiness'? Pay particular attention to any 'dual concentric' (split function) knobs to see if they've been bent – which could cause one part to catch the other, making 'live' use awkward, if not impossible.

Shake the unit (if it's small enough, otherwise gently swivel up onto one end) – is anything rattling or moving around inside? Open it up – look for any loose, partly disconnected wiring, or untidy modifications.

Re-assemble and plug in. Check that all the LEDs and lights work. Feeding a CD or tape player into the line inputs, and keeping the gain at

moderate levels, check knobs and switches for crackles, scraping sounds and other noises – using either suitable high impedance type headphones wired to plug into line outputs, or else a PA.

Expect some initial noise on controls that haven't been used recently, but if the noise persists after a few uses, that's more worrying. In older pre-used mixers (say over five years old) it's to be expected that switches and controls will need replacing, or at least cleaning, and the cost of this could be considerable.

Check connector sockets, jacks on the rear and channel strip module connections, both visually (are they badly tarnished) and physically (insert a suitable connector plug into them all to see if any are loose or stiff). If connectors are corroded, the unit may require quite expensive repairs, and there may be other, hidden damage.

Older PA gear is more mendable than most modern equipment

Are the VU needles sticking? Do they work at all? Use an FX unit and an insertable unit (EQ or compressor say) to check the PFL and aux sends and returns and inserts are all functioning OK.

It's worth working through all the channel and sub-group permutations, or as many as you can bear, if the gear is at all costly. Otherwise, you may not discover a particular and possibly crucial fault until you meet it in some particular mixing situation, some way in the future – perhaps after any warranty has run out.

With graphic EQs and mixers, you may want to call the equipment's maker and check if spare fader/slider controls are still available, and will continue to be.

Buying used multicores/snakes

If it's wound up on a drum, unravel it. Watch how every inch of the cable comes of the drum – it shouldn't be twisted. Inspect every part for signs of damage. Any grazing, cuts, nicks, kinks, deformations, softenings, melted areas, flattened sections, etc, are cause for haggling to reduce price, or repair, or rejection.

If possible, check the readings of a full test on the cable – this means using a basic digital multimeter to test the ohmage for insulation (high resistance) between all cores, for every core-to-core, and good conductance (very low resistance) from end-to-end, for every core. To be sure, do this when the cable is on the drum and also when it's laid out.

Buying used power amps

Check in the same general way as mixers (described earlier) plus watch for blown or malfunctioning channels, broken or missing controls, switches and terminals.

To test without requiring a crossover, use a two-way monitor or 'wine-bar' PA speaker (with its own internal passive crossover). To test without risking the speaker, in case the amp is faulty, connect a 40-watt car

(auto) 12V light bulb in line with the speaker wiring (either side). Then if the worst happens, the bulb will glow or blow to protect the speaker.

Take off the lid and look inside for signs of past repair work – for instance parts that don't quite fit the board, or are noticeably different to others, and fresher, more silvery solder. Some history of repairs is to be expected as an amp gets older, but should be neatly done.

Signs of possibly sub-standard, maybe inexpert or rushed work that could backfire include: white paste (used to help ease heat transfer between transistors and the heatsink metal) daubed all over; unsupported parts with bare metal legs dangling in mid air; and untidy soldering with messy blobs, textured or crystalline surfaces, and 'icicles' (pointy bits). Also watch for missing fixing screws, loose boards etc.

A heavy dust build-up suggests the amp has not been serviced much or, if the fan is speed-controlled, that the amp has been running hot a lot. Singed (darkened) areas of circuit board where it touches the heatsink also suggest this. Burnt parts, or a lingering smell of burning, suggests something has blown at some point – and it may happen again. If you find this – or if there is general corrosion or parts covered with crud – it's fair to haggle substantially over the price, if you want to take the chance at all. If you don't have access to a PA electronics expert to help you, find someone who is qualified to make an assessment and, if necessary, to do the repairs.

In all cases, before first using anything that's AC-powered:

- Check the power cable, and replace if at all deteriorated.
- Check all electrical wiring inside – a quick visual check by electrically expert eyes can save trouble later.
- Bring wiring and connectors up to modern safety standards (see Chapter 10).

Buying used PA cabs and drivers

You shouldn't buy any speakers without listening to them first. As you'll usually be spending at least £1000/$1500, a demonstration session ('dem')

If the drivers are different colours, or not the same types, the PA may be mis-matched

should be arranged – don't be afraid to ask for one.

But before you come to listen, some useful visual and mechanical preliminary checks can be made.

The first test can only be made to direct-radiating drive-units with paper/pulp (as opposed to metal) cones. It's to check that the mechanism isn't rubbing. It requires a little care – so practise on an old speaker first.

Placing your first and second fingers on either side of the centre of the cone (to avoid off-centring it), depress and release smoothly, while listening and feeling for any rubbing or scraping (you need a bit of peace and quiet for this test).

Next, check the condition of the drive-unit's edging (the surround). Is it oozing stickiness, or at all brittle? Check the cone for rips. Fortunately, all of these problems are fixable with a recone (see Chapter 9), but it obviously makes the equipment less desirable and/or valuable.

Finally, look for cracks, bends or dents in the drive-unit's chassis (metal frame) – not purely cosmetic scrape marks, but ones that might affect the drive-unit's integrity or alignment.

For example, if a chassis driver or cab has been unlucky when it's been dropped, it may be warped or bent and that may cause damage to the delicate voice-coil – possibly each time it's driven hard. The heating-up means it expands, which can cause warping – even just a few thousandths of an inch is bad enough. Or a chassis made with cast metal may have deep cracks that are largely hidden and have spread with time. Such damage could prove irreparable.

If you're buying a horn-tweeter (horn+driver in one), be aware that low-budget types don't last long. It's quite feasible for them to expire at any time after being worked hard. The same goes for piezo types.

Even with high-quality cabs, horn tweeters may sometimes die because of a manufacturing or design fault – perhaps because an onboard passive crossover has been set too low, due to a wrong-valued part being fitted, so the high-frequency driver is continually stressed.

Such features can be hard to know about in advance. Finding out more about why the PA is being sold might help you to detect them.

(It's not unknown for a pro-active user to turn such a matter to their advantage by getting the problem cured. Having bought a certain drive-unit cheaply, suddenly you're the only PA operator around that can get the good sound it's known for without spending the £100/$150 a week on replacement diaphragms that everyone else has to.)

Once powered up – through a suitably rigged crossover and amps, if these are not also part of the sale – make sure the test amps, CD player, mixer and recordings used are giving a clean sound, and that the (active) crossover, limiting, output delays, etc (see Chapter 4) are set correctly.

It's never enough just to read the manufacturer's spec sheets or even test results

Then, to check for driver problems, listen for raspy, fuzzy sounds at all volume levels, particularly very low and high.

If the sound is lifeless and dull, are the high frequency drivers dead, or just disconnected or turned down?

If the sound is hollow, ask the

same about the mid-range components.

Missing bass should be obvious enough.

A look behind the speaker grilles can also be instructive: if the drivers are different colours, or otherwise don't match, maybe they are different types – which means the PA might be a bit mis-matched. If in doubt, try listening to each cab (or driver) in turn.

Even though this may seem counter-intuitive, when listening to some drivers and not others (in a given cabinet), keep the drivers you're not listening to connected to their amplifier (don't switch them out) and leave the amp turned on. Also don't set the amp to mute (if it has this facility), but turn the gain control down.

These procedures keep the unused drivers damped, so they don't add farts and buzzings that will affect your assessment. It's also worth throwing a blanket over them to absorb any slight sounds they make.

With used **monitor speakers**, much of the above applies equally. To test them, connect up with mike, mixer and amp, and do the 'oi'/'hey' test through each monitor (see Chapter 6)

Important ergonomic matters to assess for all PA cabs are:

- ease of repair and maintenance (how many screws need to be undone to get inside, in a hurry).
- the 'pack' size (for fitting in a standard truck/HGV lorry).
- handling aids that will ease transit. Consider (for example) that four strong castors on a bass bin can cost £200 new.

buying PA gear

Used equipment to steer clear of:

- anything that you can't try and check before deciding.
- anything you're being pressured to buy.
- anything for which you can't get proof of ownership – in particular, anything you have reason to suspect is stolen. Obvious warning signs might be an unbelievably low, very 'round-figured' price, and the alleged 'owner' knowing nothing about how the gear works.

Other than the lack of recourse in the event of the gear being faulty, there are also legal risks in buying or selling stolen gear. In some countries you could face criminal prosecution for 'handling stolen goods', or taken to civil court by the original owner. At the very least, if traced, you'll usually have to hand back the equipment with no recompense.

Don't buy anything you're being pressured into buying

If you wanted to be really sure – if there's a lot of money at stake, for instance, there are a couple of procedures you could try. If buying from a company/dealer/organisation, ask at first to see the original invoice which would usually change hands when the seller bought it, then ask for a copy of it. Any discrepancies in

addresses or names (for example) should be explained in a covering letter, so it's all down in writing.

If the gear is changing hands privately, there may be no documentation – so how do you prove the seller is the true owner?

As mentioned, some knowledge on their part of how the gear works is a good start, but to be fussy you could ask to see, say, photos of the gear in use at gigs, rehearsals etc. This might be unlikely, but just going through the process of asking for this should give you strong signals as to whether the seller is genuine or not.

Listening tests

The best way to test any sound system is to set it up and play music through it – this way you not only hear the quality, but get a feel for its ease of use. This is particularly important when the gear under consideration is either: a model or make you have not used before in your present PA set-up (someone telling you it works well in their system is not a sure recipe for suitability); or, of course, if it's pre-used or second-hand.

It's never enough just to read what the manufacturer's leaflets – or even third-party 'tests', however in-depth – say about a product. Those who believe they can tell anything definite from any number of specification sheets or measurements, or from what the gear looks like are fools to themselves.

Admittedly sometimes figures or looks can be hard to ignore, but you have to go beyond this.

Quality (let alone musical quality) is too complex for any machine to judge – even the most advanced audio tests are surprisingly limited in their scope and coverage.

For example, measurements of the distortion qualities of a complete PA system are virtually unknown. Worse, many of today's audio gear reviews are accompanied by less (and arguably less good) audio testing than they used to be – if there's any at all.

There are two main methods of checking gear by listening:

- 'Single presentation' is the preferred technique. This means you take the gear away 'on trial', and use it exclusively for a week, or longer if possible. Even if your first impressions aren't good, try imagining it's the only gear there is at a mud-bound festival, and you have to get the best from it somehow.

  Afterwards, go back to the gear you were using before, spend some time re-acclimatising, and at that point you may be able to make a decision. The different may well be all too clear, but if you're unsure, repeat the test again (depending on how accommodating the seller is).
- The second option consists of either 'blind' or sighted A/B tests between different systems. This means set-ups are compared one after the other in 'controlled' conditions – the kind of procedures which are loved by academics as they seem 'objective', but are in fact by no means foolproof. There are too many subjective variables at play.

Even slight differences in volume, for instance, can alter perceived sound quality – a slightly louder system will seem to be brighter and have more bass. But then, if the frequency responses aren't identical (and few ever are), perfect matching isn't feasible anyway. To overcome any bias, try listening with one then the other system slightly louder.

Be aware that any listener's ears will start to get tired

After repeated tests, be aware that any listener's ears will start to get tired, and you may imagine hearing things that aren't there – sometimes even unknowingly compensating for tonal differences by mentally EQ'ing the sound.

In general, listening is likely to be most productive when you're in the right frame of mind to start with, and if you take regular breaks for relaxation.

In the end it's not so much about meeting arbitrary pre-determined standards as it is about gut instincts and emotional reactions to what you're hearing, and how this changes when you're listening to different systems.

The most important point about any listening tests is that you should be comparing like with like. If you plan to make regular assessments of PA gear, such as processors, mixers and

amps, it would be ideal to have a dedicated listening test room with a scaled-down version of the PA you normally use. If this isn't possible, at least try to keep notes of the gear used on any test, and where and how it was laid out and set up.

If listening tests show the sound to be wrong, bad or not quite right, consider the following:

- You may be using the system incorrectly. Have a break from it. Everyone has their bad days (wrong 'phase of the moon' or whatever). Get some sleep. Do something else. Try tomorrow when your mind is more fresh.
- Invite the maker or seller to help you check you haven't made any silly mistakes. This could be a quick run-down over the phone. If it's a brand-new design, it's even possible that the maker has got something wrong. Cabinets might have drivers with wrongly connected polarity, or defective or wrong model components fitted. It can happen.
- Atmospheric differences can affect test results – air conditions (temperature, humidity and air pressure) can have a big effect on the sound of some mikes, DJs' cartridges, and speakers in particular. While listening, try to keep the room's temperature as consistent as possible. If you know how to use them, get a thermometer, a barometer and a hygrometer (most readily obtainable from a gardening supplies shop), which indicate the air pressure and moisture content.

You should also obtain a digital voltmeter with a large, easily-visible readout, and have this connected across the power supply to the PA, so the mains voltage can be monitored. Then if the sound is 'wrong' one day, you can see if the mains voltage is down, showing that the local supply is heavily loaded, and possibly explaining the cause (having exhausted most obvious possibilities). If so, try again another time – the normal voltage might be resumed after a few minutes or hours.

(Also read the section in Chapter 10 on 'electrical safety and second-hand gear'.)

buying PA gear

gigging & touring

There's more to a successful live show than just the choice of equipment and the skills to use it well – there's also the logistics of getting the gear to the right place at the right time, and the politics and legalities of dealing with performers, promoters, venue managers, security personnel and fellow crew-members... not to mention the general public.

If record companies are to be believed, touring – once allegedly a profitable activity – makes little if any money nowadays. And PA is always cited as one of the costliest elements of touring.

It's undeniable that good, rugged sound systems can involve weighty and bulky components, and the expense incurred in moving these on and off vehicles, setting them up, tearing them down again, and transporting them around the country (or the world) can be significant.

But it's also true that over the years the pressure on tour costs, particularly in the USA, where the (strongly unionised) local crew fees have been high, has helped push forward the development of (relatively) lighter, smaller, and easier-to-rig PA systems. It's also forced touring operators to discover or develop some smoother handling methods.

Whatever level of gig or budget you're working at, there are some universal practices that are worth adopting for a hassle-free (or hassle-reduced) working life.

Smart methods of moving gear
First there's the obvious 'castored' trunks and flightcases – the ubiquitous wood-built, steel-reinforced protective boxes in which most everything, from power amps to power cables, needs to be stored and transported.

Brakes
Wheeled equipment must have at least two 'braked' castors – and these need to be kept locked at all times, except when the rack or trunk is actually being moved. It's vital to lock castors when gear is being congregated or marshalled near a ramp or stage edge, otherwise it can roll off and easily injure or maim. It may sound unlikely, but if a rack of amps (typically weighing about one-fifth to a half-tonne) does start to roll down a ramp, the momentum (roughly 'speed x weight squared') by the time it reaches the bottom can be as dangerous as a car/pedestrian accident. Even if it doesn't hit a person, the jolt when the rolling rack hits something immovable (a wall or, even worse, another rack) can prove highly damaging to the gear itself. Less roadworthy rack frames, amp innards and mixers have been critically injured by this kind of collision.

Loading on and off trucks (and trains and planes) can be tough on gear – as can the journey itself – and even the hardiest equipment needs this extra defensive layer.

Small and fragile items can also be wrapped in bubble-wrap or foam rubber first: some flightcases (and specially-designed metal 'briefcases') are lined with foam, with shapes cut out for inserting gear, such as microphones etc.

Lifting is also tiresome and potentially hazardous (we'll come back to this later), so the more gear that has castors fitted, the better – though these must be heavy-duty, high-quality, fitted with efficient brakes (at least on two of the four wheels – see sidepanel on the previous page).

Here are some more transportation essentials:

- Wheel-boards – plywood base, approximately 2x3in, fitted with large castors (each quarter-tonne-rated). Used for any gear that doesn't have castors, or where one of the castors has jammed, or whatever. The gear's own weight keeps it on the wheel-board, aided by a 'sticky rubber' mat.
  Large gear (such as mixer cases) can be transported using two adjacent boards. (Always carry a minimum of two wheel-boards.)
- Trucks (lorries, wagons, artics) should preferably have electric/ hydraulic goods-lifts fitted. This avoids the extra handling distance of ramps, the danger of roll-back, and also permits 90-degree handling, which a ramp simply doesn't allow.
- 'Hand-balled' pallet/fork-lift truck. Used with wooden pallets. Allows a large load to be wheeled across a fairly smooth and level floor. More often use in PA warehouses than venues, but found in public and institutional buildings, where they prove invaluable in certain situations.

Crew

PA crew travel with the PA and/or the rest of the crew (or sometimes even with the band, if they're very privileged). The PA crew will comprise at the very least one front-of-house (FOH) engineer, who may do the monitor mix as well. In larger shows there will be one or possibly two FOH engineers, one monitor engineer per band, and someone who looks after technical matters.

On a world tour, there may be seven people working front-of-house

---

In a really large production, such as a world tour, there will be maybe seven people working FOH (at least if you're in the megastar realm) – some of whom will just be looking after one part of the system – then two crew on monitors, and three 'system technicians'.

A touring PA crew are paid weekly, or even monthly, or at the tour's end. They may also need to receive daily allowances ('per diems').

Local crew (for 'casual' manual help with lifting gear etc – sometimes known as 'humpers') are hired for each show – and it may be safer to pay them *after* the show, just in case they're a bit too casual and don't stick around…

Useful stuff to have on tour

- Mobile phones.
- Staffed telephone line, fax line, fax machine, and laptop for email.
- Two crew radios to site security. Spare battery packs & solar/mains chargers.
- PA crew vehicle or marquee, and crew-only porta-toilet.
- Copy of licences for radio mikes.
- Copies (not originals) of manuals for all gear.
- Copy of makers' instructions for fitting diaphragms.
- If taking gear abroad, you may need 'carnets' detailing every item, and proof of ownership.
- Driving licences, plus passports, visas etc as required.
- Climbers' head torch (eg Petzl) – leaves hands free for work in dark corners or outdoors at night.
- 'Swiss Army' knife or similar.
- Ear plugs (there's more about these in Appendix C).
- 'Breathable' waterproof jacket so you can dry-out inside it.
- Hip flask filled with suitable restorative.
- Hammock and relevant fixings.
- First aid kit and emergency procedure chart (recovery position etc).

PA hire

Even PA systems that are put together and used mostly for particular artists are hired out to third parties during quiet periods. (It all helps towards the purchase cost and upkeep.)

The enquiry to hire may come from a band, their own engineer, their management or their record company.

> # The client may want to hear the PA in action at a gig you are doing

If you're in charge of the PA, have an equipment list ready to send out. The client may also want to hear the PA in action at a gig you are doing (even if you've been recommended to them – bear in mind that with PA, as with performing, you're only as good as your last show).

Naturally, to be representative, they ought to hear the PA in the same sort of venue size as they'll be wanting to play in. They may also (quite sensibly) need to see if they can get on with whoever's doing the mixing.

With smaller systems it's common for the band to supply their own FOH (and monitor) mixing engineer. With larger rigs, particularly, demands by the hiring artistes or their engineer(s) to have total technical control should be resisted, unless or until there is a proven track record.

Band managers may resist giving total control to the PA company for the sound, but this approach is likely to be the best one if the PA operator is competent. The usual half-way house is that the PA company supply a 'babysitter' – their own FOH engineer who can take control of the system, and otherwise provide guidance.

Choosing a PA to hire

If you're listening to a PA with a view to hiring it, you might want to make some notes, so comparisons can be made with others you hear, and the most suitable rig chosen – as long as the person making the notes has good listening skills, and the style of music used is suitable for the PA, and familiar to the listener.

Notes could be made based on the PA response at each frequency range (sub, LF, MF, HF etc), and given marks out of five, say, or judged on each of various criteria.

For instance: level at each frequency (too little, too much, OK); clarity, clean-ness and musicality. To be thorough, this should be done from various listening points in the auditorium.

You might also rate things like microphone choice; FX choice; desk (mixer/board) facilities; desk EQ; ease of use; smoothness of changeover (in multi-act or multi-set gigs). And any other specific tests you may devise.

Hire charges

Typically, gear is hired out at a rate that's a percentage of its replacement value, per day, usually with increasing discounts for longer periods, or a series of dates. Plus crew costs and 'consumables'. Other direct expenses that may be claimed for are negotiable.

Contracts should be read carefully to ensure you are not made liable for something you wouldn't normally expect to pay for. Make sure everyone's clear whether the price quoted is before or after taxes.

For a tour spanning several weeks or even months, haulage fuel may form a large part of your costs. Considering the potentially unstable price of fuels, it's wise to add a provision that any fuel increases above a certain threshold can be levied on top.

The price can sometimes be brought down by cutting some carefully selected corners – for example:

- hiring fewer effects and toys. Maybe the band can borrow or supply some themselves? It's worth mentioning to the client that rather than spending the

---

Tip from the top – Lost?
Here's an idea if you completely lose your way on-route to a gig in a strange town or country – you can't find the venue and you need to get there quick... Drive your truck into any large petrol/gas filling station, and ask at the cash desk about local taxi-cab firms. You want their most experienced driver. Order the taxi, and when it arrives get the driver to lead you to the venue via the quickest, least congested route – you just follow behind in the truck.

Don't forget to leave a generous tip...

money on a 'flavour of the month' FX they could spend it on a good engineer who could probably get the desired effect using existing gear, and make the sound better all-round.

- fewer bass bins (they add the most cost in haulage fuel).
- using a more local crew.
- reducing PA power and/or venue size in towns where the band is 'less popular' so a large PA is not wasted when 50 per cent of the seats are empty. (Some venues can be reduced in size by moving partitions, tiered seating, etc.)
- analysing what the competition are supplying, to see why their offer may be cheaper than yours. That cheap deal may include minimal FOH and monitoring gear, with everything else charged as 'extras' – including engineer.

A definite booking should be secured with a deposit, and preferably written (or at least verbal) mutual confirmation. Payment for 'one-off' jobs is normally made in cash, on arrival of the PA system, and before it's even unloaded.

This is only fair, since in reality even the nicest promoters have disappeared into the night, leaving everyone unpaid, and venues are well-known for unexpectedly going into liquidation.

Promoters can disappear into the night...

For regular work with reliable customers, payment would be negotiated. Usually some will be upfront, some weekly during the rehearsals and gigs, and some after the event/tour has finished.

Support bands

It's important to establish from the start whether the support band is paying for its share of any PA hire directly to you, or whether the main act is paying for the whole thing. It's obviously wise to check with both parties and get it in writing.

Support bands are sometimes viewed as a necessary evil by most everyone else involved in a gig or tour (the less sympathetic might say an 'unnecessary evil'.) But if you've ever been in a new, breaking or even un-signed band, you'll know what it's like from the other side.

Support bands often feel they get a bum deal – particularly from sound and lighting engineers. They may sometimes have a point...

Tips from the top – **Health**

It's always a good idea for at least one member of the crew to be trained in first aid techniques. It may as well be you... In the meantime, there are self-help and precautionary steps that everyone involved in PA work can take. First of all, here are some quick tips on preventing back strain when lifting and moving heavy gear – or 'avoiding ergonomic hazards', as they say in the US of A:

● First of all, if the object looks like it's too heavy for one person to lift, just get some help. Forget macho – how macho is it to be laid out in a hospital bed in traction?

● If you are going to tackle it yourself, think of the following word – **BACKUP**. It stands for the following:

Back straight – don't curve your spine
Avoid stretching – keep the object close
Clutch firmly – get a good secure grip
Knees bent – helps with balance
Use your legs – let them take the strain
Putting down – do it the same way

If you've wrenched something, bashed something, cut something, or you're just generally feeling poorly, think on this: Hospitals are no friends of minor complaints, and in some countries treatment is uncertain and expensive. Or you might be stuck on a festival site, feeling ill, but too badly needed to leave. Or say you witness a fellow crew-member lying injured, and there's no one else to help them...
Assistance could be at hand, in the form of a book like *The Family Guide to Homeopathy* by Dr Andrew Lockie, which has some sound advice on first aid and 'bodily disorder' treatment, using homeopathic remedies where appropriate. The remedies listed can be safely self-prescribed, and are low-cost. A basic first aid kit of about 20 types of 'remedy' pills, one tincture and five creams covers most situations – from burns, crush injuries, weird food poisoning, sprains, smog fumes, and all manner of other minor troubles that stop you giving 100 per cent.
Of course, if the injuries are plainly serious, or first aid doesn't ease matters fairly quickly, or symptoms worsen, immediate hospitalisation is advisable. This is one of the good things about this book – it identifies and distinguishes serious, possibly life-threatening conditions from ones that you can treat and recover from quickly.

The hard fact is that the audience and the crew are almost entirely there for the main act – they're the stars and/or employers.

Sound & lighting people are obliged to save the best for last, to keep a few tricks up their sleeves. So the support band inevitably comes off second best.

(It's not unknown for the support act to blow the headline artist off the stage, but it's usually in spite of the mix, rather than because of it.)

You can look at it two ways. If the support band are keen to forge a serious, successful career, they've got to behave professionally and diplomatically, and realise their (hopefully temporary) lowly position in the scheme of things.

Being nice to engineers and stage crew, helping them rather than hindering them, buying them the occasional beer etc, can smooth things along very well – and is much more likely to result in an acceptable mix, and a re-booking.

Word gets around in such a close-knit, pretty incestuous industry.

Similarly the engineer might just as easily treat the support band with some respect, maybe hand on some sage advice and encouragement, and deliver a good sound for them into the bargain (albeit not as sparkling as the main act's).

As well as this being the 'decent' thing to do, there's also the prospect that the up-and-coming support band may well remember that particular engineer and, when they're in a position to do so, employ them themselves at some later stage.

Diplomacy & negotiations

Almost invariably, sound (ie the PA) takes second place to production and even lighting at a show. Despite the obvious fact that without sound there is no gig. Gigs have gone ahead without stages, without lighting, and without artiste hair-management... but never without a PA (except perhaps in the case of a non-vocal band in a tiny venue). The PA's status should be assured. But as ever, the less secure parties tend to be the pushiest...

Before a tour starts is the best time to ensure the band and their management understand the central role the PA plays, and that under-funding it may not be in their best interest in terms of a good show.

A classic example of the kind of three-way power struggles that can occur on tour is where rigging points for PA flying are limited by pre-determined lighting and staging requirements. The set and lighting designers will claim they concocted their design months earlier and any changes will incur heavy extra production costs. Yet the PA's points should have had equal priority, since they can make or break the sound.

The result can be that your top PA gear underperforms (and the FOH sound requires endless tuning) throughout an entire tour, because it has to be hung from the wrong acoustic positions.

For instance at London's Wembley Arena (a venue whose future is now sadly in grave doubt) it was well known that the best speaker position was a central overhead cluster. Yet set designers would often insist on putting, say, a 20-foot wide cardboard 'moon' or other stage gizmo in that vital central position.

General and mutual solutions to this sort of problem need discussing in plenty of time, away from the heat of the show, when PA engineers and owners find themselves in relaxing places with their production and lighting management friends.

Sound level limits

If noise levels set by local authorities are exceeded at a gig, the promoter or venue owner may have to pay a fine. If you're lucky, this may be negotiable – if for example the PA's FOH mix engineer and other crew have gone out of their way to co-operate, and particularly if a substantial proportion of the excessive levels have been caused by particular artistes and their engineer. Don't count on it, though.

Also, in many countries (such as EU member states) if the sound level in a workplace (averaged over several minutes and called 'Leq') exceeds 85dB-SPL, technically employers should take action to provide basic protection for their workers. Present laws don't exempt or increase the threshold where music is concerned.

On the other hand, like many laws, it isn't always heavily enforced in the live sound world. It also only affects employees (not self-employed contractors, volunteers and enthusiasts), who should be able to find out more from their management.

In terms of hardware, complying with the law isn't very expensive. Buy everyone ear defenders and arrange for enormous red, flashing lamps to light all over the backstage areas if the sound level on stage exceeds 90dB-SPL (Leq).

In the UK, USA, and many other countries, there are a host of laws about making noise, especially at night. (This has become more of a big issue in recent times, although it's certainly not a modern phenomenon – the earliest surviving law records in London, dating back 600 years, are about noisy neighbours, busy manufacturing or partying late at night. Plus ca change…)

The laws about nuisance and disturbance may involve confiscation of equipment, sometimes even for a first-time offence.

To be on the safe side, especially with outdoor events:

- keep as far away from built-up areas as possible – distance is the best and cheapest volume reducer.
- reduce volume after midnight, then further at 2am; for all-night parties, don't increase until 9.30am.
- if an official challenges you, be prepared to hold a discussion about social responsibility. Negotiate. You may be able to reach a compromise on volume and EQ (low bass is often a problem) until the sound is acceptable to you and the neighbours.

(There's more about sound and hearing protection in Appendix C.)

Truck weight

The legal weight limits of transport vehicles are easily exceeded. A truck can be technically overloaded with PA gear before it is stacked above waist height. And with pressure on tour costs, one vehicle

may be expected to do what is not strictly legal.

The legal weight limits of transport vehicles are easily exceeded

On many main roads the police randomly but routinely stop and check the weights of larger transport vehicles. Other than any possible criminal proceedings or penalty 'points' on your driving licence, the bad news is that they will also require that the excess weight is removed before the vehicle goes on its way. This means that another vehicle will have to be brought to collect the discarded gear – all adding to the time and cost involved, and something you

---

Guarding gear at the gig

Usually the venue owner or the promoter is responsible for providing security for the PA gear while it's on their land. You should make an agreement about where responsibility for the security of the PA is delineated – clearing up any 'what if' circumstances.

Small valuable items of gear, particularly microphones, are easily stolen. Mikes are not only relatively easy to pocket, but there are more potential buyers than for other parts of your PA – they're attractive items in their own right, and prize trophies to some artists' fans.

If the stage is left unguarded before –

or especially after – a performance, mikes can disappear from stands very quickly, particularly if the stage is easily accessible. It's not even unknown for them to be stolen while in use – especially if the audience surge onto the stage, or a mike-wielding singer jumps into the crowd.

To save the hassle and expense of lost mikes, there are several lines of defence you could adopt:

- Make sure venue security personnel are aware of the risk to all portable items – they may not always realise, for instance, that mikes need to be guarded at least as much as a guitar or a DJ's records.

- Crew should remove all mikes from the stage immediately the set (or the encore) ends – giving priority to microphones near the front of the stage (usually the vocal mikes).
- Especially-prized microphones can be fitted with anti-theft devices – from simple 'postcoding' or 'zip-coding' with ultra-violet-sensitive pens, to more elaborate radio trackers – or even a remote-controlled release of coloured liquid exploding from within the mike casing to mark and identify the thief… (Are we getting a bit extreme here? It's a thought, anyway.)

gigging & touring

can do without on the way to meet a deadline.

Security & insurance

A small PA can range in value from £1000/$1500 – for a used semi-pro set-up – to easily £10,000/$15,000 for a modest high-quality set-up, then upwards, towards £250,000/$375,000 for a festival or stadium PA.

PA equipment requires specialist insurance – though only a few companies offer this kind of cover. As with all insurance policies, you'll feel it's money down the drain, no matter what you're charged – unless you need to make a claim. But can you afford the risk of being uninsured, should disaster strike? Insurance could save your livelihood. It also relieves a lot of stress and strain – both worrying about what might happen, or trying to go through legal proceedings to prove 'blame' if something does happen.

And while you're insuring the gear, ask about 'public liability insurance' – if someone has an accident at a gig, they might claim it was your fault for positioning your gear where it was, and you could be sued. For a very large sum. You need to be covered – no question about it.

If you feel insurance is too expensive to bother with, it could be a sign that the PA isn't being used enough – or maybe it's worth more than you're charging for it. (If you only use it occasionally, you might be able to

negotiate a special insurance rate for several named days per year.)

A major reason for being insured is to protect you in the event of theft. Stolen equipment isn't often recovered. Even if you know who has it, you may not be able to get it back. Insurance can help ease the pain – though it's not a licence to be careless. You must take 'due care'.

Whether or not you choose to insure, you should take a variety of practical steps to discourage and counteract theft:

Gear security checklist

● If any gear has a serial number, or any other identifiable feature (ask the maker if you're not sure), record it when you first acquire the gear. Keep all records on computer, backed up on removable disks.

This is also the time to insure the gear (when you first get it) – not two weeks later, after it's been stolen. For insurance purposes you need to decide on a value you want to recover in case of loss (which may be higher than the buying price, as replacement costs may be greater for newer models).

● Update the inventory regularly – every year at least. If insured, lodge a copy with your insurer. Make sure that updated amounts (of insurance) are confirmed in each year's new policy.

● Photograph the equipment – out of the rack, with shots from enough angles to identify it.

● Open up the gear and stick name labels in unobtrusive (and non-heated) places – invisible inks may be used.

● Avoid any labels (or other signs, like gear maker's stickers) on the outside of your van/truck or gear store-house, which might suggest what's inside. Choose an uninteresting, non-descriptive trading name. PA stores need no local publicity at all.

● If possible, try to live near your gear. Failing this, store it next to reliable friends who are always at home. Larger companies have been known to store equipment in farm warehouses next to occupied farmhouses in the countryside, rather than in theft-prone industrial 'units'.

● Gear storage areas should ideally have no windows, except for tough, translucent sky-lights (light but no vision). Where the value warrants it, all doors should be fortified and double locked, top and bottom. Multiple doors are sensible. Distributed anti-ram-raid pillars are best placed behind shuttered doors. Generally, keeping the security features (like locks, chains and pillars) indoors will reduce the signs that there is anything of interest there. Similarly, any security cameras should be covert.

● Once gear leaves the building it should be guarded at all times (insurance policies may specify this). Gear in an unguarded van/truck (even left

The six stages of show production – a cynic's guide...
1. Wild enthusiasm
2. Total confusion
3. Utter despair
4. The search for the guilty
5. The persecution of the innocent
6. The reward and promotion of the incompetent

for a few seconds) provides a quick-witted thief a 'few grands-worth' of gear – villains have vans too, remember. If a truck filled with PA gear has to be parked overnight, someone will have to sleep inside.

Other insurance no-no's

Apart from the 'leaving-the-truck-unattended' sin, other excuses that insurance companies quote to avoid paying-out on claims include:

- *Fraudulent conversion* – you lend the gear to someone who disappears with it.
- *Under-valuation* – if the value of what you've insured is clearly too low, then the claim on any individual item(s) will be reduced in proportion. Second-hand gear will need careful appraisal and agreement with the insurer, especially if it may go up in value (if it's 'vintage' or collectable, say).
- *Inherent vice* – refers to gradual deterioration. If something about to disintegrate fails, then you can't claim.
- *Electrical or mechanical derangement* – refers to damaged gear that's been wrongly wired or misused. Using a bass bin underwater is not covered, nor is connecting a mixer's power cable to a too-high voltage supply.

If in doubt about exceptions in an insurance policy, you're always free to ask, and get the exact limits set out in plain language. Despite some evidence to the contrary, insurance companies are not secret societies.

Someone else's insurance?

Sometimes a claim can be made on another organisation's or individual's insurance. In this case it shouldn't matter if you are insured or not.

For example, say a part of your PA is sent away for repair, and is stolen from the repairer. Their insurance should cover you. Similarly, equipment might be damaged in transit to you – and hopefully those paying for the transit are insured.

In these cases, contact the company concerned, get their insurer's details, then talk with the insurer to make a claim. If the other party turns out not to be insured at all, of course, then you're stuffed.

PA gear requires specialist insurance, offered by only a few companies

Fire

Fire is less common than theft, but more likely to be devastating. The risk of fire can be reduced if you:

- Turn off all un-needed electrical power when the PA store is vacated. Special wiring keeps a few critical circuits going – eg alarm panel, exit lamps, small house lights.
- Use heating that is safe – such as electrical storage, electrical underfloor, or electrical oil-filled radiators. Also

eco-methods that store heat.
- Never leave oily or greasy rags about in a heap. Spread out/hang on a rail.
- Have all electrical wiring inspected annually. Check all terminal screws are tight, through to the supply intake.
- If there are signs that rodents have entered your building, all wiring should be inspected for damage. Many fires in warehouses are electrical and caused by hidden rodent damage to wiring. Stripped of insulation, the wiring arcs, eventually causing a fire which typically spreads to nearby combustibles.

Smoking is sometimes banned at gigs and tours – though realistically this has to be balanced against the risks induced by having a stressed, nicotine-deprived crew.

It doesn't happen very often, but if modern PA equipment catches fire, toxic fumes may be produced. Older materials just burned merrily – wood smoke isn't so harmful. But the smoke from MDF (medium-density fibreboard) used for some cabs is highly toxic (due to the plasticised glue used to bond the wood fibre).

It's often unsafe to attempt to use a fire extinguisher on such a fire. If you're not sure, get out of the way and call the fire service.

sound
& hearing

This section explores how sound is constructed and described – the different frequencies and dynamics involved, and the terminology used by sound engineers and other PA professionals to characterise specific types of sound; plus comparative sound pressure levels and the impact (good and bad) these can have on our ears and long-term hearing.

Of all our senses, hearing seems to be able to operate over the widest range – sometimes to an almost inexplicable extent. So it's important for all sound system users to be aware that music's subtler qualities and communication value may be restricted or even prevented when a PA system damages or 'twists' the musical signal.

Compared to the average figures reeled-off in acoustics and electronics text books, some people's perceptive abilities seem to extend up to ten times further than expected. For example, some individuals are sensitive to 4Hz... an ultra-low frequency often described as 'subsonic' (inaudible except through bone conduction), and which is one-fifth of the frequency where normal hearing ceases in most people.

It has also been noticed that some musicians can detect the difference between tones only 0.1 per cent apart up to at least 10kHz – this implies the human ear-and-brain combination is sometimes capable of resolving timing differences of around a tenth of a millionth of a second.

Frequency ranges

In terms of the sound spectrum, we usually refer to three basic frequency bands: low-frequency (LF), mid (MF) and high (HF).

Bass energy in music (or bottom-end) is the content with frequencies below 300Hz (Hertz, or cycles per second), and 20Hz is considered the lowest limit of normal hearing. To a greater extent than with mid-range sounds, the useful dynamic range of bass frequencies is compressed naturally by the ear.

Below 150Hz, bass becomes increasingly visceral. At the high sound pressure levels (SPLs) at which heavy metal or hard rock is performed (110dB+), kick-drum frequencies centered around 120Hz can be felt in the solar

It's possible for sub-bass signals to be felt before they're heard

plexus. These are literally 'hard' and many listeners may find them offensive or at least unpleasant.

Below 100Hz, bass softens. Reggae, funk and house music

make full use of the 'pleasure' region centered on 80Hz. PA system sound levels in this range have often been observed peaking at up to 135-145dB (which is 10 to 15dB beyond the

Music can have low HF power but high HF energy

supposed pain threshold, without any harm or even physical discomfort – see sidepanel). Indeed many listeners describe the experience as cathartic, or giving rise to raised consciousness.

These lowermost audible frequencies are also the ones most strongly experienced when hearing underwater (an environment not unlike that in the womb, as it happens).

Below 40Hz bass becomes more tactile again, and by 16Hz it is no longer audible through the ears but sensed solely through bone conduction. This kind of frequency range can alter the way we perceive music, since sound is transmitted much faster through solids (earth, floor, feet, skeleton) than through the air – so it's possible for sub-bass signals to be felt before they're heard (you might experience this in thunderstorms sometimes).

In most music there is little explicit, large signal sub-sonic musical content: it's rarely produced by acoustic instruments – with the exception of very large pipe organs, giant gongs, and long horns, and in

the sub-harmonics of the lowest notes on a bass guitar. It's much more easily produced by an increasing variety of electronic equipment, and is commonly used for dance music 'sub-bass'.

At frequencies below the fundamentals of musical instruments, say below 15Hz, there's still plenty of sub-bass present in the real world – even if it is just the rumble of stageboards, underground trains, or air conditioning. These sounds may be best preserved for a recording, and are best avoided live, to save using up valuable PA power on sonic 'garbage'.

Unfortunately the abrupt filtering needed to remove undesirable subsonic content slightly delays the wanted bass signals.

Mid-range is anything between 300Hz and 4-5kHz, and mid itself is generally split into low-mid, high-mid, and whatever's in-between.

High frequencies (HF, treble, or top) begin at around 5kHz (5000Hz). Higher still, peoples' hearing rapidly becomes insensitive: depending on your genetic make-up, diet, health, age, and cumulative exposure to non-musical percussive sounds (particularly hammering and gunshots), the upper hearing limit at quite high SPLs (100dB) typically varies between 12 and 20kHz.

Above 3 to 5kHz the average power levels of nearly all kinds of music, integrated over a minute or longer, are less than the levels at lower frequencies, and reduce further with ascending frequency. As a rule of thumb, the power in the mid-to-low bass regions is at least ten times (20dB) more than the power at 10kHz.

A great body of music revolves around relatively abrupt change. Sounds stop and start all the time. Rhythmic sounds may have durations under one second, and the HF components of these can have momentary (but repeated) levels as high as the largest bass signals.

In this way, music can have low high-frequency power, but high HF energy.

Music's HF dynamics are most apparent in live work. Capturing HF dynamics up to 20kHz is aided by capacitor microphones (which can have a far smoother, higher-extending, and less compressed response at HF than most dynamic/ moving-coil kinds). And also by active microphone splitters which buffer (strengthen) signals from mikes, preventing subsequent losses in, and sound-smearing resonances caused by, the stage-to-FOH mix-position cabling. (See Chapter 1 for more on mikes and splitters.)

While our conscious hearing stops at 20kHz, higher 'ultrasonic' frequencies in music, up to at least 80kHz, can be

Our conscious hearing stops at 20kHz

perceived by the human brain (some animals, of course, seem to hear such frequencies much more readily).

In pioneering listening tests carried out by recording mix-console designer Rupert Neve and producer Philip Newell, among others, when frequencies

sound & hearing

above 20kHz are filtered out of a music signal (or 'programme'), sensitive listeners notice a lack of vitality.

More recently it's been demonstrated 'objectively' – in the sense that specific neural activity and chemical production has been measured – that the subliminal perception of the ultrasonic sounds associated with music enhances pleasure.

Some digital equipment limits PA high-end frequency response in a uniformly hard and unnatural way. By contrast the response of some microphones and pickups continues (to a greater or lesser degree) well above 20kHz.

The sound of music
One of the key qualities of music (as it has developed in Western culture at least) is tone. Some tones, called fundamentals, are harmonically (for which read mathematically) connected to others with different but related frequencies.

Together, fundamentals and harmonics, and their phase relations, along with the envelope (the 'shape' of the sound developed by the averaged amplitude) create a timbre – the musical (or otherwise) quality of the note or tone we hear.

Tones which are not harmonically related may be discordant. If they are harmonically related, but adversely so (usually odd harmonics above the 5th), they are 'dissonant'. When a tone changes in intensity, its 'frequency' (as perceived by the ear) changes – this is known in musical terms as 'pitch'.

Sounds that have no dominant, identifiable tones are 'atonal'. They are akin to noise bursts. Continuous sounds, both tonal and atonal, get boring after a while.

The tonal waveform changes over a short period (usually measured in milliseconds) but each subsequent cycle is identical to the first. Continuous atonal sound is considerably more interesting – as exciting as a waterfall can be, say.

Music's most vital component is its dynamic – it creates richness and adds to emotional impact. Tonal and atonal sounds

---

Continuous sounds, tonal or atonal, get boring after a while

---

fade and increase, stop and start, and change in pitch and frequency in diverse ways, creating wave-patterns (as can be seen on an oscilloscope – see Chapter 9) which rarely repeat exactly, and appear visually chaotic.

The overall amplitude (size or loudness) pattern is called the 'envelope'. Timing creates the music's meaning, with a precise schedule of the beginning (attack) and build up of each tonal and atonal building block, and its sustain (levelling off), decay and release.

In most locations, sound is reflected off nearby surfaces, causing multiple early reflections

or reverberation (reverb). This added complexity makes the sound appear richer. Complex distortions, both gross and subtle, caused by microphones, speakers, electronics and cables can cause deviations in what the musically adept and experienced ear expects.

Tonal qualities can be unduly emphasised or retracted, timing thrown out of sync, subtle dynamic contrasts and 'edges' blurred, and spacial qualities bizarrely warped or flattened.

Sound terminology

As an adjunct to the full technical Glossary on p219, this section lists some words commonly used in the description of sound:

Asymmetry – all sounds comprise alternate compressions and rarefactions of the air (or other physical medium) – in other words pressure cycles back and forth.

For a completely pure tone, the net change in air pressure is nil (assuming an average of whole numbers of cycles). For most musical sounds, though, the waveshape that describes the pressure moment-by-moment is skewed or lopsided at various times, leaning more in one direction than the other. This is called asymmetry.

If the music is in a well-sealed room it would have the effect of very slightly varying the atmospheric pressure over periods of several seconds.

When a music programme (the technical term for the original signal) is converted into an electrical waveform, asymmetry causes DC voltage shifts, which can upset the operation of some PA gear.

Transients – most kinds of music in its raw, live state has unpredictable bursts which can be 6, 10 or 20dB higher than the average, and last varying lengths of time. Fast 'edges', quicker than an SPL meter takes to respond, can be 20 to 40dB above the average short-term levels. The ear barely registers these as loudness, yet notices when they are missing, and certainly notices something nasty is happening when these edges cause the signal chain to hiccup for a longer period.

In most PA systems, transients are rescaled with compressor-limiters, which squash rather than remove the effect. The difference between the average level (or root-mean-square, commonly known as RMS) and the highest peak levels of a programme is called peak-to-mean ratio (PMR), where 'mean' is math-speak for RMS. (The term 'crest factor' has a similar definition but is best kept for steady waves, like AC power.) PMR is initially determined by the genre of music.

Some approximate examples of this would be:
- Orchestral, bebop jazz: 20-30dB
- Live rock, most genres: 10-20dB
- House music, digital: 10dB
- Muzak, lifeless 'mall' music: 4-8dB

Variability – music is not predictable. There is no algorithm (as yet), and computers are still incapable of differentiating between signals which are music and those which are noise.

Lots of adjectives have been employed, even invented, over many centuries to try to capture the essence of particular aural experiences – with varying degrees of success.
Here's a selection of some of these words and what they mean (roughly) in audio terms – and how they might be usefully employed when mixing (see Chapter 8).

First of all, words that seek to define certain frequencies and tones (from high to low):

Sheen – very high treble, above 16kHz: often absent, or heard as hiss/digital noise

Tizzy – excess around 12 to 16kHz: usually over-emphasising cymbals' high harmonics

Airy/smooth/open – apparently effortless high treble: seeming to extend further than the music

Closed-in – treble lacking above 10kHz: almost the opposite of airy

Enclosed – dull, coloured, airless

Dull – general lack of treble

Hard/metallic/brittle – excess of high, metallic-sounding harmonics, usually around 8 to 16kHz

Bright/brilliant/glassy – excess around 4 to 8kHz

Sibilance – excessive amounts of 5 to 7kHz

Aggressive – preponderance of mid-high energy (3 to 6kHz), often phasey and distorted

Crisp – plenty of clean 3-4kHz

Presence – the frequencies around 2kHz

Muffled – where high frequencies are reducing rapidly above 2kHz

Articulation – audibility of the inner detail of complex sounds, particularly those in the main vocal range (300Hz-3kHz)

Nasal – 1kHz emphasis

Recessed – general lack of mid-range

Barky/woody – characteristic mid-bass resonances

Honky – excess around 600-800Hz: like a Cockney saying 'oi', or improperly used mid/HF horn speakers

Lean – slight, gentle reduction below 500Hz: or very clean, transparent bass

Chesty – excess in the 200 to 400Hz area: especially with male vocals

Boxy – excess around 250-450Hz: as if inside a cardboard box

Rich – a downtilt in level above 300Hz: also, a slight excess of reverb

---

How music confounds science

You could say music operates in more dimensions than science can accurately measure. More than being simply soundwaves moving in physical space and time, there's the harder-to-quantify effects that pitch and rhythm changes can have on an individual listener – and the even more complex and unpredictable effects it can have on several individuals in different parts of the same venue, with different physiological and psychological backgrounds and make-ups.

Tables, charts and waves on an oscilloscope are helpful tools for certain audio engineering tasks, but when it comes to the whole picture, the sound of music can only really be gauged by an experienced and sensitive pair of ears – and almost by definition there will be no optimum means of music reproduction that's perfectly suited to everyone, despite all the technological 'advances' made to date.

Dark – sound that tilts down from the bass upwards

Punchy – around 120 to 160Hz: a high-definition area

Thin – overall lack of bass

Balls/ballsy/gutsy – low bass that is visceral, ie can be felt

Boof-boof – around 80 to 90Hz: soft bass area

Chunky – 80 to 90Hz: 'sample' bass with added harmonic definition

Gutless – absence of low bass

This next batch contains some more general terms used to describe the sonic or dynamic properties of music, particularly live music:

Ambience – to do with mood, feeling, atmosphere – subtle, low-level, often non-musical background sounds, mainly subliminally appreciated

Analytical – when sound equipment seems to reveal excess of the 'stitchwork' in music. Sometimes describes when a system has distortions that unduly emphasise detail or 'edges'

Attack – the speed and crispness with which a note/tone/signal begins

Boomy – poor bass damping: a bad loudspeaker-amplifier combination

Clear – plenty of contrast

Clinical – suggests sound that is clean, bright, sharp, detailed; but may be used as mildly pejorative, perhaps implying emotional qualities are lacking or held-back

Congested – see Smeared and Thickened, which are facets of the same effect

Contrast – clarity, differences in tone: as in visual art/photography

Detail – more spacial version of Dynamic

Dry – sound tending to lack reverberation

Dynamic contrast(s) – subtle changes in level or pitch embedded among much larger changes

Dynamic range – in PA-speak, the 'amplitude performance envelope' of a sound system; in musicians' parlance, the programme's intensity range

Etched – finely detailed

Focus – sharpness of detail: may vary across the soundfield, in all three dimensions

Euphonic – pleasantly distorted at the expense of accuracy

Fast – incisiveness of attack, particularly of bass fundamentals; but as bass doesn't move fast, by definition, most likely a reflection of rapid damping, proper synchronisation between the fundamental and harmonics, and correct reproduction of all associated harmonics

Fuzzy – spikey yet soft texture caused by high distortion and compression

Glare – distorted mid-treble; also tonal imbalance or forwardness

Grainy – excess texture: a kind of distortion, usually in the high mid-range.

Gritty – like grainy, but harsher and coarser

Grunge – like gritty, but more muffled ('grunge' music is closer to being 'gritty' music)

Hardness – fatiguing 'wood block'-type of mid-range emphasis

Harsh – dissonant and/or discordant: unpleasant

Hologramatic – when coherent and correctly focused sound (or light) enables higher dimensions to unfold

Image/imaging – ability to portray width, depth, and sometimes height

Incisive – conveying the 'slicing' sound of close-miked snares, like a sonic machete or knife: indicative of good attack synchronisation; like Fast & Slam

Layering – sounds having a precise depth in a soundfield, with the implication of many depths or infinite gradation

Lifeless – superficially perfect, anodyne reproduction conveying no emotion or interest: commonly caused by forcing equipment or system to manifest a perfect measured frequency response without regard for factors affecting space or dynamics

Loose – badly damped bass

Lush – see Rich

Micro-dynamics – lifelike energy (transients) in small, low-level sounds

Muddy – see Smeared; especially applied to bass

Muffled – no contrast: opposite of clear

One-note bass – poor damping of major resonance(s) in low bass

Pace – ability to make music seem to unravel at the pace (or beats-per-minute, BPM) it was recorded at, rather than slower

Phasey – symptomatic of a frequency response that undulates like a comb. Co-exists with a manic, zig-zag phase response, literally 'phasing' our hearing system

Pinpoint – when the image is very stable and finely etched, like some metal sculpture

Punchy – similar to Slam, but can have a pejorative element of 'One-note bass'

Rich(ness) – lots of coherent reverb: more usually applied to programme (music) rather than equipment

Rounded – loss of attack transients, due to poor damping, poor HF response, or slew limiting

Rhythm – infectious vibe inherent in a live show, which makes people want to move and dance

Slam – convincing, correctly-synchronised attack for a fundamental in the 125Hz area

Slow – rhythm seems less fast than it should be

Smeared – when an otherwise sharp image seems to be portrayed through butter-smeared glass: caused by excess incoherent reverb, too much harmonic and intermodulation distortion and/or timing errors

Solid – well-damped bass

Sound-stage – the space between and around a truly stereo-driven, dual-arrayed PA in which a stereo (Greek for 'solid') image appears

Squashed – seeming absence of most dynamic contrasts: may be caused by hard limiting

Thickened – reduced dynamic contrasts: can be caused by compression or soft limiting, or more subtly by any path component, from microphones to resistors

Timing – time is a another kind of 'spacial' dimension: the timing between sound components at different frequencies coming from one or more instruments may be unlike

the original sound – compare bass to mid, bass to treble, etc; delays of milliseconds or less can be audible

Transient – abrupt, short-lived events in music: skilled ears can resolve differences in attack slopes and harmonic synchronisation down to tens of microseconds

Transparent – when you feel you're hearing just the music, not the PA equipment: a sense of there being 'nothing in the way'

Woolly – see Loose

Music & your ears

The bad news is that loud sound can damage the ears. The good news is that music may have a lower risk of doing this compared to non-organised sounds (otherwise known as noise), even at the same sound pressure levels (SPLs). In theory, at least.

The theory is that the ear can take 15 to 20dB higher levels from music than from 'noise', particularly if it's non-stressful – ie enjoyed by, or at least not upsetting to, the listener.

But beware: these figures are based on an ideal sound system set up in a perfect acoustic environment. Unfortunately, in the real world – where PAs and venues are rarely perfect – damage can and does occur in

some individuals exposed to loud music, especially over time.

Since professionals in the music industry – whether they're musicians or sound engineers – rely on reliable hearing for their livelihoods, they may be naturally disinclined to own up in public to any ear problems: but an increasing number have privately admitted their concerns.

Tests over the past 30 years on a link between hearing damage and loud music have provided mixed results – often contradictory, and overall surprisingly inconclusive.

It appears likely that certain people simply have a lower hearing-damage threshold than others – so while some can go gigging and clubbing for years

Some people may have a low hearing-damage threshold

with minimal effect, others can show signs of tinnitus (continual ringing in the ears) and other ear problems after only months of exposure to loud music. And sometimes this is just the start of an irreversible deterioration in

---

Comparative sound levels (dB SPLs)

0dB – quietest audible sound
20dB – quiet recording studio
60dB – normal speech (at arm's length)
70dB – quiet music (at arm's length)
80dB – average acoustic instrument (at arm's length)

90dB – large vehicle passing close
100dB – loud vocals (at six inches)
120dB – aircraft engine (six metres)
125dB – toms at close-miking distance
130dB – pain threshold
135dB – bass drum, close-miking distance
140dB – explosion
Note that every 6dB rise in SPLs

represents a doubling of sound pressure; adding 10dB is roughly a three-fold increase; a 20dB jump means ten times higher. Which means adding 60dB is an increase in SPLs of 1,000 times (10^3)... So the sound pressure when you get to 140dB is one million times greater than it was at 20dB in that quiet studio.

sound & hearing

hearing (which worsens with age in most people anyway).

Unfortunately there's presently no way of knowing in advance who is going to be vulnerable in this way. So caution may be the best option if you want to be sure of avoiding harm.

Measuring sound levels

Sound pressure levels (SPL) are measured in dB SPL. These dBs (which are 'code' for a range of pressure levels) are absolute 'units'. dB SPLs can be weighted – as is done in equipment noise measurements: 'A'-weighted is centred on 1kHz, with most of the bass and treble filtered away.

This sort of reading is designed to approximately simulate the sensitivity of hearing to mid-range sounds. It will naturally work in favour of loud bassy and toppy music, if keeping down the dB SPL is the reason for monitoring the sound levels.

Bad quality sound and damaged gear contribute to ear trauma

'C'-weighted readings don't employ filtering, and these come far closer to giving an objectively correct 'flat response' reading. But if used for SPL monitoring, C-weighted readings work against sound levels being high at any frequency. The C-weighted reading is however usually the one to use for reading signals in the room, eg

for room EQ, using pink noise (see Chapter 9).

The maximum feasible working range (for average human ears) is from about -10 to +140dB SPL. That's a huge, huge pressure variation, of 30 million to one – this is why dBs are handy, as they can easily cover such vast number ranges, keeping to friendly, easy-to-work-with two and three-digit numbers.

Causes of ear trauma

Apart from a biological predisposition to hearing damage (a lower-than-usual threshold), most ear trauma can be put down to the following:

- Bad quality sound – nasty harmonics (even if they measure small SPLs with single tones). For example, any equipment with a 'spiky' rather than 'smooth' distortion residue (test gear is needed to see this). Some poor-quality digital gear also wrecks waveform reconstruction in deeply un-musical and potentially damaging ways.
- Damaged equipment (see also Chapter 9).
- Abused equipment – eg clipping, too heavy a load.
- Peaks in the PA system at aurally sensitive frequencies, between 1 and 5kHz. Caused by drive-units, cabinet interactions and complex acoustics. May be called 'spiky colouration'. It seems some people's ear passages create big response peaks from mid frequencies, to over 22kHz, making them far more sensitive.
- 'Ringing' in metal-diaphragmed high-frequency

(HF) horn-drivers. Metal cones and diaphragms generate more unpleasant sounds than other types (paper cones, phenolic or plastic-cloth diaphragms), and some people are much more affected than others (as mentioned above).

- Generation of highly dissonant sub-harmonics (1kHz tone creates a sub-harmonic 'tone' at one fifth of this frequency = 200Hz, say) – made by (mainly) compression drivers. These are not natural harmonic sounds and can be highly unpleasant, even in 'subliminal' doses.
- Sudden (attacking) solo sounds – individual sounds that rise to high levels in thousandths of a second, or less. For example, if a drummer suddenly strikes the snare drum hard, just once, while you're standing near, with no other music going on, it can be unpleasant – though when that identical sound is played within the context of a piece of music, it's nice enough. (The noise of a gun going off at close range is far more sonically stressful than even the loudest heavy-metal drum kits – as it's got even faster attack and higher peak sound levels.)

Getting too close to a loud drum kit is generally more stressful than approaching other loud instruments. First of all, percussive noises – where something is being hammered to make the sound – are the most stressful type to the ear. They combine sharp attack, sprays of harmonics and spiky peaks, and can also seem less powerful than they really are (and also produce a misleadingly low dB SPL reading on average-responding meters).

- In terms of prevalence and potential for doing harm, worldwide public enemy number one is bad venue acoustics.

Briefly, most buildings are 'phased' (sonically confused) by high-energy music. Most are bad enough with less 'spiky' classical or folk playing.

Bad acoustic effects involve massive 'comb-filter' colouration – this means having the sound level rise 25dB for a few Hz, then dip 25dB for a few more Hz, and so on, over most of the musical range. The result is aurally tiring at the very least.

Even the worst graphic-EQ-abusing inexperienced engineer cannot wreck sound as efficiently as an acoustically-unfriendly room that's driven too hard.

It's not so much the size of the room as the shape – nice-looking but overdone features like deep balconies are bad news, but so are boring boxy spaces, not to mention certain surface materials (an excess of harder materials like metal, glass and glossy plaster hardly help absorb reverberant high mid and treble), the walls' thickness (the thicker the better – 12-to-20ft-thick castle walls are suggested for a serious venue), and solidity (older-style constructions are usually best), and so on.

Any areas of undamped metal can cause truly dreadful sounds – like a ripped metal drive-unit.

It's a sobering thought that some of the best buildings for high energy music appear distinctly 'low tech' and often date back to pre-history – for instance tents (marquees, big tops etc), earth and stone buildings (quite open ones are good).

Percussive noises are the most stressful to the ear

The conclusions are obvious: all public venue architects would do well to be humbled by the terrible damage done to the enjoyment of music (especially in the 20th century) by blind belief in the 'progress' of modern building materials.

In so-called 'pre-civilised' times, most people would have experienced loud music (such as massed drums and horns) at tribal ceremonies and the like, and probably enjoyed better sound quality with it. It's only in recent centuries that architects have forgotten about

meeting such acoustic needs. In general, the less 'synthetic' the environment for music, the better.

Even non-structural vibrating objects at a venue can be 'set off' by music. Any areas of metal (sheet, grid and mesh) – for instance security barriers – are regular causes of dissonant sounds (it's no accident that good-sounding metal cymbals are hand-made by craftsmen using centuries-old skills). Metal doors may be deadened by opening them, and replacing them with a security guard if needed (assuming 'sound-escape' is not a problem).

Dance parties in modern warehouses, made almost entirely of undamped corrugated metal, can be sonically unpleasant for the same reason. If you have to use these venues, hang canvas sheets (flame-retardant-treated) along the walls and ceiling if possible – this will help absorb at least some of the more metallic-sounding harmonics produced by the panel vibrations.

Signs of ear stress

In approximate order of appearance, these can range from a queasy 'sea-sick' feeling, or an inability to focus on the music (the inner-ear balance mechanism may be affected); to buzzing, crackling,

---

Earplugs

There are many kinds of earplugs and ear defenders – some almost useless for PA work. Others are worse than no use – they can actually exacerbate damage. The key point is that the correct, most comfortable and useful type may not be the first and most obvious you'll find. Strictly speaking, ear defenders and plugs should be obtained from a hearing (audiology) clinic that specialises in musical clients, and the best earplugs are custom-made for the wearer so they fit the individual's ear cavity exactly.

For help on choosing earplugs, and more information on hearing protection in general, try the RNID (the UK's Royal National Institute for the Deaf) (*www.rnid.org.uk*), or a specifically music-oriented hearing advice group like HEAR (*www.hearnet.com*).

sound & hearing

rasping sounds, or a tickling sensation in the ears; to temporary deafness (threshold shift – where the hearing level drops for a while before gradually coming back up); and finally, and most alarmingly, pain.

If your ears are constantly bombarded, tiny hair cells in the inner ear begin to die

Pain in the ears is a loud, flashing warning sign. It says, "leave at once or else protect your ears". If your PA system creates pain (or other signs of discomfort), something is wrong. It's not necessarily simply 'too loud', but the sound certainly needs adjusting in some way (see Chapter 8 for more on this). If that's not in your control, distance yourself from the sound as soon as possible.

If your ears are stressed in this way, you could use earplugs (see sidepanel) or other measures to reduce the sound levels they're subjected to, at least for a while, to give them a rest.

If PA engineering is a part-time occupation, try to avoid day jobs that involve regular noise exposure, or exposure you can't control. It doesn't have to be loud – just persistent. (Even quite quiet computer cooling fans can trigger tinnitus – sometimes music played over the top of it actually stops the effect.)

Be 'boring' and wear ear defenders when mowing the grass or using power tools – it all helps reduce your overall exposure and possible long-term damage.

Tiredness, jet-lag, having a cold (flu, etc), or otherwise feeling under-the-weather can also make the ears more sensitive to trauma, so be more prepared to protect them. If your job involves making important sound quality assessments, it's best not to even attempt this until you feel better.

Too loud and too long

Excessive aural stress causes a release of body chemicals which are meant to protect your hearing by shutting down the tiny hair cells in the inner ear. This is what causes 'temporary threshold shift' – basically short-lived deafness. After the loud event, the chemical is withdrawn, though the hair cells still need some time to de-tox and recover.

If your ears are constantly bombarded, the cells begin to die. If enough of them die, your hearing sensitivity is permanently impaired by a certain degree.

As ears are normally most sensitive between 1 and 6kHz, damage tends to occur first at these frequencies. The first sign of this may be that you find it hard to hear people speak (voices are at the lower end of this range).

In the UK, visiting a health-service hospital audiologist costs nothing. Elsewhere, if PA is a regular occupation, it would be worth spending some money for an annual hearing check (it's your most important tool, after all).

Tinnitus – ringing in the ears

– is a common after-effect of loud sound exposure (especially aurally distressing PAs), and can be very upsetting for anyone, whether musically inclined or not. The 'ringing' (the actual sound varies from person to person – it can be metallic clanking, rustling, or something similar to feedback), can be at any frequency (bass, mid, top) and can last for any length of time.

There is no 'cure' – though the sufferer can employ techniques to help ignore the sound as much as possible. It may also come and go at random – even long-term sufferers have known it turn off one day, as if a switch has been found by a dozy sound-engineer in your ear.

Sometimes tinnitus has no sonic connection at all – it may be genetic, or brought on by stress, or by some other medical condition (even some drugs can exacerbate or trigger it). But everyone who uses their ears for a living (or even for recreation) should be aware of the facts and the risks.

Pain in the ears is a loud, flashing warning sign

glossary

An exhaustively comprehensive A-Z of PA and sound-related terminology, with around 600 technical words and phrases applicable to the PA industry – from textbook electronic definitions to working abbreviations and slang expressions commonly used by sound engineers, musicians and DJs.

A

AC: alternating current – used in the supply of electric power to most homes and venues (which in the UK is called 'mains' electricity), as opposed to DC (direct current) normally obtained from batteries. Could also be called 'alternating voltage'. Varies from zero to full in one direction, then back in the other direction (the opposite polarity) 50 times per second (50Hz) in the UK and Europe, and 60 times/sec (60Hz) in US. Can cause interference to sensitive audio signals (this can sometimes be heard as low hum interference on sensitive audio signals). Electricity suppliers use AC because it's readily converted, using a transformer, to very high voltages for efficient distribution over long distances, then back to lower voltages for our use. Again, voltages vary from one territory to another (see 'voltage').

acoustic: (i) the natural (non-electrically aided) sound produced by a musical instrument, or enhanced by the shape or construction of a room; (ii) guitarists' abbreviation for 'acoustic guitar'.

acoustic lens: unlikely-looking set of slanted slats (or 'louvre plates') at the front of a PA horn, which disperse high frequencies over a wide angle for good coverage. Rarely used in frontline PA today.

active: powered – not necessarily amplified, but using electronics to assist or improve functioning.

active crossover: see crossover.

ADAT: a type of multi-track digital audio tape.

ADSR: stands for attack, decay, sustain, release – the elements that make up a standard volume 'envelope'.

ADT: artificial (or automatic) double tracking – used to reinforce an existing signal, for instance making one singer sound like two.

AES: Audio Engineering Society (based in New York) – made a limited contribution to music PA, but has developed several international standards.

aerial: a metal object designed/shaped to emit or receive radio waves – called 'antenna' in the US.

glossary

AFL: after-fade(r) listen – hearing the signal after its level has been adjusted by the channel fader. Also called 'post-fader listen'.

A-gauge: referring to a quarter-inch stereo (TRS/3-pole) plug (or 'jackplug' in the UK) with normal-sized tip, as used on many headphones.

alternating current: see AC.

alternator: alternating current generator.

ambience: the general sound quality or atmosphere created by a particular room or venue, or by signal processors designed to recreate acoustic qualities (eg reverb).

ambisonics: robust multi-level, multi-channel sound system/methodology, invented by Gerzon in the UK in the 1970s.

ammeter: a meter that measures or indicates current, in amperes (amps).

amp: (i) short for ampere – a unit of current, the symbol for which is I; (ii) short for amplifier – could be a power amp or backline instrument amplifier or line amp: a device for increasing the level of a signal).

amp rack: sturdy frame or rack for mounting power amps – usually deeper than normal instrument racks, and often very heavy.

amplification: making a signal bigger (may refer to voltage,

analogous to signal level and loudness, or current).

amplitude: academic term for size or magnitude – particularly of a signal voltage or level.

analog (UK analogue): 'traditional' electronics, where soundwaves are turned into analogous electrical waves, not chopped up and processed mathematically, as with digital audio.

anodised (US anodized): having a hard, protective layer of aluminium oxide over an aluminium surface – anodised surfaces are often black but can also be brightly coloured, and make a good heat conductor while acting as an electrical insulator.

antenna: a metal object designed to emit or receive radio waves – called an 'aerial' in the UK.

anti-clip: circuitry that prevents amplifier overload (see 'clip').

anti-phase: (i) alias for the 'cold' side (negative, or return wire) of a pair of balanced signal wires; (ii) a signal source where the phase (or more correctly the polarity) has been reversed.

anti-surge: a 'delayed' fuse (body marked 'T') that withstands brief current surges without breaking. Note that it doesn't *prevent* current surges.

arc/arcing: (i) continuous electrical spark, long used for powerful spot-lighting; (ii) same sort of spark, but can be far

smaller and occurring inside connectors, cables, components of equipment due to mechanical or electrical failings. Can easily cause a fire, or serious burns, and also liberates toxic gases, including ozone. Turn off the power if you ever see or suspect an arc (if you hear crackling, sparking or smell burning).

array: a group of speaker cabinets arranged to work in mutual support – often piled vertically or splayed out in one or two planes.

artic: short for 'articulated lorry', antiquated British term for a large juggernaut truck.

artist(e): handy all-encompassing name for performers, singers /vocalists, musicians, groups/bands, announcers/comperes, MCs/callers, DJs, rappers/poets, drama queens, etc.

asymmetric: a quality of music signals with strong timbre, particularly vocals, where one or other of the alternating polarity halves ('upper & lower') are different in level.

attack: the speed at which a sound (or filter, or 'envelope') reaches its maximum level – for instance percussive sounds usually have fast attacks, while smooth, liquid sounds have slow attacks (see also ADSR).

attenuation: reducing a signal's level.

attenuator: electronic circuitry that reduces level, usually in fixed steps of useful round-figure amounts, like -10dB, -20dB. Also the knob or switch that controls the setting.

aux: short for auxiliary – a mixer section and control which redirects part of a channel signal elsewhere (to add effects, or for monitoring etc).

aux rack: a mount and enclosure for effects (FX) and related auxiliary gear.

aux return: inputs on the mixer used for adding back the signal from/with the FX.

aux send: an output to FX and other places.

A-weighting: (i) a standard sort of high and low-pass filter that's regularly mis-used for equipment specification testing, as it can make performance figures look better; (ii) an identical filter, used for reading a limited range of acoustic sound levels.

AWG: American Wire Gauge – designates the diameters of the conductors in all kinds of cables. Similar to British SWG, using a set of mainly even numbers between 0 and 40 instead of mm².

B

back contamination: when two signals mixed together affect each others' sources, so a pure signal is no longer available from one – it was a problem in primitive passive mixing schemes used before 'virtual

earth' mixing electronics was introduced into PA consoles in the late 1960s. Still occurs when mikes have to be passively mixed.

back EMF: a counteracting ('reactive') signal produced by loudspeaker drivers (see EMF).

back-end: the amps and speakers in a PA set-up (see also 'front-end')

backline: (i) musical instrument amps (usually for guitars), placed in a line across the back of the stage, or stacked up in a wall or crescent for visual effect; (ii) in modern stage sets, can also include racks of sampler channels and their sub-mixers; (iii) anything on the stage.

baffle: the front panel or baseboard of a speaker cabinet, onto which direct-radiating drivers and smaller horn flares are mounted.

balance: (i) tonal 'symmetry' of sound – correct amount of all frequencies present; (ii) 'old school' BBC term for the 'mix' – the blend of the different vocal and instrumental sounds; (iii) left/right side equality, as in hi-fi usage.

balanced: signals or in/out connections where there's a pair of conductors (hot and cold) that are separate from earth/ground/shield, carrying opposing versions of the same signal at exactly the same level, and also having near identical impedances compared to ground. Greatly helps to avoid noise problems in compatible circuitry and equipment.

In a balanced instrument signal cable (lead, cord) a pair of identical insulated wires operate as the hot and cold signal conductors, while a separate ground/earth shield protects the audio signal from noise interference. Longer cable lengths can also be used compared with unbalanced leads.

bandwidth: (i) in analog terms, the frequency range over which some device or system usefully works, measured in Hz and kHz for PA; (ii) in digital terms, the rate (measured in MHz or GHz) at which the processor can 'clock' (push through) streams of data.

bantam: a solid, mini (phone/jack) plug of very high quality, used for dense patchbays on some mixing consoles.

bench repairs: fixes and maintenance done in a workshop, rather than at a gig or on the road ('field' repairs).

B-gauge: a smaller-tipped quarter-inch plug, sometimes called 'military phone plug' or 'GPO plug' – often used for professional patchbays/jackfields.

bi-amp(ed): any system using a two-way active crossover – which usually means that, for each complete speaker unit, two separate amplifiers are used, one to drive the bass speaker and one to drive the mid/high speaker.

bias: a critical 'tune-up' setting of an amp, tape machine, or

glossary

other equipment, generally involving some auxiliary voltage or current that helps the circuitry to work properly.

bipolar: short for bipolar junction transistor (BJT) – the 'ordinary' species of transistor, as opposed to the other major type, the field-effect transistor (see transistor, FET, JFET and MOSFET).

BiFET: a 'hybrid' integrated circuit that uses J-FETs at the input stage but bipolar parts elsewhere (see FET).

bin: (i) a bass speaker horn – so-called because some early types looked like rubbish/trash bins when seen in transit on their backs; (ii) a storage space for some types of computer data, used in audio and acoustic analysers.

BJT: bipolar junction transistor

BNC: rugged, reliable high-quality coaxial cable connector widely used for audio test gear, RF (radio frequencies) and video, as well as unbalanced digital links. Has a bayonet latch, like a light bulb. (There's disagreement over what the letters BNC stand for – some say 'bayonet nut connector', while others say 'bayonet Neil-Concelmann' (after the US designers.)

board: slang term for a mixing console, or desk.

boom (arm): a rotatable extension on a microphone stand, for added reach.

boot: (i) to start up/switch on a

system; (ii) a thick, soft, plastic (somewhat condom-like) object surrounding high-voltage electrical connections, such as power switches and fuses inside gear. Also used over the backs of multicore connectors, and temporarily over plugs, to keep out rain, dirt, etc.

bottomed: when a speaker's cone is overdriven so that moving parts clash with fixed parts – eg the cone bangs the magnet's pole piece. Commonly due to over-powered amps and/or excess low bass boost, but also caused when drive-unit suspensions weaken with age.

box: slang term for a large PA speaker cabinet or bin.

BPM: beats per minute – the tempo of the music.

BPO: British Post Office – original source of many electronic equipment standards, such as the 19-inch rack.

breaker: an electrical circuit breaker, effectively a re-usable, re-settable fuse, protecting against overcurrent, fire or electrocution, and commonly used in PA set-ups instead of (or as well as) fuses.

breakdown: (i) process of dismantling the PA and stage set-up after a gig; (ii) electrical term describing failure of insulation or parts meant to sustain high voltage, which can involve arcing, and may be a fire risk.

break-in: see breakout.

breakout: one or more cables in

a loom/snake/multicore cable brought out to a plug for access and use. (See also fantailed and tail.)

brick: slang word for a 'power strip' or multi-way plug-board.

bridge, bridging: (i) a way of joining two amp channels (of a stereo amp) so the output 'swing' is doubled and power delivery is increased (particularly into higher impedance speakers); (ii) an AC to DC rectifier (a rectifier passes current forwards only) assembly comprising two or more diodes; (iii) connecting additional equipment to/across a given line level feed – as when 'daisy-chaining' is used.

BS: (i) British Standard – UK system of standardisation for electrical equipment, for compatibility and safety, often shown by a 'kite-mark' and a number (similar to US 'UL' mark); (ii) US term for something that comes out of the back end of cattle and some people's mouths.

bullet: a high-frequency horn with a bullet-shaped 'phase plug' (which helps soundwaves flow in phase, improving sound quality and efficiency).

buffer: an amplifier or circuit stage inside equipment, mainly intended to isolate the preceding signal source from loading, and to provide a lower impedance drive that's more powerful and won't droop or distort when loaded heavily Actual signal level (voltage) may be unchanged or slightly increased.

glossary

bus/buss: (i) in audio, a mix-buss is an elongated common point or 'conduit' into which many signals can be 'poured' or grouped without overloading, and without risk of 'back-contamination'; in a computer/digital context, MIDI, SCSI & USB are examples of connections that have 'bus' properties; (ii) in electrical systems, a bus is an elongated main distribution conductor (usually a solid bar) shared by several outgoing conductors.

BV, bv: short for backing vocals.

C

Cannon: alternative name for an 'XLR' plug or socket, after its American inventors (see XLR).

cab: a speaker cabinet or enclosure – may contain drivers for one or more frequency ranges.

cable: (i) another name for a (speaker/instrument/power) lead – the sheathed connecting wires, with or without connectors; (ii) in 'flying' means wire rope.

capacitor microphone: also commonly known (not technically accurately) as a condenser mike. Generally high-quality, if relatively fragile, and requires 'phantom power' to operate. It contains two thin metal plates, one of which (the diaphragm) is moved by soundwaves to create a change in electric charge (capacitance). The modulating voltage is fed to a head-amp and transformer before being output as a high-level signal with low impedance.

capsule: the guts of a microphone – its delicate inner workings, as opposed to robust exterior casing designed for handling.

cans: slang for headphones.

cardioid: description of the 'heart-shaped' response pattern of a unidirectional microphone (see 'polar pattern').

cascade: a linking of two or more devices or circuits where the output of one feeds directly to the input of the next.

CCW: counter-clockwise (UK anti-clockwise) – usually means 'turning down' a control knob.

CD: (i) compact disc – digital media often used to provide pre-gig background music, or backing tracks to aid performance (also useful for sound-checking the PA); (ii) constant directivity – used of speaker horns (see Chapter 4).

centre tap (US center tap): an output provided at the mid-point of a power supply or transformer, usually designated zero volts (0V).

CE: a mark applied to equipment sold in EU (European Union) countries, theoretically indicating that the equipment meets (largely unspecified) Euro-regulations.

ch: abbreviation for channel – could be of a splitter, mixer, crossover, power amp, etc.

channel: an input on a mixing desk, or amp, or other gear, and/or the controls relating to it.

channel separation: the extent to which signals in separate channels are kept apart. Opposite of 'crosstalk' – so if crosstalk is low, it means separation is high, and vice-versa.

changeover: stripping down a finished set-up and rigging and soundchecking the next act at a multi-act show or festival.

charge: the ability of elementary particles to attract or repel other particles, depending on whether their charge is negative or positive (electrons and protons are essentially microscopic magnets). The symbol for charge is Q, though it's measured in units called coulombs, the symbol for which is C. The charge (in a copper wire, for example) can be forced to move in a certain way by attaching the wire to a battery or some other kind of electrical 'pump'. This process is known as 'electro-magnetic induction'. The resulting flow of charge is called current.

chip: tiny slice of semiconductor material (usually silicon crystal) on which an 'integrated circuit' (IC) is built (see IC).

choke: a special type of inductor (coil) that's designed to work with audio or radio

glossary

frequencies while also handling high DC currents – which would more normally degrade performance.

Cinch plug: German name for a phono/RCA plug (after a major connector maker).

class A amp: purist (usually power) amplifier giving minimal distortion but low efficiency – they run very hot for the small power they give.

class B amp: amp that uses the least standby power with acceptable performance.

class A-B amp: amp that operates in Class A for most signal excursions, and in class B, with slightly lower fidelity, for occasional peaks.

class D amp: lightweight amp that 'slices up' the sound, reducing it to timing of on/off pulses, and called 'digital' for marketing. Elegant on paper, cheap to make, but still not suiting real-world use after 40-years-worth of attempts.

class G and H amp: power amps that save on heat dissipation, weight and cost by varying their headroom to match the signal's immediate demands – but the switching involved can affect the sound quality.

clevis: see shackle (flying equipment).

clip: abrupt flattening (chopping off) of the peaks of a signal, when the equipment can go no further – creates aggressive, hard sound,

removes most sonic detail, hurts nearly everyone's ears and can quickly blow drivers.

cluster: a group of horns or speaker cabs, normally HF, and usually flown, arranged to act as a single sound source.

CMR: common mode rejection – a measure of hum and noise reduction (as in balanced connections which use opposite-polarity signals to cancel out noise).

CMRR: common mode rejection ratio – amount of hum and noise reduction compared to some reference.

CMV: common mode voltage – the hum and noise enemy itself.

coil: also called an inductor, choke or reactor – literally a coiled-up wire, it concentrates the magnetic field caused by audio signals, RF or other AC in the conductor. The field then reacts against itself changing – hence a coil 'resists' and so filters-out fast changes in applied voltages, meaning high frequencies. Widely used for passive crossovers and some equalisers to separate low and high frequencies, and to keep RF noise out of audio inputs and outputs.

cold: the negative, low, or return side of a balanced line, or balanced input or output.

colouration (coloration in US): tonal (frequency response) and timbral problems and aberrations, as well as ordinary distortion – usually quite a complex effect caused by speakers and room acoustics,

rather than a gentle roll-off of treble, say.

combo: (i) an instrument amp with integrated speaker cab; (ii) old-fashioned/obsolete term for a band or group of musicians.

common mode – external signals (buzz, noises) that you don't want to leak into your PA system – see CMR, CMV.

compandor: a device or circuit that combines compression and expansion.

compensation: adjusting some internal parameter to tune-up or smooth out the operation of a circuit or apparatus – for instance to stop unwanted RF oscillations.

complementary: equal but opposite matching parts. Usually used in PA to describe two genders of transistors or a matching pair of crossover filters. Can also apply to plugs & sockets or well-matched low and high drivers and/or horns.

comp-lim: short for compressor-limiter – a processor that can perform both compression and limiting.

components: (i) loudspeaker parts, ie drive-units, horns, adaptors, etc; (ii) electronic and electrical building-block items such as diodes, coils, capacitors, relays, resistors, etc

compression: squeezing the dynamic range by reducing peaks and troughs in sound level – making loud sounds quieter and/or quieter sounds louder. An effect often used on

glossary

either individual signals (tightening up vocal or basslines) or whole mixes (allowing the overall level to be increased without clipping/ distortion).

compression driver: a dedicated, highly efficient high-frequency driver designed for horns (horn flares), which can look rather like lifting-weights and are often as heavy. The shape of the horn unit, and the physical compression of the sound, means the sound can be thrown further, though with a narrower spread.

compressor: sound processor that can be set up for smoothing dynamic range to minimise sudden leaps in volume, so overall perceived loudness can be increased without clipping.

concert pitch: standardised instrument tuning used in most western music (at least since 1960), where the A above middle C has a fundamental frequency of 440Hz – which can be measured using an electronic tuner or checked against a tuning fork.

concert sound: alternative term for PA.

condenser mike: popularly used though technically less accurate name for a capacitor microphone.

conductor: something that allows electricity to pass easily through it (in other words it allows its electrons to move around freely from atom to atom), presenting low

resistance. Good conductors include metal and water.

cone: front, moving part of a loudspeaker drive-unit – normally round, usually black and made of paper-maché or plastic, or else silvery and made of metal foil. It vibrates very precisely to turn electrical signals back into soundwaves.

console: or mix console – another name for a mixer, board or mixing desk.

consonant: musically pleasing.

control gear: the more tweakable elements of the sound system – like the front-of-house and monitor mixers, FX and drive racks, and their ilk – as opposed to the speaker cabs, amps and cabling.

convection: natural cooling method, where hot air rises, drawing in cooler air to replace it.

cooling tunnel: a 'heat sink' in the form of a chimney or a tunnel down which cooling air is blown.

cord: US alternative word for instrument or electrical lead/cable.

coulomb (C): unit of electric charge – one coulomb-per-second is a current of one ampere.

CPC: circuit protective conductor – electrician-speak for electrical safety earth conductor.

CPS: cycles per second – same

as Hertz, which is the unit of frequency.

crest factor: also called 'peak factor', or peak-to-mean ratio (PMR) – the difference between the peak and average (RMS) levels in signals; 3dB for a pure sinewave, 10 to 20dB for most music, or below 10dB when the sound is heavily compressed.

crossover: (i) a unit that splits a signal into two or more complementary frequency ranges – for instance sending low frequency sounds to the bass cabinets, and remaining mid and trebly sounds to the mid and HF horns. An active or 'electronic' crossover uses powered electronics to provide refined filters and other signal processing capabilities, producing a better result than passive crossovers, which are simpler and 'powered' from the signal itself; (ii) central region of a signal (as seen on an oscilloscope) where a wave passes through zero and changes polarity (alternates) – something that happens many tens, hundreds or thousands of times a second with sound signals. Some circuits and equipment can develop distortion in this area; (iii) cross-fader on a DJ mixer; (iv) a form or style of music that straddles genres.

crossover distortion: a horribly graunchy kind of distortion caused by bad or sick amplifier circuits. Note: has nothing to do with speaker crossovers.

cross-patching: (i) changing

glossary

the channel used by one mixer (say FOH) for a given signal source (probably because of a failure somewhere). Usually intended to avoid affecting the channel connections of other mixers (eg monitoring, recording) that are connected to the same source; (ii) a mistaken patch where instrument sources/lines get crossed over.

crosstalk: signal leakage from one channel to others (eg on a mixer).

CSA: cross-sectional area – size of a conductor (especially speaker, power). A larger CSA provides a higher current rating, in a fairly proportionate way.

current: the rate of movement of electrical charge (also described as coulombs per second). Current itself is measured in amperes (symbol I). Ohm's law says that the current flowing through a circuit or device is equal to the voltage across it divided by its resistance ($I = V \div R$).

current draw: current taken by PA or other gear from the main power supply.

CW: clockwise – usually equates to 'turning up' a control knob.

cycle: one whole wave or single vibration of a signal, from maximum in one direction, through zero, to maximum in the other, and back to zero. Can be audio, RF or AC electrical. One cycle per second is expressed as 1Hz.

D

DAC: abbreviation for digital-to-analog converter.

daisy chain/ing: several units (eg amps, speakers) cabled together in a chain. Usually it is the drive signal, but it could be (mains) power.

damping: the opposite of resonance, ringing or reverb – a deadening-down effect used in speakers, mikes, acoustics and electronic circuits, to get better sound quality by stabilising or restricting spurious effects.

damping factor: (i) a figure of merit for a power amp, saying how good it is (or should be) at damping the speaker cone's rapid back and forth motion (ie absorbing 'back EMF'); (ii) also defined as speaker load impedance divided by power amp output impedance; (iii) another term for 'Q' in parametric EQs.

DAT: digital audio tape – may be used to provide backing tracks at gigs, or to record the event straight to a master.

dB: deciBel – unit used to describe relative levels (of sound, voltages etc), or absolute levels if used with certain suffixes that specify a reference point (see dB SPL, dBu etc).

dBA: a 'weighted' sound level reading that emphasises mid frequencies, and progressively ignores low bass and HF. Useful for some jobs, but wrongly used for others.

dBC: a sound level reading that shows nearly all musical frequencies equally.

dBm: a level or scale used in older audio equipment. 0dBm = 0.775 volts (775mv) and also 1 milliwatt (1mW) of power into a set loading resistance of 600 ohms, based on an obsolete concept borrowed from telephone technology – though you'll still see dBm mistakenly used instead of dBu in PA gear that has never used 600-ohm loading.

dBr: r is for relative, a dB scale like dBu, but where 0dBr is any level you like, or that is useful to refer from, such as the clip (overload) level of a power amp.

dB SPL: the everyday scale of sound pressure levels, where 0dB SPL is the lower threshold of human hearing.

dBu: the everyday dB scale used in PA, where the reference at 0dBu = 775mV (0.75V), for any impedance – this is the most commonly used deciBel unit in the UK and Europe.

dBV: used where the reference is 0dBu = 1volt (1000mV), for any impedance – a variation that's not widely adopted.

DC: direct current – in theory a rock-steady voltage, as from an ideal battery. In practice, it may vary in time (with ripple, noise, discharge) but should maintain a constant polarity.

DC-coupled: signal coupling in the PA signal path which permits frequencies down to zero Hz to 'pass', and which also has nil phase shift or signal delay at low frequencies.

DC fault: where an output bears any significant DC voltage, which can cause thumps, bangs or even damage, when connections are made.

DC offset: a DC voltage appearing at (usually) an output – usually benign, becomes a DC fault if not.

DC servo: circuitry used to prevent DC offsets developing, by automatic compensation.

DDL: short for digital delay line, a signal processor which creates delay or echo effects.

decay: gradual drop in level of a sound or signal – for some sounds it starts as soon as the 'attack' has peaked.

de-esser: a dynamics processor that reduces 'sibilance' on vocal signals, or compresses the HF part of a signal.

delay: (i) deliberate delaying ('retarding') of signal to some speaker cabs to help compensate for distance (at large venues or outdoors), and so improve the resulting overall sound; (ii) an effect added to an instrument or voice to create either a thick doubled-up effect, or a more pronounced echo; (iii) unwanted processing delay in digital electronics, otherwise known as 'latency', which can be maddeningly long. Can also occur in analog electronics, but usually only on the edge of being audible as it's measured in millionths of a second.

delay tower: speaker cabs placed towards back rows of auditorium, or downfield, to help compensate for distance without creating an echo – as sound travels through air at only about 660mph (far slower than electrical signals through wires).

deoxidise: clean/restore electrical contacts.

derate: to use below maximum ratings, to increase reliability and/or safety.

desk: short for mixing desk – another name for the mixer.

device: a key component or unit – may be a transistor, IC, horn driver or special effect box.

DI: stands for direct injection – a means of isolating, adjusting and balancing a line-level instrument signal (from keyboards, guitars etc) so it can be connected to the PA's stagebox at a suitable level and without creating buzzes.

diaphragm: (i) the actual 'cone' of a compression driver (like a small, inverted standard speaker cone), made of either very thin, tearable metal foil, or 'phenolic impregnated material', and shaped rather like the contraceptive of the same name; (ii) the complete replacement assembly, including voice-coil. (Old diaphragms are often used as ashtrays or paperclip holders.)

differential: opposing 'push-pull' signals, outputs and receiving inputs. When these

are also identical in size and have equal impedance to ground (earth), these become balanced signals, outputs and receiving inputs.

digital: system of transmitting or storing information (including sound) as a series of numbers. Since 1980 has been gradually replacing 'traditional' analog systems in many music-related applications, despite early concerns about loss of sound quality, particularly at extremities of frequency ranges, and problems with unwanted delays (latency). Advantages include accuracy, memory and automation facilities, and a clean, low-hiss delivery.

dim: a switch on some mixers that can drop the output to low levels – useful when setting up.

DIN: (i) acronym for Germany's Industrial Standards Organisation; (ii) a family of fiddly, multi-pin connectors developed about 50 years ago, following a German industrial standard, and nowadays mostly used for MIDI connections; (iii) British expression for a loud, ugly noise.

diode: an electronic device, valve or semiconductor which acts as a one-way current valve. 'Rectifiers' are diodes designed to handle high currents.

directivity: (i) the angular sound dispersion of a drive unit or whole cab, which usually must be specified (in degrees horizontal and vertical) at different frequencies; (ii) a map of 'reception strength', for a mike or pickup device. In all cases the change in sound level is progressive, so 'the line is drawn' usually where the level has dropped by 6dB = by 50 per cent.

directivity index: the ratio (difference) of sound level from a speaker, front to rear.

discrete: (i) separate mixes; (ii) 'visible' electronics, where separate component parts are used – as opposed to silicon chips, where different, interconnected parts are created as one microscopic block.

dispersion: the direction and distance that a sound is 'thrown' by a speaker system – alias 'coverage'. A less technical description than directivity.

dissipation: power wastage emitted as heat.

distortion: (i) corruption or deviation from the original, preceding or intended signal, however introduced into the system; (ii) 'fuzz' or 'overdrive' effect deliberately created by using an FX unit (usually on electric guitar).

distro: slang abbreviation for 'distribution', as in electrical power around a PA or venue.

diversity: describing a radio system that switches between one or more aerials or input channels to maintain good reception. Widely used for stage radio mikes.

DMM: digital multi-meter – for reading AC & DC volts and amperes, resistance in ohms, and possibly other quantities,

with digits rather than a 'swing-needle'.

DNR: shorthand for 'dynamic range'.

dolly: trolley on wheels for moving heavy gear.

drain: (i) another name for the bare wire that makes connection with the foil shield inside some signal cables; (ii) one of the two main terminals on a MOSFET; (iii) US slang for an earthing/grounding point or terminal.

drive-unit/driver: (i) the 'loudspeaker' inside a cabinet (box) or behind a horn flare; (ii) a transistor in the stage before the final output; (iii) any device or circuit that feeds or controls another circuit.

drum-fill: (i) a monitor cab dedicated to the drummer's needs; (ii) a short, fancy bit of drumming between main patterns.

dry: slang for signal without reverb (or any other effect).

DSP: digital signal processing, the digital 'engine' behind digital crossovers, processors and effects (FX).

ducking: where one signal, often music, automatically dips in volume when another is introduced. Used by DJs and MCs to drop the music level when they want to speak.

dummy load: used to test amps at high power levels without filling the area with loud sound.

DVM: digital volt meter – may also be capable of other tests.

dynamic mike: a common, rugged, versatile low-cost kind of microphone that uses 'dynamo' (motor) principles. (*Not* related to signal dynamics.) It's also known as a 'moving-coil' mike, because a small coil inside is vibrated by soundwaves and, thanks to electro-magnetic induction, creates an electrical signal.

dynamic range: difference in dBs between the maximum output level before clip (say +20dBu) and the noise floor, where signals disappear (say -80dBu) – in this example it's 100dB.

dynamics: expression in music through a range of volume (intensity) shifts.

E

earth (in US called 'ground'): (i) in power wiring, the safety wire (usually green, sometimes green & yellow-striped), connected to the earth at some point (known by UK electricians as CPC – the circuit protective conductor); (ii) the thing connecting a conductor to the physical earth (soil, sand, rock) beneath us – which could be a building frame or a large metal stake dug into damp soil, or even, in unfortunate circumstances, a person; (iii) any point taken to be at zero electrical potential (zero/0 volts, 0V).

earth lift (US, ground lift): a fairly safe means of curing 'earth/ground loop' problems (far better than fiddling with earth/ground wires in plugs). This is a switch or linking arrangement that either temporarily disconnects or places an impedance between an earthed equipment chassis and the signal 0V. Useful or essential when systems are unbalanced, or even sometimes if they are balanced.

earth loop (ground loop): a problem caused by circular or multiple-looped connection of ground conductors – if two or more pieces of earthed equipment are cabled together it can result in audible mains hum or buzzing.

earth/ground shield: see shield.

echo: repetitions of a signal, with gradual reduction in volume, recreating the reflective acoustics of, for example, a large cave, canyon or stone building.

effects (FX): (i) creative audio signal treatments; (ii) the analog or digital signal processing units used to achieve these results (often called effects/FX units).

effects loop: connections from and back to the mixer, which allow FX to operate on selected sound signals.

E-fields: electric fields that can transmit hum, noise or crosstalk signals from one wire to others. This type of field is easily shielded with thin metal or even just conductive paint.

EIA: Electronic Industries Association, notable only because EIA speaker test standard was developed into the AES's drive-unit power rating test standard.

electret microphone: lesser type of capacitor mike – doesn't require phantom power, but uses a battery.

electric: (i) occasional UK slang shorthand for 'electricity'; (ii) guitarists' shorthand for 'electric guitar'.

electro-acoustics: (i) academic name for key parts of PA, where electronics and acoustics come together, meaning inter-disciplinary work with mikes, mike amps, power amps, speakers, cables, cans and acoustics; (ii) acoustic guitars which have built-in pickups and sometimes pre-amps.

electronic crossover: misnomer for 'active crossover'.

EMC: electro-magnetic compatibility – standard testing for electronic (including PA) and electrical equipment to ensure it will (theoretically) work OK with other gear, or at least won't cause serious damage or interference.

EMF: electromotive force – the property that a generator or battery has which makes a current flow. Measured in volts.

EMI: electro-magnetic interference – electrical noise pollution caused by any sort of electrical or magnetic phenomena, both natural (static, lightning) and, more often, human-made. May manifest as buzz, crackles, static.

EQ: short for equalisation (US equalization) – filters, resonators and other EQ (either analog electronic circuits or digital realisations) which alter tone/tonal/timbral qualities by cutting and/or boosting certain frequencies. Ranges from just treble and bass controls with limited adjustment, operating at fixed frequencies (mixers usually have at least three, with a mid control), or complex – for instance a variable 'parametric' EQ. (See also 'graphic'). Some EQ is required for standard correction purposes, and is never altered nor accessible: for instance 'RIAA' EQ for playing vinyl, 'NAB' EQ (and others) for some reel-to-reel tapes.

excursion: the movement of speaker drive-units.

expander: (i) device for increasing the dynamic range of a sound signal (instrument or overall mix), so the loud bits get louder and/or the quiet bits quieter – mostly the opposite of a compressor; (ii) a keyboardless synthesiser module.

F

fader: slider volume control, mainly on mixers – used rather than rotary controls to enable several to be moved at once.

glossary

fantailed: one cable 'split out' to several destinations (see breakout).

FCW: fully clockwise – used on schematics and in technical instructions to describe the maximum setting of a control knob.

feedback: (i) acoustic feedback – howl or squeal through the FOH PA or monitors; (ii) more generally, *negative* 'feedback' is the universal control system in nature, as well as electrical engineering – as something goes wrong, it's pulled back the other way; 'positive feedback' is the usually destructive opposite type of feedback that uncritically reinforces its stimulus signal, building it up until the system saturates, or oscillates – as in the familiar howl or shriekback through the speakers.

female: referring to a connector designed to receive the pins/prongs of a male plug or socket.

FET: field effect transistor – may be J-FET or MOSFET type. (As the 'T' in FET already says it, there's no need to say 'MOSFET transistor'.) See 'transistor'.

figure-eight (UK, figure-of-eight): a microphone response pattern where the pickup sensitivity of the mike is the same at the front as at the back, but lower at the sides.

fill: cabs or horns used to 'fill in' an area with sound which wouldn't be covered otherwise.

filter: a circuit or digital process that allows parts of a signal at certain frequencies to pass through unchanged, while rejecting or reducing others.

five-way: a system which splits the sound into five frequency ranges, usually sub-bass, bass, low-mid, high-mid, and HF.

flare: a moulded or cast horn.

flex: another (archaic) word for power cable/lead/cord.

floating: electronic connections which either aren't earthed/grounded, or don't need an earth/ground or other common 'reference' point.

floor monitor: alternative (US) name for a 'wedge', a stage-mounted monitor cab.

floor stand: a stubby mike stand for miking up something low-down (guitar amp/kick drum etc).

flown: see 'fly'.

fly: to hang cabs or horns (usually above the stage).

FM: frequency modulation – as used to transmit signals in VHF and UHF radio/wireless microphone and guitar systems.

FOH: see 'front-of-house'.

FOH engineer: person who mixes the 'out-front' sound that the audience mainly hear.

foldback: BBC and 'old school' British term for the monitor system.

folded horn (or bin): one that curls round, keeps it compact, suits bass only.

formant: academic word for the harmonic (timbral) patterns that characterise different instruments or vocals.

four-way: usually means a system with separate bass, low-mid, high-mid, and HF elements.

frequency: the rate (in pulses or cycles per second, or Hertz/Hz) of a sound or electrical (or any) vibration, also known as a wave, pulse or any other 'waveshape' that's repeated at some regular rate. In PA, significant frequencies range from 5 cycles per second (5Hz, lowest sub bass) to over 20,000 times/sec (20kHz, the audible/ultrasonic threshold). In music, what we hear as pitch is partly 'frequency sensation'. See also hertz.

frequency response: the way in which, and the amounts by which, sound (acoustic) and signal (electrical) levels vary across the audible range of frequencies, and beyond.

front-end: the input sections of a PA mixer's channels.

front-of-house: see FOH.

full-range cabs: speaker cabs that contain an integrated set of drivers to cover the entire frequency range, or all but the sub-bass. Can help achieve compactness.

FX: shorthand for 'effects', also known more formally as signal processors – boxes that can be used to creatively/ artistically alter sound.

G

gain: amount of increase or change in signal level. When dBs are used, increased gain is shown as '+dB', reduction is shown '-dB', and no change as 0dB.

gain structure: the 'map' of gains shows how much levels are boosted and/or cut by throughout the PA's or a mixer's signal path, and the cumulative results for typical fader positions.

gate: (i) a sound dynamics processing device that can 'clean up' signal feeds – achieves this by muting the feed when the wanted signal is absent or too small to be useful (so, for instance, muting low-level background noise during quiet passages). Can be triggered by the wanted signal or others. Can also be used to create sound effects; (ii) a signal that triggers another device.

generator: used for producing DC power at a gig (usually outdoors) where 'normal' electricity supply is unavailable or not required – sometimes affectionately called a 'gennie'.

gig: a live musical event.

gooseneck: a flexible-stemmed microphone or lamp-holder.

graphic: short for graphic equaliser – a type of EQ with individual faders (from four or five to 30 or more) for adjusting the levels of various pre-determined frequency bands. The physical fader positions result in a visual 'graphic' representation of the adjustments made.

ground: another (US) word for 'earth' – but also used to distinguish other common points (0V, zero volt) in audio and PA gear, that are separated from true earth.

grounded bridge: a rarely-used industrial-grade power amp technique that allows bridged channels to be bridged again to achieve very high voltage swings.

ground lift: see earth lift.

ground loop: see earth loop.

ground shield: see shield – it's rare for a shield not to be grounded.

group: (i) a bunch of channels or signals brought together on a mixing desk for easy joint manipulation; (ii) the fader which controls such a group; (iii) a band of musicians/performers.

group delay: frequency-dependent delays in a signal, such as the bass 'boooph' sound of a kick drum 'hit' arriving after the higher-frequency 'edges'.

H

handball: US (military) slang for team act of moving heavy PA boxes by hand rather than with a forklift-truck.

handling noise: the unwanted noise made by a microphone and its cable when physically moved around during performance.

hash: US slang for noise, RFI, buzzes.

hard short: electrician's term for when a solid short circuit is caused by purposeful wiring in error, rather than (say) a few wire strands accidentally and lightly touching.

harmonic: spurious added 'timbres' or 'notes' that are often musically related (by intervals) to the sounds causing it. The basis of all musical instruments, but also randomly created by PA systems – not always very musically.

harmonic distortion: 'ordinary' distortion occurring in analog (audio) electronics and speakers and mikes, which involves harmonics being produced.

headline: the main (or most publicised) act on a bill.

headroom: the amount of available space (in dB) above the normal operating level before serious distortion sets in – should be at least 15 to 18dB in professional systems.

head-set/head-worn: a hands-free mike system, with headband and mini boom-arm in front of the mouth.

heat exchanger: a more

glossary

descriptive term for heatsinks, which 'exchange' heat with the surrounding air.

heat sink: means of diverting unwanted, excess and potentially destructive levels of heat from electronics that handle high power, particularly in a compact space.

hertz (Hz): cycles per second (CPS), the unit of 'frequency'. (Named after Heinrich Hertz, the 19th-century German physicist, a pioneer of transmitting signals using radio waves.) There is a direct relationship between hertz and audible pitch: basically the higher the fundamental frequency of a note, the higher its pitch. For instance, 440Hz is the fundamental frequency of the note A (above middle C) at 'concert pitch', while an A an octave higher has a fundamental frequency of 880Hz (ie double). See 'octave'.

HF: high frequency (also referred to as treble, or top end) – in sonic terms usually refers to the frequency range above about 5kHz.

h-field: magnetic field.

HGV: UK abbreviation for heavy goods vehicle – big trucks used for carrying large PAs, lights and sets.

hi-packs/high packs: slang for (usually) integrated MF+HF cabs – the upper part of a packaged full-range sound system.

high-pass filter (HPF): a circuit

or digital process that allows parts of a signal at higher frequencies to pass through unchanged, while rejecting or reducing those at all lower frequencies. Example: removing unwanted bass.

high-Z: high impedance – as in low-grade mikes, pickups and some electric pianos. Worth avoiding or, if must be used, connect with short low-capacitance cable. (See Z).

hiss: white noise, or any random noise where high frequencies dominate.

HMP: high melting point – as in solder used for very hot parts.

horn: (i) a flared tube placed in front of a (often HF) drive-unit to magnify the sound, and control directivity; (ii) the whole cabinet the flared tube is mounted in.

hot: the live, or positive, side of a balanced line, input or output.

house: (i) the venue (ii) type of dance music, invented in the US in the mid-1980s (from 'warehouse party').

house console: alternate term for FOH mixer.

howlround: another word for 'positive' acoustic feedback – when a PA system feeds on itself – which can result in squeal, growl etc, depending on which frequency feeds back first.

howlround threshold: the sound level above which

howlround (feedback) begins.

HPF: see high-pass filter.

HRC: high rupture capacity – type of heavy-duty ceramic, sand-filled fuse used in high-current situations where normal glass fuses may explode.

HSE (Health & Safety Executive): official UK body with powers to stop a gig if something contravenes safety regulations.

HT: high tension – another term for high voltage, typically above 50 to 100 volts in PA terms, but above 600V to 1500V in venue electricians' territory.

hum: low-frequency noise, usually associated with AC power frequency interference (around 50/60Hz).

humping: UK term for lifting, carrying and stacking PA gear.

hybrid: a circuit or other unit that uses mixed technologies (eg analog and digital).

I

I: symbol for ampere, the unit used to measure current.

IC: see integrated circuit.

IEC: the long-established International Electrotechnical Commission (1907) – best known for the low-tech, plasticky mains plugs (called 'Euro-connectors' or

'Europlugs' in UK) which appear on (and regularly fall out of) low-power PA gear, and are the bane of most PA FX and drive racks.

IMD: intermodulation distortion – same cause as harmonic distortion, but the resulting spurious notes/sounds created are not necessarily at all musically related. Higher levels of IMD may be heard as ugly 'grunge' or 'grit'.

impedance: (symbol Z) more complex than resistance, but also measured in ohms (Ω). Everything offers some opposition to a current or signal flow – cables, components, mikes, amps, speakers... In a DC system this would by easily calculable using Ohm's law: resistance = voltage divided by current (R = V÷I). In AC systems, resistance is augmented by reactance, which depends on the capacitance and inductance in the circuit, which in turn varies with the signal frequencies involved. Also, devices in the signal chain will often have an input (or load) impedance and a separate output (source) impedance, depending on whether they are receiving or sending the signal.

inductance: a property met by alternating current flowing in a wire, especially when concentrated in a coil. Involves magnetic energy storage, the opposite of capacitance.

inductor: another (generic) name for a 'coil'.

in-ear monitoring: where each musician's monitor is an ear-piece, instead of going to foldback/monitor speakers – allowing performers to hear themselves more clearly as well as curtailing feedback/howlround problems.

infill: a monitor fill that points inwards from inside of stage left & right PA stacks.

infrasonic: academic term for frequencies below normal audibility (about 20Hz) – also called (less accurately) subsonic.

input channel: the channel 'strip' generally found en-masse on the left portion of a mixer – it deals with individual sound sources, rather than groups.

input impedance: the impedance or loading presented at the inputs of PA and other signal-handling equipment.

inrush current: a high, short-lived current surge in the power supply, when one or more large transformers and/or capacitors are first energised – occurs when larger power supplies and amplifiers are first switched on. Can cause lights to dim.

insert: a two-way socket ('jack') that allows a signal processor to be 'patched' (connected) into a mixer's channel.

instability: self-oscillation in circuitry inside equipment, caused by poor design, manufacturing defect, a bad repair, or wearout – may be at RF, hence not audible, but can still harm PA gear.

instrument amp: (i) musician's amp/pre-amp, which may be just a 'head' or a combo; (ii) an academic name for a balanced input stage, or amplifier used there.

insulator: a material that's a very poor conductor of electricity, and can be used to protect humans or separate conductors. Good insulators include plastics, glass, rubber and dry wood.

integrated circuit: a whole circuit of many different electronic parts etched in layers on a pin-head-sized piece of silicon (or other semiconductor).

inverter: (i) a power conversion device, changing direct current to alternating current (some types are unsuitable for PA use); (ii) circuitry or switchable setting that reverses the polarity of a signal.

ISO: International Standards Organisation – only known in PA-land because it long ago agreed on sets of standard centre frequencies that regular graphic equalisers should have.

itinerary: the list of places on a tour.

J

jack: (i) (in the UK) short for 'jackplug' – a widely-used connector (originally made for telephone switchboards): there are several types, the commonest are quarter-inch diameter, although 'mini-jacks' appear on some portable and low-tech gear. Basic mono-type TS (tip & sleeve) and two-pole/stereo-type TRS (tip, ring & sleeve) connectors are used mostly by musicians, but also appear on some PA gear (eg mixer send/returns) – but they're not liked by sound engineers because of unreliability and general unsuitability; (ii) in the US, 'jack' usually refers to the socket into which such plugs are inserted.

jackfield: also called a patchbay – a row of jack sockets, either standard or 'Bantam' type, which may be used in PA for FX patching.

J-FET: junction field-effect transistor – unlike a MOSFET, not used for power handling, but as a specialist type of transistor in mixers, pre-amps, radio mike circuits, etc, either as a discrete part or hidden in the inputs of some widely used ICs.

jitter: digital timing inaccuracy – major cause of poor sonic quality (hard sound) in digital gear.

joule: unit of energy (symbol J) – 100 joules per second is 100 watts.

K

k: short for kilo – as a prefix it means one thousand, as in kW (1000 watts), kHz (1000 hertz), kΩ (1000 ohms).

key: the trigger input of a 'gate', used for sophisticated triggering – may be frequency-sensitive as well as level-sensitive.

keys: abbreviation for keyboard instruments, including synths.

kick drum: slang for bass drum.

knee: area on a curve where relations between two parameters make their main change of direction or 'slope' – in PA, usually refers to a compressor's compression ratio: for example 'hard-knee' compression kicks in more dramatically, while 'soft-knee' is smoother in effect.

L

L&R, L/R: left and right – stereo channels.

lavalier: a miniature suspended (or clip-on) microphone.

LCD: liquid crystal display – a digital display screen.

LD: abbreviation for lighting designer.

lead: (i) UK word for cable – rhymes with 'deed'; (ii) shorthand for 'lead guitar' – the main guitar within a group, which plays most of the solos and/or riffs.

LED: light-emitting diode – used for simple alpha-numeric displays and on/off indicators.

leg: (i) one part of a lead (or an electronic circuit) that splits into two or more sections; (ii) one part (section) of an in-tour schedule.

leq: sound level used by environmental 'noise-control' officials that's averaged over several minutes. This permits brief high level passages in the music, above a sound level limit, provided quieter sections follow, to 'pay back'.

level: the amplitude of a signal – in music and PA gear this usually equates to volume, intensity and higher dB figures.

LF: low frequency (bass) – say below 500Hz, down to 20Hz.

limiter: a dynamics processor, unit or circuitry that largely prevents a signal going above a preset level.

linearity: academic term for absence of distortion in audio circuitry. Parts and circuits that cause distortion are said to be non-linear.

load: (i) something that absorbs electrical power, or into which an electronic signal source is driven (for example, a speaker, or a mixer's input seen by a mike; (ii) the total impedance of the several speakers connected to a power amp etc; (iii) the energy consumption of equipment, represented by the current draw at the power intake.

log: short for logarithmic, as used in mathematics, and for calculating dBs.

long-throw: a 'horn' speaker cabinet radiating a narrow beam of sound which projects further than a wider spray.

loom: a bound-together bundle of cables (or wires inside gear), so they can be handled as one entity (see also 'multicore/snake').

low packs (lo packs): slang for LF cabinets or bins that are part of a packaged system.

low-pass filter (LPF): filters-out frequencies above the cut-off frequency, allowing those below to pass through. Potential complement to a high-pass filter.

loudspeaker: a transducer that converts electrical signals into acoustic soundwaves (ie the reverse of a microphone). The most common type of speaker used in PA systems is a moving-coil (just as the most common mike is a dynamic type), though niche speaker types include those using electrostatic or ribbon mechanisms. The word 'loudspeaker' can be used to refer either to the complete speaker unit (including cabinet/case), or just the drive-units, or drivers and horns, depending on context. It's sometimes abbreviated to LS, or more often just 'speaker'.

LPF: see low-pass filter.

LVD: low-voltage directive – umbrella term for supposedly harmonised EU safety regulations for voltage above 50V, up to 1500V (low is rather a misnomer).

M

m: short for milli – a thousandth part; used as a prefix, as in mA (milliampere, a thousandth of an amp), mV (a thousandth of a volt).

M: mega – a million; used as a prefix, as in MHz (megahertz, a million hertz).

mains: UK term for AC power, as delivered by the electricity supplier to wall outlets in homes, venues etc.

male: referring to a connector with one or more external pins or 'prongs' designed to fit into a mating (reciprocal) 'female' connector with corresponding chambers.

MBS: minimum break strength (of wire cables etc).

MCB: miniature circuit breaker – an electrical power switchboard type that's rated to break fault currents up to 6000A. (See RCD)

MF: mid frequency – a wide-ranging term covering sound frequencies between around 500Hz to 5kHz (also subdivided into high-mid, low-mid, etc).

MI: abbreviation for 'musical instrument' – refers to equipment aimed at musicians, as opposed to professional audio engineers, and to the industry supplying this equipment.

mica: a natural electrical insulator (really a rock related

to slate) that also conducts heat well, used in many power amps and power supplies to connect live-bodied transistors to earthed heatsinks.

micro (μ): a millionth part – used for describing very small electronic component values.

microphony: when vibration generates electrical noise or signals – may or may not accidentally create musical effects. Common in valve/tube instrument amps.

mid-band: middle of the human hearing range, where the ear is most sensitive – typically 500Hz to 5kHz, centring on about 1.5kHz.

MIDI: Musical Instrument Digital Interface – industry-standard for communication of control signals and data between electronic musical equipment.

MIDI port: MIDI connection socket (or MIDI jack in US) – the MIDI Out of the master/controller device is linked up to the MIDI In of the 'slave' unit; unless it's one of a chain of slaves, in which case MIDI Thru sockets are used to link the slaves.

MIDI timecode (MTC): a means of synchronising MIDI-based instruments with each other or with computers.

mid-range: referring to the mid-band sound frequencies.

mids: abbreviation for mid-range speaker cabs or frequencies.

glossary

MIL-spec: rigorous specifications for electronic parts devised over the past 50 years by US Dept of Defense – similar to 'DEF' (defence contractor) standards in UK. PA and modern electronic equipment reliability has greatly benefited from these programs.

mix buss: see buss.

mixer: (i) the mixing desk, board, or console – where all the sound signals are collected and processed, as required. Can be anything from four channels to several dozen, depending on the size of the line-up and/or venue; (ii) the FOH (or monitor) sound engineer.

mm²: (UK, Europe) a conductor's area – determines approximate current capacity and exact resistance per unit length. Signal wires are typically 0.2mm², speaker wires typically 2 to 6mm², and most PA power wires range from 0.5 up to 50mm².

momentary ('mom'): push button switch that only operates while pushed, like a doorbell – used for talkback microphones and other controls.

monitor: a specialised speaker pointed at the artist(s) so they can 'monitor' their own performances.

monitor engineer: person who creates and manages the musicians' monitoring mixes, making sure they can hear what they need in order to play comfortably.

monitor mix: a mix customised for the requirements of a particular performer, usually sent from a dedicated mixer operated by a monitor engineer at the side of the stage.

mono: short for monophonic – (i) in audio electronics, only one final output or mix (also 'monaural'); (ii) monophonic, as in a synthesiser that can only be played one note at a time.

monolithic IC: an integrated circuit created all-in-one (monolith) from a tiny slice of pure silicon.

MOSFET (MOS-FET): metal-oxide semiconductor field-effect transistor – a versatile type of transistor that can be made to handle high power, and is widely used in power amps and switching ('electronic') power supplies.

motherboard: the main 'base' PCB (printed circuit board) inside some equipment into which smaller boards are mounted or can be plugged.

moving-coil: see dynamic.

multicore: a fat sheath neatly containing many cables (either balanced pairs for mike/line or speaker leads) connecting the FOH mixer to the stage. Also called a 'loom' or 'snake'.

multimeter: useful measuring tool that reads DC and AC volts, amperes, resistance in ohms, etc, over wide ranges.

mu-metal: a magnetic shielding metal – a special kind of steel, expensive to form, so only useful for mass use (audio transformers) or brute-force solutions.

music power: a different method of showing power delivery capability figures, more sensible for music, but mis-used by discredited makers in the past, as it made amps appear more powerful than others rated by steadier methods (such as so called 'RMS' or 'true average' power). Also creatively utilised to make speakers appear to handle more power – a 250w RMS (average) rating is often called '500W music' (or 'programme') power.

musicality: long used by musicians before re-employed by hi-fi enthusiasts in late 1970s to describe sound qualities of gear – easy to hear, yet as hard to describe as music itself.

multi: short for 'multicore cable' – plural: multis. In US called 'snakes'.

M3, 4, 5...M6, M10 (etc): series of standard metric (machine) screw sizes widely used in UK, Japanese and European PA gear.

N

n/c: no connection – an unused contact in/on a plug or socket etc.

negative feedback: see feedback.

neon: an indicator light that uses red/orange neon gas (or related 'noble' gas if other colours are required), so avoiding the need for a fragile filament – works on high-voltage AC or DC.

nested delays: a sequence of delays, where each subsequent one is timed from the one before. Then if the first has to be altered, all the later ones keep in step with it, and so a cascade of alterations isn't required.

noise-gate: see gate.

non-linear: see linearity.

octave: (i) an interval (change of frequency) with a ratio of 2 to 1 – or 1 to 2. When a sound's frequency is doubled, it and the pitch heard go up by one octave, and when it's halved, it and the pitch drop an octave. Two octaves upwards means a quadrupling (x4) of frequency. Like dBs, this is a ratiometric scale (steps of x2 rather than +2) that suits human sound perception; (ii) musically, a distance of 12 semitones.

off-mike: when vocalists forget to keep their mouths in front of, or even anywhere near, the microphone.

offset: (i) residual or fault-caused DC voltage at the output, or even input, of audio equipment; (ii) angling or positioning of equipment (PA cabs, cartridges on tone arms, etc).

ohm (Ω): unit of electrical resistance/impedance.

ohmage: slang expression for amount of resistance, as in: 'that speaker has the right ohmage'.

Ohm's law: states $V = IR$, where V is voltage, I is current, and R resistance. Also $I = V \div R$ and $R = V \div I$.

open: for instance when a mike is plugged in and the channel fader up and ready to receive a signal.

O-ring: a rubber 'o' shaped ring, used as a seal or anti-undo (friction) lock.

oscilloscope: short for cathode ray oscilloscope – a key visual-based test tool that shows all kinds of signals and waves, from DC and audio to RF waves, on a screen. Analog types work in real time, digital types don't and are inferior for non-steady music type waveforms. Abbreviated to 'scope.

out-front: slang term for the main FOH PA (mix position).

outfill: monitor fills that point outwards from the outside of the L & R stage PA stacks.

out-of-phase: when the polarity (or 'phasing') of a signal conflicts with (and tries to cancel) another signal, causing partial or full cancellation if combined. See 'polarity'.

P

P: see power.

PA: (i) originally derived from the phrase 'public address' – in its abbreviated form it applies to concert sound systems of all sizes; (ii) pro audio – the industry of professional sound engineers and equipment makers/suppliers (iii) personal appearance – a celebrity booking, at which the artists may or may not perform (or more likely mime) their latest hit single; (iv) personal assistant. (It can cause much confusion if you mix up any of these definitions...)

pad: switchable attenuator, used to drop ('pad out') excess level.

pan: the left & right 'balance' control on a mixer.

pan pot: short for 'panoramic control potentiometer' – the knob fader that rotates between two paths, apportioning signals to the left and right.

parametric: short for parametric equaliser – an EQ that allows the most precise control over a sound's tonal quality, through three main adjustable parameters: gain (boost/cut), frequency, and Q (which controls the breadth of the affected frequency range).

parallel mode: two or more amp channels connected together in parallel (+ to +, – to –) to double the current (rather than voltage) capability, for driving low impedances (low loads). For expert use only, as gains must be closely matched.

parasitic(s): something unwanted in a circuit – stray capacitance, inductance or combinations of such, causing variations from proper performance.

parasitic cone: a mini speaker cone in the centre of some large-coned drive-units which gives some treble output, used when a tweeter can't be afforded.

passive: non-powered circuitry.

patch: (i) to connect (into a circuit, for example), or join two sockets with a cable/lead; (ii) a preset sound called up on an FX unit, synth, etc.

patchbay: another name for a jackfield.

patchman, patchperson: PA crew member at festivals or multi-event, whose sole job is to ensure that all the on-stage signals are correctly connected and to plan and manage the logistics of this, including dealing with sudden billing changes – such as finding mike lines for 22 extra conga players.

PCB: printed circuit board – the thin insulating sheet covered with conducting interconnections and electronic components that all 'solid state' and even most valve/tube equipment circuitry is built upon.

peak: (i) the highest value – the transitory peak level of a signal, seen as a wave's tip on an oscilloscope – as opposed to the highest 'RMS' value, which is averaged out over time. These can differ by a factor of 1.5 to ten times in sound signals; (ii) the highest point of a frequency response or EQ curve, where boost is greatest; (iii) the optimum point of some 'upward' tuning process, such as

the best aerial/antenna position.

peaking out: see 'clipping'.

peak limiting: a more academic term for 'clipping'.

peak-to-mean-ratio (PMR): see crest factor.

peak-to-peak (P-P): term used in 'scope measurements, where the peak values of the signal (in each direction, +/-) are clearly visible. For a pure tone signal only, a P-P value (usually a voltage) is twice the peak value, and 2.8 times the RMS (average) value. See also 'crest factor'.

pepper pot: the dispersion control part of a specialised high-frequency horn, a type of acoustic lens used especially in wedges – circular front comprises plates covered with holes, hence the name.

PFC: power factor correction – feature of large or mass-produced electronic power supplies (for a mixer or computer etc) and in high-power amps, which forces tidier mains current draw, thus reducing mains pollution and saving energy.

PFL: pre-fade(r) listen – a button on the mixer which enables the selected channel's signal to be monitored, regardless of whether the channel's fader is up or down.

phantom power: 48-volt DC power (from the mixer or mike splitter) shares the balanced mike connection's three wires, one side (+48V) being passed

onto both the mike's balanced signal conductors, the other (0V) is connected to the shield connection directly, without the need for any extra wiring. Dynamic (moving-coil) mikes don't use it, and the power, being the same on each signal wire, stays invisible (hence 'phantom'), while capacitor mikes can readily access it. 'Doesn't mix' with unbalanced mikes.

phase: the timing of two signals relative to one another, expressed in degrees. A complete waveform cycle of a particular frequency is 360 degrees, so if two identical signals at the same frequency are 180 degrees out-of-phase, they'll cancel each other out. Can be compared to, but shouldn't be confused with 'polarity'.

phase plug: a device that blocks the centre of a horn's throat, forcing the soundwaves to take a particular path, usually out around the sides, and helping them 'into phase'.

phase shift: (i) where the timing of different frequencies is gradually shifted – most audible at low frequencies; (ii) common name for polarity flip or changeover – a 180-degree reversal of signal 'direction' (a bit like the negative of a photograph).

phasiness: describing the highly-coloured sound from two or more speaker cabs interacting to create an unpleasantly spikey 'comb filtered' response – so called because the frequency response is seen to have numerous close-spaced peaks and troughs, like the teeth on a comb.

phone plug: US term for plug with a quarter-inch diameter barrel (known in UK as a 'jackplug').

phono plug: UK name for an RCA plug – an unbalanced, unprofessional connector, widely used on hi-fi gear and budget home recording gear etc for its compactness and low cost (originally designed for connecting RF signals in the 1940s). Also called a 'Cinch plug'.

piezo: short for piezoelectric – in PA terms, a low-power HF driver technology that transduces audio in sound by exciting/vibrating a quartz crystal.

pig: slang term for large, awkward-to-handle power transformer or Variac (variable transformer).

pink noise: random noise (sounds like a waterfall) that is similar in character to music, with most of the power in the low end. Used for setting up PAs (along with a spectrum analyser – called real time analysis), and for the endurance-testing of amps and speakers.

pk: shorthand for peak.

plug: a connector on the end of a cable, cord, lead or device, which couples with a socket.

PMR: peak-to-mean ratio – see 'crest factor'.

PO: short for power output – as in capability or location of.

polarity: (i) the opposing negative or positive 'halves' that all analog and some digital signals have (the alternation back and forth is visible at low bass as the pulsating of a speaker cone); (ii) the arbitrary labelling of opposite (180 degrees different) sides of balanced signal and also speaker connections as 'hot' and 'cold' and '+' and '−'. Commonly called phase, but true phase (shifts) can assume any value in degrees, whereas polarity only 'knows' 0 (zero) and 180 degrees. A full cycle, from positive to negative and back again, happens over 360 degrees, so if two identical signals have opposite polarities they'll be exactly 180 degrees out of step with one another – which has the same effect as being directly 'out-of-phase', in other words they'll cancel one another out. This is why polarity and phase are often used synonymously, though they're only interchangeable when the signals are musically simple ones, such as sine waves.

polar pattern/diagram/plot: nothing to do with signal polarity – this shows the 3D response pattern of a speaker or microphone in the horizontal or vertical plane.

popshield: see windshield.

pot: short for potentiometer – a kind of electronic tap (faucet in US) or flow-valve which lies behind most variable controls (volume knobs, etc) on analog gear. Not the knob itself,

though it's sometimes mistakenly used that way.

power: (symbol P) the rate at which energy is expended, wasted, transferred or absorbed, measured in watts and defined in electrical terms as voltage multiplied by current (VxI).

power bridging: an older term for bridging two power amp channels.

power density: (i) the density of a signal – see 'crest factor'; (ii) an amplifier or amp rack's power output capability expressed in terms of some form of spatial volume, such as 'watts-per-rack-unit'.

power factor: a measure of the amount of wasteful 'borrowed' energy in AC power circuits.

PPM: (i) peak program meter – original and best peak reading 'needle meter' to BBC spec, rather oddly scaled 0-7, though others exist with own specifications and dB scales; (ii) parts per million – used when measuring distortion, noise or other signal contamination.

pre-amp: short for pre-amplifier – a circuit designed to boost low-level signals to a standard level and EQ them before they're sent towards the power amp stage for full amplification.

pre-fade: a signal feed taken from a mixer before the volume fader.

PROM: programmable read-only memory.

glossary

programme (US program): (i) a music signal; (ii) a rating based on music signals, as opposed to artificial test signals.

proximity effect: (i) in microphones, the bass boost and popping that occur when you get up close; (ii) with closely-spaced speaker cables, high current peaks cause the cable's resistance to increase 'dynamically'.

PSU: power supply unit – usually changes AC mains into precise DC supply (or supplies) for powering electronic equipment; preferably a separate unit, not built into the gear it powers, in order to keep noisy electrical circuitry away from sensitive audio circuitry.

pumping: common un-musical effect of a wrongly set-up compressor or limiter, where the level is quickly varying in time with the bass beat or content.

punt kit: items and equipment used to deal with PA emergencies.

push-pull: using two identical power devices (in power amps) or drive-units (in cabs) working oppositely, meaning 180 degrees 'out-of-phase' with one another. Done to provide more power and/or lower distortion. In a push-pull driver the diaphragm is attracted to one side at the same time as being repelled by the other, producing a more linear, less distorted response.

PVC: poly-vinyl chloride – a plastic available in many grades

of suppleness, used for insulating most electrical conductors and cable sheaths. Flammable and, if heated, emits toxic fumes.

Q

Q: 'sharpness' of a 'bell type' (frequency-specific) equaliser, corresponding to the resonant frequency of a resonator. High Q (stronger filter resonance) reduces the affected signal to a narrow frequency range around a selected point; low Q (less filter resonance) means the affected range is wider around the central, peak point.

quad: (i) a four-channel device or set-up, such as a chip or sound system; (ii) a British (hi-fi) amp maker.

quarter-inch: diameter of commonly-used plug/connector, known as a jackplug in the UK and phone plug in the US. May be 'mono' (TS, tip & sleeve connections), or 'stereo' (TRS, tip, ring & sleeve). In metric areas, it's 6.35mm.

quasi-complementary: amp output stage that uses 'same gender' output transistors, tricking one of them to act as the opposite gender, or otherwise perform to get the symmetrical output capability that music signals need.

quiescent: standby mode, tickover, idling. Another name for bias (biasing current), used to help power amplifiers work with low distortion.

R

R: stands for resistance – measured in ohms (Ω).

rackmount: gear that's designed to be screwed into standard 19in-wide equipment racks.

radial: a type of mid and high-frequency horn, with a rectangular format and a slightly 'lip' or 'slit'-like appearance. In some types the lips are swollen, and are known as 'baby bums' ('baby butts' in the US).

radio microphone: or wireless mike – a microphone that combines a traditional capsule with a wireless transmitter, sending signals to a remote receiver without the need for connecting cables.

ramp fill: a monitor cab (usually a wedge type) for the sort of performers who have to move up to higher parts of the stage set during their act.

RCA plug/jack: US term for what's called a phono plug/socket in Britain – named after the US company that first developed it in the late 1930s, specifically for VHF (RF) use. Long superseded technically, it's still a budget audio favourite, but a pest to professionals.

RCCB: residual current circuit breaker – see RCD.

RCD: residual current device – a life-saving circuit breaker that operates when more than a small current leakage (to somewhere it shouldn't) is detected, preventing some (not all) causes of fire and/or electrocution.

RCBO: Euro-term for a combined RCD & MCB (mini circuit breaker) which provides combined protection against electrocution, fire and overcurrent in a space-saving form.

receiver: a device that accepts signals and re-converts them if necessary (as with a radio/wireless microphone set-up).

rectifier: (i) a diode (solid state or valve/tube) that's dedicated to converting AC power to DC, as it passes current in a forward direction only; (ii) a solid pack of four or more rectifiers – see 'bridge'.

residue: unwanted 'background' signals or noise – below/behind the wanted signal.

resistance: an impediment to the flow of electrical current through a wire in a DC circuit (or soundwaves through the air or other material). In AC circuits it is combined with reactance to create impedance. Ohm's law dictates that resistance can be calculated by dividing the voltage across a device by the current running through it ($R = V \div I$).

resolution: (i) a measure of the smallest increment of change offered by a PA, measuring system or control etc; (ii) sonic coherence or detail.

resonance: (i) the vibration (intentional or otherwise) of anything in the musical environment, which can have an impact on what an audience ultimately hears. Natural, sustaining oscillation can occur in musical instruments, speakers, cabs and rooms, and in circuits (such as EQs). Also synthesised by drum machines, keyboards, synths etc. Wanted in musical instruments, sometimes wanted a bit in EQs and rooms, unwanted most everywhere else. The diaphragm of a transducer (as in a microphone or speaker) has a resonant frequency at which it naturally vibrates. The sound that such equipment ultimately produces can be controlled by selective damping ('resistance') of these natural vibrations; (ii) a filter control used in synthesis which amplifies the cut-off frequency.

returns multi: a separate multicore cable/snake used to convey the drive rack's output back to the out-front PA in each wing or overhead, usually via the stage.

reverb: short for reverberation – (i) the natural reflections of sound off hard surfaces, creating a room's distinctive acoustic ambience; (ii) an electronic (digital, analog or electro-acoustic) device that replicates (usually) a number of different room environments.

RF: short for radio frequency – ranges from below 100kHz to over 200GHz (gigahertz): in other words it's too high to be audible by humans, so can be sent through the air without being an acoustic nuisance. Similar in principle to audio signals in wires, but higher frequencies used for all sorts of wireless communications.

RFI: radio frequency interference – noise or audio upset caused by radio transmissions being picked up by other circuits. RF 'pollution' is everywhere but strongest in urban areas.

RF burnout: when RF in power amps (either picked up or through internal 'RF oscillation') causes overheating or burning.

RF oscillation: when gear self-oscillates at frequencies above audio (RF) – this is very 'leaky' and affects sound quality.

rhythm: (i) the beat or tempo of a piece of music as (often) provided by the drum part; (ii) guitarists' shorthand for a guitar which plays mostly chords, rather than riffs or solos, thus emphasising the rhythm of the music.

rhythm section: the instruments that provide the rhythmic backing to a piece of music – usually means at least the drums (and/or percussion) and bassline, but may also include chordal guitar, piano etc.

ribbon mike: uses a piece of very fine (ribbon-like) metal foil suspended in a magnetic field. Good at capturing nuances of sound especially vocals – though delicacy makes it less than ideal for knockabout or outdoor gigs.

glossary

rig: (i) as a noun, colloquial term for a sound (or lighting) system; (ii) a truck (in US); (iii) as a verb, to set up a sound system, or other stage gear.

rigging points: on a flown PA cabinet, this is where the hoist and safety chains are fitted.

ringing: a resonant sound that's on the verge of continuous oscillation – can precede full-blown howlround/feedback.

ripple: (i) a sawtooth wave with a buzzy sound and a 100/120Hz fundamental frequency – occurs in most AC/DC power supplies; (ii) the reducing wiggly trace of a wave that's suffering from ringing.

riser: a raised section of stage.

RMS: root-mean-square – a true average measurement of an ever-changing music signal, be it expressed in volts, amperes, or dB etc. Simple averaging of a complex signal gives wrong results. Full definition is 'the square root of the average value of the squares of the instantaneous values over a complete cycle.'

rolling riser: a riser on castors, used to mount and then move stage-sets rapidly into place at multi-set events where a revolving stage isn't used.

roll-off: the rate at which a filter attenuates a signal (in dB per octave, dB/oct), after passing the cut-off 'corner'. Comes in multiples of six: 6, 12, 18, 24dB/oct.

routing: the paths taken by the different signals inside the mixer, into sub-groups, sends, outputs, etc (rhymes with 'tooting' in UK, 'outing' in US).

RTA: real-time analyser, also called audio spectrum analyser – used to test single components or entire PA systems, often using a pink noise test signal.

rubbing: aftermath of a speaker's voice-coil being overheated. Swollen adhesive rubs against the sides of the magnet, only a few thousandths of an inch away.

rumble: low-frequency vibration and noise, usually physical or mechanical rather than electronic – eg. traffic, trains, thunder, air conditioning fans.

runner: (i) someone who carries items or messages (if intercom fails) between the stage and FOH mix position (or elsewhere); (ii) guide rail on some racked gear, allowing equipment to be easily slid out.

running order: (i) the sequence of acts on a multi-act bill; (ii) the list of songs expected to be performed by an act – also called a set-list.

S

sampler: digital audio recording device used to capture and/or reproduce sections of music, sometimes manipulated by time-stretching or pitch-shifting. Popular with many current bands/acts/DJs.

saturation: (i) when a sound system is driven into compression and/or clipping; (ii) when transistors reach their 'end stops', passing the most current that can be forced; (iii) when a magnetic core (inside inductors, transformers) is likewise 'flooded out'.

schematic: electronic circuit drawing (diagram/plan).

'scope: (i) see oscilloscope; (ii) 'scope out' – means sorting out, evaluating or investigating a problem in a PA rig.

screen: see shield.

scrim: thin cloth that set designers may use to hide PA cabs – can have a costly effect on sound quality depending on material and types of ink/paint used. Early liaison essential.

second breakdown: a weakness in bipolar transistors used in high-power amps, which causes fatal meltdown.

sensitivity: amount of input voltage needed to achieve a given output level – low sensitivity means more volts are needed. Generally PA amps have low sensitivity, to keep unwanted noises down. For speakers (only), sensitivity can also be a partial measure of efficiency.

separation: (i) sonic/electrical isolation between, say, left and right stereo signal paths, or adjacent mixer channels, thus minimising 'crosstalk'; (ii) physical separation of instruments and performers, to minimise 'spill'.

set: (i) an act's turn on-stage (ii) the physical staging, risers, backcloth, etc.

set-list: the songs, or repertoire, being performed – often (poorly) handwritten, usually delivered to the mix engineer at the last minute (inadvisedly)...

shackle: a U-shaped fitting, closed with a pin (used when flying cabs etc).

shield: braided or twisted metal wire sheath which helps protects an audio signal from external noise interference, draining away RF noise and other unwanted signals and fielding them to earth – or at least neutralising their effect. Also called a screen.

shock-mount: a device for reducing the transmission of shock and vibration – generally a section of rubber with threaded (metal) couplings attached to each end – much as used in vehicles for engine mounting.

short: 'short circuit' – a connection that has near zero resistance. Usually unwanted, causes malfunction, and is potentially destructive in power circuits.

shunt: (i) term for a device connected 'in parallel' – where the current flowing in a circuit is divided between two or more elements; (ii) a large resistor, used in-line to read current indirectly by observing voltage drop.

shunting: placing some electrical component (wire, switch, capacitor, resistor, etc) across another so the total current is shared between pathways.

sibilance: harsh, high-frequency distortion on vocals (sounds like bad lisping), often due to poor mike technique, placement, or mixing. Also caused by HF driver's distortion, or after diaphragm damage.

sidefill: specialised monitor (foldback) cab, placed to the left and right of the stage to fill these areas.

sig gen: short for signal (tone) generator – used to inject sine, square or other waves into equipment for testing.

sightline(s): visibility of the act on-stage for the paying audience, as against being obscured by the lights, PA gear, etc. It should be remembered that what the crew/engineers need to see is not the same as what the audience wants to see.

signal delay: see so-called 'time delay'.

single-pole: (i) a circuit or device with only one side, or element – a single switch etc; (ii) a crossover with a basic slope of -6dB per octave – which means signal voltage halves for every doubling in frequency (up or down), a natural relationship.

six-way: a system that splits the sound into six frequency ranges, generally sub-bass, bass,

low-mid, high-mid, low-HF, and high/super-HF.

slant plates: see acoustic lens.

slave: (i) a circuit or device under the control of another; (ii) a power amp 'slaved' (daisy-chained) off another to add power without any independent controls; (iii) a low-budget musicians' instrument ('MI') power amp.

slew rate: the maximum rate of change in a signal – slew rate is high for high-frequency signals and very high for radio-frequency signals of a given size. Often mixed up with 'slew limit'.

slew limit: the maximum rate of change in a signal that an amp or other signal path device can handle, without losing control and creating distortion.

sling: an assembly of wire ropes/cables that connects the cab load to the lifting point when flying.

SMPTE: stands for Society of Motion Picture & Television Engineers – refers to the timecode created by said body, used to keep audio tracks in sync, either with each other or with video.

SMT: surface-mount technology – micro-miniature circuitry and parts, first pioneered in the 1970s.

snake: slang term for multicore or loom – a sheathed bundle of cables/leads.

snapshot: a saved image of a

glossary

mixer set-up for future re-use.

softener: sheathing materials used to protect the load or wire rope/cable from damage during lifting, or when in place. May be disposable.

solid state: where all the active (amplifying, rectifying) components of a circuit are solid (transistors, chips etc).

solo: button on a mixer that isolates one channel and feeds it to a monitor – usually only for engineer's use, and could be disastrous if fed to main mix by mistake (some desks have a 'Solo In Place' button which does exactly that); (ii) a highlighted instrumental from one player – may or may not require boosting from mixer; (iii) a lone performer/performance.

soundcheck: a chance for the mixer to be set up, each instrument's EQ to be tuned-up, and howls eliminated. Note, it is not supposed to be a band rehearsal...

S/PDIF: the Sony/Philips standard for a digital connection on CD players etc – uses either ordinary phono (RCA) or superior Canare or BNC connectors.

speaker: see loudspeaker.

Speakon: proprietary locking plug & socket specifically designed by Neutrik for connecting high-power speakers.

spec: short for specification – the technical and statistical data

that describes how a device functions.

spill/spillage: unwanted external sounds picked-up by a microphone.

spike: another name for transient.

SPL: sound pressure level (in deciBels). Measure of sound intensity, analogous to the voltage level of the PA signal.

splitter (box): a dividing device that ideally isolates signals going off to different places – for instance enabling signals from the stage (mikes, pickups, DI) to be fed to more than one mixer, with minimal interaction.

stack: several speaker cabinets (either backline or PA) positioned on top of each other.

stagebox: a junction box at the stage end of the multicore cable/snake/loom, usually with XLR sockets into which mike, pickup and DI leads are plugged.

star/star earth: a central point where many earth (0V) wires are joined on a 'round table' (equality) principle – may comprise a stack of crimped eyelets (ring terminals) or star-shaped multiple solder-tags (tabs), solidly connected to a main earth/ground point or 'buss'.

state of the art/SOTA: the 'latest thing' in an area of technology that's fast developing.

stereo: short for stereophonic –

(i) using left and right channels to reproduce higher dimensional sound (nearer to 3D at best) than single-channel mono(phonic); (ii) any two-channel equipment, whether used for left-and-right signals or not.

stiff: slang for a low impedance – one which can be heavily loaded without sagging;

stray/strays: unwanted side-effects in electronic circuits or components.

stretch frame: an extra section of mixer that can be linked by a 'buss' cable to extend the number of input channels.

sub: normally used as shorthand for a low-bass speaker (a sub-woofer).

sub-group: another name for one of the groups on the mixing desk.

sub-harmonic: inverse of a harmonic – an unwanted spurious frequency made when some speakers distort, that's at an integer fraction (one-third, one-tenth etc) of the original frequency. Creates disturbing effects, like on a horror movie soundtrack.

subjective: based on personal perception, opinion and experience, without the aid of test instruments.

sub-mixer: a secondary mixer, often an older or more junior unit with fewer channels – usually used for additional subsidiary signals, like pre-recorded feeds, percussion, etc.

glossary

subsonic: frequencies below normal audibility (about 20Hz) – should really be called 'infrasonic'. Made much use of in dance music.

supply: short for power supply.

support: the 'junior' act or other performer who 'opens' a show or plays before the headline act – and is sometimes (sadly) treated to a more cursory PA service. Also called the 'opener'.

surge: (i) an inrush – a temporarily high current drawn, for example, when large transformers (mainly inside PA power amps and mixer power supplies) are first turned on; (ii) a temporarily high power supply voltage. Both can be subdued by using a 'surge protector' circuit or device.

sweep: (i) a special sine wave test signal, with a smoothly increasing or decreasing frequency – used to check response over a range of frequencies; (ii) sweep (or swept) EQ is a simple 'parametric-type' equalisation, where particular focal frequencies can be selected rather than a pre-set 'brush', but without the added precision of a Q control.

swing: or voltage-swing – the full capability of an amplifier, relating to the way music or a test signal moves from a peak point in one direction to a peak in the opposite – rather like a musically driven pendulum. This property is exaggerated in high-power amps.

SWL: safe working load (in terms of weight).

T

tail: (i) short, free (loose) conductor or short cable section; (ii) prepared ends of a cable or wires; (iii) a finished note or event in the signal, fading away.

tail-out: see breakout.

tandem: US jargon for a mode where two or more amp channels are driven equally by one source.

talkback: slang for one-way FOH-to-stage intercom link, built-into some mixers.

TDS: time delay spectometry – using 'echoes' to test audio gear performance.

THD: total harmonic distortion.

thermal: to do with heat.

thermalled out: when excess temperatures cause shutdown, sometimes automatically re-setting.

thermal runaway: when a power amp's output stage loses control, and the power transistors suffer catastrophic destruction through accelerative overheating.

thimble: grooved metal fitting to protect the eye (fastening loop) of a wire rope when flying.

third man: not a spy, but PA crew member who acts as the monitor engineer's right-hand person and looks after the PA gear on stage. Positions and removes mikes, lays cables, organises local crew, stacks and tears down PA cabs, etc.

three-phase: standard AC electricity is sent around the country in this form – uses three live terminals and sometimes also neutral. One of the three phases is normally supplied to domestic/small consumers; larger industrial premises are supplied with all three phases and will distribute the power internally according to requirements. Has considerably higher voltages between lives (415V in UK) than any live-to-neutral (240v in UK), so it is more dangerous than single-phase power.

three-way: see 'tri-amped'.

tie clip: a miniature clip-on microphone.

timbre: tonal quality or colour of a sound.

time delay: a misnomer, should really be 'signal delay'. At a large or outdoor venue, where speaker stacks are a long way apart, part of the signal can be delayed to avoid the disconcerting repeat echo effect otherwise heard (see delay tower).

time domain: seeing musical signal events played out over time, rather than against frequency.

tolerance: maximum variation

glossary

in PA gear or mass-produced objects – rather than pretend perfection. Tolerance quantifies the maximum imperfection allowed for, example: ± xdB.

tone burst: an indoor PA and speaker test method which uses swept sine-wave tones in staged bursts to avoid disturbed readings, as would be caused by exciting the rooms' acoustics with continuous tones.

tone generator: see signal generator.

top end: treble or HF (high-frequency).

t-nut: used to hold (chassis) drive units in place (in wooden cabs only) – bites into the wood, holding itself in place once first tightened up.

trace: (i) the moving dot of light on a 'scope screen; (ii) a PCB copper track.

transducer: a device that changes one form of energy into another – for instance a microphone changes soundwaves into electrical signals; a speaker does the reverse.

transformer: a fundamental electrical/electronic component that can change voltages and impedances up or down, and provides isolation, balanced connections, etc. without affecting frequency – as used in an AC adaptor, which transforms high-voltage AC mains current into low-voltage DC for powering some electrical gear.

transformer switch: typically a switch that flips over the cross fader, used by DJs when 'scratching'.

transient: a sudden, steep, short-lived change or burst in any signal or waveform, may be musical or not. Examples: switching click, record scratch, cymbal hit.

Transit: a medium-sized van/small truck – mass-market British-made Ford model, traditionally used to transport small PAs (and bands) to and from gigs.

transistor: a basic semiconductor part which can amplify signals – invented in the 1940s as a smaller, more reliable alternative to the valve/tube. There are two main types of transistor: bi-polar junction transistor (BJT) or field-effect transistor (FET or MOSFET).

transmitter: something that sends a signal – could be converted to RF first, as with a radio (wireless) microphone.

tri-amp: a three or six-channel amp, for tri-amping.

tri-amped: a system using active crossovers with three-band outputs and separate amps per channel.

trunk: a large, sturdy case, used to keep cables, spares, tools and related accessory items.

TRMS: true root-mean-square – spec of a test meter, indicates that true RMS (not estimated) values of AC voltage and current will be displayed.

TRS: (tip, ring, sleeve) describing the various connective parts of a 'stereo' or balanced quarter-inch (phone/jack) plug. The tip is the knobbly end, the ring is the part between the two insulating stripes, and the sleeve is the rest of the metal barrel.

truss: frame or scaffolding built to support elements of the stage rig (PA or lighting).

TS: (tip, sleeve) – the connective parts of a standard mono (or unbalanced) quarter-inch plug (see TRS, phone, jack).

tube: US word for valve – short for electron(ic) vacuum tube.

tweeter: a high-frequency speaker.

twisted pair: audio cable pair arranged (by twisting) to reject hum and RF noise fields, when used with balanced outputs and inputs.

two-way: (i) a speaker system comprising just LF (low) and HF (high-frequency) drive units, or a crossover for such; (ii) a switch that has two positions where connections are made.

tx: generic abbreviation for 'transmitter' or 'transmission'.

U

U: as in 1U, 2U, 3U – a standardised measurement of rack unit height: 1U = 44.45mm (or 1.75in).

glossary

U, Uo: strangely useless international/Euro standard electrical designation for power supply voltage, seen on Germanic electrical connectors & apparatus.

UHF: ultra high frequency – radio frequencies long used for TV, also available for stage radio/wireless mikes. Equipment in this range works over 'line of sight', reduced by thick solid walls, and needs only small aerials/antennae.

UL: mark used on products in the US to indicate they have been tested for 'safety' – name comes from the testing system originators, Underwriters Laboratories Inc.

unbalanced: (i) a single-sided circuit or connection – usually as in a cable/lead that has a sole central signal wire and uses the ground shield to double-up as the second signal wire; (ii) a disabled balanced circuit.

undamped – see ringing, damping.

unity gain: a gain of x1, or no change – in other words, what goes in comes out the same level.

UPS: uninterruptible power supply – valued for computers used at gigs, and rather vital when power is sourced from unreliable site generators.

VA: volt-ampere – a measurement of power in a DC

circuit, expressed in watts. In an AC circuit, the power calculation has to take account of reactance, which reduces the true available power. In most PA gear, true wattage is typically around 60 per cent of the VA rating; or looking at it the other way, VA is usually about 166 per cent of the given wattage rating.

valve: British word for a vacuum tube.

Variac: a trademarked variable AC transformer. Not a safety transformer.

VCA: voltage controlled amplifier – allows gain to be controlled remotely. Also readily permits control ganging of channels/groups etc, and computer control and recall in PA mixers.

Veroboard: multipurpose copper-striped circuit board with grid of holes, used for rig repairs and equipment customisation. Called 'Perf board' in US.

V-FET: a type of MOSFET – old terminology.

V-I limiter: dynamic output protection in many power amps against stressful speaker loads – V-I stands for voltage & current.

VHF: very high frequency – could be RF or audio above 10kHz. Frequencies in this band are available for stage radio mikes.

volt (V): unit of electrical 'pressure' or electromotive

force/energy. Ohm's law states: voltage = current multiplied by resistance ($V = IR$) for DC circuits.

voltage: the amount of energy that's carried by a charge as it flows in a circuit or wire.

voltage swing: see swing.

vox: shorthand for vocals.

VU: originally 'volume unit' – a swing-needle meter responding to average signal levels, ignoring peaks and spikes but giving a good idea of subjective loudness.

watt (W): unit of power – the rate that energy is transferred (or work is done) over time, equal to a certain amount of horse-power, or joules per second (and named after James Watt, British pioneer of steam power). In electrical terms, $W=V \times I$ (where V is voltage and I is current), and since Ohm's law says $I = V \div R$, the W can also be calculated as $V \times V \div R$, or $V^2 \div R$ (in other words, watts = voltage squared divided by resistance). See also VA.

wavelength: the distance between identical points on a wave cycle.

wedge: a monitor cab that's placed on the stage floor, and is 'wedge' shaped to help direct the sound directly at the performer's ears. Also called 'floor monitor'.

glossary

W-bin: type of bass speaker enclosure in approximate W shape.

wet: with an effect added (especially reverb).

wheelboard: a skateboard-type device used to help move big cabs and trunks.

white noise: naturally produced random noise – sounds more trebly/hissy than pink noise.

windshield: a protective cover, built into most stage mikes or may be a removable (spongy) accessory – transmits most sound while preventing damage and unwanted noise like wind and explosive breath from close-up vocal action. Also called 'popshield'.

wing: one side of the stage, or the area beyond – usually where the PA speakers and amps are located.

wireless: a way of communicating without using connecting cables – usually using radio-waves (RF) or invisible, infra-red light.

woofer: a low-frequency speaker.

WPC: watts per channel.

X

XLR (or XLR3): a three-pin plug or socket, taking its name from its connections (eXternal, Live, Return) – the de facto standard audio connector for

mike, line and also many speaker level jobs. Also called a Cannon connector.

XOR: shorthand for 'crossover'.

X-talk: short for crosstalk.

Y

Y-lead: a splitter (or combiner) lead/cable – most common type is female XLR to two male XLRs, to drive two inputs, but may also be quarter-inch types.

Z

Z: symbol for impedance – measured in ohms, for which the symbol is Ω, or omega, the last letter of the Greek alphabet (hence the Z in English).

zero level: not 'nothing', but an arbitrary (0dBr) or standard (eg +4dBu in most PA) starting-point from which all other levels above and below are measured.

Z-in: shorthand for input impedance – the 'load' presented to signals 'looking into' gear.

Z-lead: a 'punt' cable that helps solve a variety of problems and emergency situations, when signals have to be accessed or diverted or abnormally connected. Comprises two female and two male line XLRs, all wired together with short lead lengths in a 'Z' shape.

Z-out, Zo: output impedance – the resistance seen looking 'up' an output of an audio device. A low output resistance means more drive 'power'.

Zobel: a network of resistors and capacitors used in the output stages of amps to ensure stability, meaning to prevent RF oscillation.

zero volt: the name given to reference ('ground') connections, busses and wires in electronic circuitry – also called 0V (in UK pronounced 'oh volt')

zero volt switch(ing) (ZVS): a 'mains cycle-synchronised' method used in some lighting and other AC powered gear that's regularly switching on/off – ideally should reduce 'hash' and clicks over the PA.

index

index

index